Springer Proceedings in Physics **328**

Indexed by Scopus

The series Springer Proceedings in Physics, founded in 1984, is devoted to timely reports of state-of-the-art developments in physics and related sciences. Typically based on material presented at conferences, workshops and similar scientific meetings, volumes published in this series will constitute a comprehensive up to date source of reference on a field or subfield of relevance in contemporary physics. Proposals must include the following:

- Name, place and date of the scientific meeting
- A link to the committees (local organization, international advisors etc.)
- Scientific description of the meeting
- List of invited/plenary speakers
- An estimate of the planned proceedings book parameters (number of pages/articles, requested number of bulk copies, submission deadline).

Please contact:

For Americas and Europe: Dr. Zachary Evenson; zachary.evenson@springer.com
For Asia, Australia and New Zealand: Dr. Loyola DSilva; loyola.dsilva@springer.com

Aldrin Antony · Saji K J · Priya M. J.
Editors

Proceedings of The 4th International Conference on Optoelectronic and Nano Materials for Advanced Technology

icONMAT 2025, 11–14 February, Cochin, India

Springer

Editors
Aldrin Antony
Department of Physics
Cochin University of Science and Technology
Cochin, Kerala, India

Saji K J
International School of Photonics
Cochin University of Science and Technology
Cochin, Kerala, India

Priya M. J.
Department of Physics
Cochin University of Science and Technology
Cochin, Kerala, India

ISSN 0930-8989 ISSN 1867-4941 (electronic)
Springer Proceedings in Physics
ISBN 978-981-95-2696-3 ISBN 978-981-95-2697-0 (eBook)
https://doi.org/10.1007/978-981-95-2697-0

This Springer imprint is published by the registered company Springer Nature Singapore Pte Ltd.
The registered company address is: 152 Beach Road, #21-01/04 Gateway East, Singapore 189721, Singapore

Preface

It is with great enthusiasm we present the proceedings of the 4th International Conference on Optoelectronics and Nanomaterials for Advanced Technology (iCONMAT), jointly organized by the Department of Physics and the Centre of Excellence in Advanced Materials (CAM), Cochin University of Science and Technology (CUSAT), held at Kochi during February 11 to 14, 2025. This volume comprises 30 peer-reviewed contributed papers that have been selected and presented at the conference.

The purpose of the conference was to explore the rapid advancements in optoelectronics, nanomaterials, energy harvesting, and thin-film technologies fields that have led to revolutionary breakthroughs with significant industrial and societal impacts. To facilitate in-depth exploration, icONMAT 2025 was organized into seven key symposia: Nanomaterials and Devices, Energy Harvesting & Storage, Thin Films & Coatings, Biomaterials & Engineering, Ultrafast Laser Technologies, 2D Materials & Devices, and Quantum Materials. These thematic areas covered a spectrum from fundamental research to applied technologies, promoting interdisciplinary collaborations and addressing global scientific challenges. In addition to these sessions, the conference featured technical training sessions led by the professionals. These sessions provided participants with hands-on experience in advanced fabrication techniques, high-precision characterization methods, and real-world applications of emerging technologies.

icONMAT 2025 brought together an exceptional assembly of around 300 participants, including distinguished scientists, researchers from premier institutions and industry leaders across India and worldwide. With over 38 invited talks, 5 technical talks, 52 oral/flash presentations, and 110 poster presentations, the event highlighted the growing interest in interdisciplinary research in optoelectronics and nanomaterials. Each contributed paper included in this volume were peer reviewed before being accepted for publication in the proceedings.

We hope the findings and ideas shared in these proceedings will encourage continued innovation and foster collaboration in the field of optoelectronics and nanomaterials.

July 2025

Aldrin Antony
Saji K J
Priya M. J.

Acknowledgement

We extend our heartfelt gratitude to all those who contributed to the successful conduct of the icONMAT 2025 and the publication of this proceedings.

We gratefully acknowledge the generous financial support received from Anusandhan National Research Foundation (ANRF), Department of Biotechnology (DBT), Defence Research and Development Organisation (DRDO), Board of Research in Nuclear Sciences (BRNS), Council of Scientific & Industrial Research (CSIR), and our industrial sponsors for the smooth execution of the event. Special thanks to our keynote speakers, invited speakers, and session chairs for sharing their valuable insights and enriching the academic sessions with their expertise.

We sincerely thank all the contributing authors for their high-quality submissions and the reviewers for their meticulous evaluation and constructive feedback that helped enhance the standard of the papers included in this volume.

We extend our special appreciation to the research scholars of Nanophotonic and Optoelectronic Devices Laboratory (NPOED), Intense Laser-matter Interactions Group (ILAI), and International School of Photonics, CUSAT, for their unwavering support, hard work, and dedication from the initial planning stages through to the successful completion of these proceedings. We sincerely thank the Conference Chairs Prof. M. K. Jayaraj, Prof. Joaquim Puigdollers, Prof. Gin Jose, Dr. Riju C Issac, Dr. Aldrin Antony, Dr. Saji K J, organizing committee, and student volunteers, for their tireless efforts and invaluable contributions in making icONMAT 2025 a resounding success.

Finally, we thank Springer Nature for partnering with icONMAT 2025 and for their support in publishing this volume. We look forward to continued collaboration in the future editions of the conference

Aldrin Antony
Saji K J
Priya M. J.

Acknowledgement

Contents

Sm^{3+} Activated Phosphor-in-Glass: A Novel Orange-Red Emitter for Inorganic Colour Conversion in LEDs

Jeena Rose Jose[1,2], M. S. Amogh[1], Ashika J. Ashok[1], Gin Jose[2], and P. R. Biju[1](✉)

[1] School of Pure and Applied Physics, Mahatma Gandhi University, Kottayam, Kerala, India
prb.mgu@gmail.com

[2] School of Chemical and Process Engineering, University of Leeds, Leeds, UK

Abstract. A set of orange-red emitting Phosphor-in-Glass (PiG) samples based on Sm^{3+} activated $SrTi(PO_4)_2$ phosphor (STP Sm) with varying phosphor content (x = 3, 5, 7, and 9 wt%) were fabricated via a one-step melt-quenching method. Structural studies, including HR-TEM, SAED, and FTIR, confirmed the successful integration of polycrystalline phosphor powder into the amorphous glass matrix. The optical and luminescent characteristics of the prepared PiGs were analysed using UV-Visible transmittance and photoluminescence spectra. PL emission spectra revealed intense emission for all PiGs under 401 nm excitation, with a prominent emission peak at 595 nm corresponding to the $^4G_{5/2}$ to $^6H_{7/2}$ transition of Sm^{3+} ions. Chromaticity studies demonstrated that the prepared samples exhibited colour coordinates of (0.571, 0.422), resulting in orange-red emission with a colour purity of 98.2%. These results highlighted the STP Sm PiGs as promising inorganic colour converters for lighting and display applications.

Keywords: Phosphor- in- glass · $SrTi(PO_4)_2$:Sm^{3+} phosphor · TZN glass

1 Introduction

Phosphor-converted LEDs (pc-LEDs) have gained attention for their longer life span, cost-effectiveness, eco-friendliness, and high energy efficiency, making them popular in automotive, display, and plant lighting applications [1–3]. Commercial white LEDs typically combine a yellow phosphor, such as YAG:Ce^{3+}, with a blue LED chip by means of organic resin to realise white light emission. However, issues like low thermal conductivity, UV intolerance, and poor chemical and thermal stability of organic resins reduce the quality of emitted light and the life span [4, 5]. Moreover, the viscous nature of organic resins leads to uneven phosphor dispersion, resulting in blurring of emitted light and chromaticity shifts over time. To address these drawbacks of phosphor-embedded organic resins in pc-LEDs and withstand LED junction temperatures (~150 °C), inorganic light-emitting materials with better heat resistance, along with luminescent stability, are needed.

In order to overcome the drawbacks of phosphor-embedded organic resin, researchers have explored inorganic colour converters such as phosphor glass ceramics, bulk glass

A. Antony et al. (Eds.): icONMAT 2025, SPPHY 328, pp. 1–9, 2025.
https://doi.org/10.1007/978-981-95-2697-0_1

phosphors, and phosphor-in-glass (PiG) systems [6–10]. Among these, PiG, a simple mixture of luminescent phosphor and transparent glass, has emerged as a promising candidate due to its low production cost, high luminescence efficiency, reduced phosphor usage, and ease of fabrication. The use of a transparent glass matrix in place of organic resin significantly enhances thermal and chemical stability while maintaining the luminescence efficiency of phosphor. Unlike ceramic phosphor plates and phosphor glass ceramics, which are limited to specific types of phosphors, PiG offers greater flexibility in phosphor selection based on the properties of the host glass. Additionally, its optical performance can be tailored by adjusting the phosphor concentration, PiG thickness, and sintering conditions. Y.K. Lee et al. demonstrated the use of Pb-free silicate glass with commercial YAG:Ce^{3+} phosphor, achieving excellent thermal and colour stability [11]. Since then, various PiG systems using silicate and tellurite glasses combined with YAG: Ce^{3+} have shown promising luminescent characteristics for solid-state lighting applications [12–14].

Building on insights from previous studies, this work focuses on the fabrication and detailed structural, optical, and luminescent characterisation of PiG samples with varying phosphor content, to evaluate their potential as inorganic colour converters for LED applications. High-purity orange-red emitting $SrTi(PO_4)_2$:Sm^{3+} is selected as the luminescent phosphor, owing to its excellent luminescent properties as well as the inherent advantages of phosphate-based phosphors, including environmental friendliness and low production cost [15, 16]. Tellurite-based glass is chosen as the host material for incorporating the phosphor due to its low melting point, which helps to prevent the degradation of phosphor during synthesis, and its high refractive index, which minimises refractive index mismatch at the phosphor-glass interface in the PiG system.

2 Experimental

2.1 Synthesis of STP Sm PiG

Orange-red emitting $SrTi(PO_4)_2$:Sm^{3+} phosphor with its optimum Sm^{3+} doping concentration (0.05 mol) was synthesised via high-temperature solid state reaction method as detailed in the previous study [15]. A set of PiG materials based on this phosphor and TZN glass (70 TeO_2 - 20 ZnO - 10 Na_2O) with 3, 5, 7 and 9 wt% of phosphor content by a relatively simple one-step melt quenching method. High-purity TeO_2, ZnO, and Na_2CO_3 were accurately weighed according to the molar composition of the mother glass (TZN) and thoroughly mixed with the appropriate amount of phosphor powder for 30 min using an agate motor. The resulting mixture was then sintered at 730 °C for 15 min, followed by pouring the melt into a brass mould preheated at 230 °C. The samples were annealed for 8 h to relieve internal stress. Further, the PiGs were polished into the desired shape for various characterisations. For ease of notation, the PiG samples were labelled as x% STP Sm PiG, where x represents the phosphor content (x = 3, 5, 7, and 9 wt%).

2.2 Characterization Techniques

The microstructural properties of the STP Sm PiG were investigated using high-resolution transmission electron microscopy (HR-TEM) performed with JEOL JEM

2100. Thermal characteristics were determined using a TA-Q20–2047 differential scanning calorimeter (DSC). The structural nature of the samples was analysed by X-ray diffraction (XRD) using a Rigaku Miniflex 600 with Cu-Kα radiation ($\lambda = 1.5406$ Å). Functional groups were identified through Fourier transform infrared (FTIR) spectroscopy using a PerkinElmer Spectrum 2 spectrometer. Optical transmittance was measured using a Varian Cary 5000 UV–Vis–NIR spectrophotometer. Photoluminescence analysis, including PL excitation and emission spectra and lifetime, were characterised using a Horiba Fluorolog-3 spectrofluorometer equipped with a phosphorimeter.

3 Results and Discussion

Fig. 1 gives the X-ray diffraction pattern of 9% STP Sm PiG in the 20–80 diffraction angle range. The XRD patterns of all PiG samples are the same, irrespective of the incorporated phosphor concentrations. The broad diffused peak around the angle of 30° can be attributed to the amorphous tellurite host glass. The absence of any characteristic peak corresponding to Sm^{3+} doped $SrTi(PO_4)_2$ phosphor can be attributed to the relatively low concentration of incorporated phosphor [17].

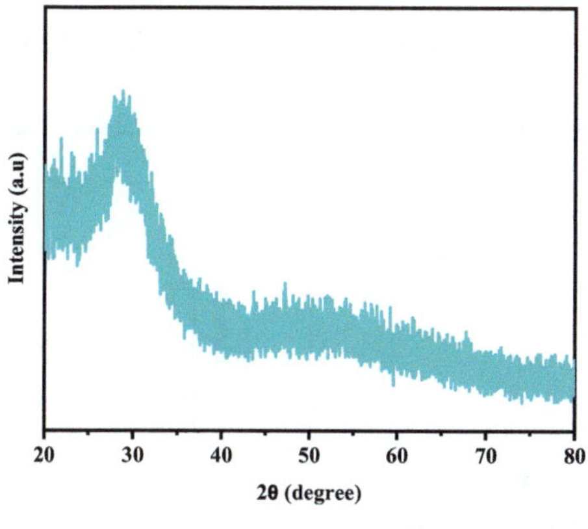

Fig. 1. X-Ray diffraction pattern of 9% STP Sm PiG.

The presence of well-defined lattice fringes in the HR-TEM image displayed in Fig. 2 (a) confirms the distribution of polycrystalline $SrTi(PO_4)_2$:Sm^{3+} phosphor in the nanometer range. The selected area electron diffraction (SAED) pattern of the prepared PiG, shown in Fig. 2(b), exhibits bright spotted rings along with diffuse halos, which can be ascribed to the polycrystalline nature of the incorporated phosphor and the amorphous nature of the glass host, respectively. These observations further confirm the coexistence of phosphor and glass phases, supporting the successful formation of the phosphor in glass.

The FTIR transmittance spectrum of 9% STP Sm PiG is shown in Fig. 3(a). The spectrum displays multiple bands associated with the tellurite-based mother glass and the embedded phosphor. The IR bands observed around 988 and 1030 cm^{-1} originate

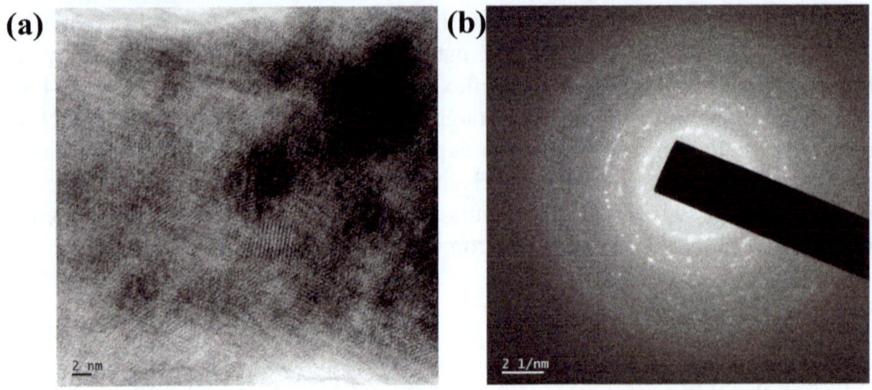

Fig. 2. (a) HR- TEM in 2 scale and (b) SAED pattern of 9% STP Sm PiG

from the incorporated $SrTi(PO_4)_2$:Sm^{3+} phosphor and can be attributed to the symmetric and asymmetric stretching vibrations of the $(PO_4)^{3-}$ anionic group, respectively [15, 16]. Vibrations of the continuous TeO_4 network in the tellurite glass contribute to a band at 580 cm^{-1}, related to the connectivity in the tellurite glass network [18]. The presence of vibrational bands corresponding to both the phosphor and the glass matrix further confirms the successful incorporation of the phosphor into the glass.

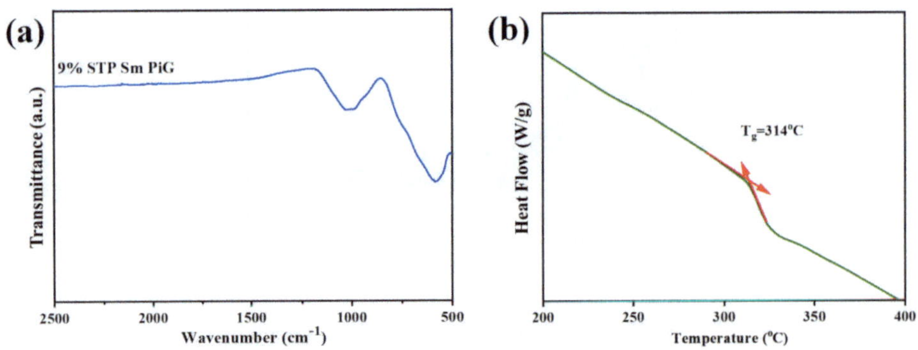

Fig. 3. (a) FTIR spectrum and (b) DSC plot of 9% STP Sm PiG

The thermal behaviour of the prepared PiG was analysed using the DSC (differential scanning calorimetry). Variation of heat flow with changing temperature of the representative sample is given in Fig. 3 (b). The glass transition temperature (T_g) of the 9% STP Sm PiG is estimated to be 314 °C, based on the exothermic hump observed in the thermal profile. The higher glass transition temperature of the prepared PiG compared to that of the pure TZN glass (287 °C) can be attributed to the increased rigidity of the glass network induced by the incorporation of the phosphate-based phosphor [18, 19]. Moreover, the stronger P-O bonds in the incorporated phosphor, compared to the weaker Te-O bonds in the host glass, may contribute to the increased transition temperature observed in the prepared PiG system [18].

Fig. 4(a) gives the UV-Visible transmittance spectra of the prepared STP Sm PiGs for varying phosphor concentrations. The spectra reveal a slight decrease in transparency as the phosphor content increases. This reduction in transparency might be due to the formation of more non-bridging oxygen (NBOs) with increasing phosphor content [20, 21]. To further investigate the effect of phosphor addition on the optical properties of the glass matrix, the optical band gap energies were calculated using Tauc's relation given by [17];

$$\alpha(\nu) = \beta^2 \frac{\left(h\nu - E_g\right)^{1/2}}{h\nu} \tag{1}$$

Where $\alpha(\nu)$ and E_g represent the absorption coefficient corresponding to the frequency ν and the optical band gap energy, respectively.

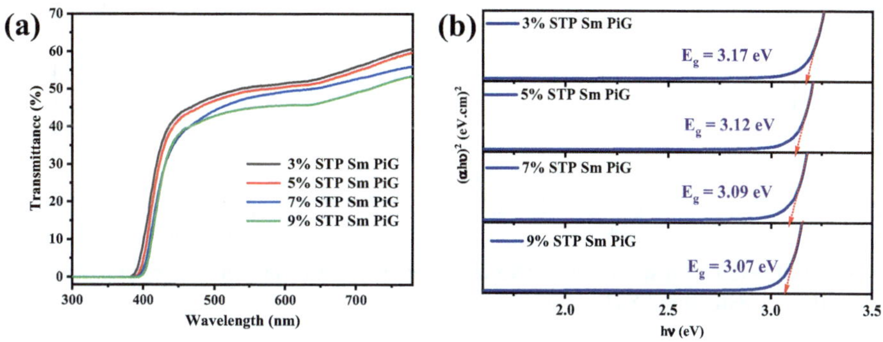

Fig. 4. (a) UV- Visible transmittance spectra and (b) Tauc's plot of STP Sm PiGs with varying phosphor concentrations.

Fig. 4(b) depicts the Tauc's plot for the prepared STP Sm PiG for different phosphor concentrations along with the estimated optical band gap energy. The estimated E_g value exhibits a gradual decrease from 3.17 to 3.07 eV with increasing phosphor concentration from 3 to 9 wt%. This reduction in band gap energy can be attributed to the formation of NBOs resulting from the incorporation of phosphate-based Sm^{3+}-doped $SrTi(PO_4)_2$ phosphor into the tellurite-based host glass [22]. This incorporation induces the creation of a significant number of NBOs, which disrupt the glass network. Additionally, the formation of surface dangling bonds on the phosphor introduces localised states within the band structure, further contributing to the decrease in band gap energy [21]. This reduced band gap energy with the phosphor incorporation can be correlated with the improved crystallinity of the PiG matrix.

Furthermore, to explore the photoluminescence performance of the STP Sm PiG, PL excitation and emission spectra were recorded for a series of samples with varying phosphor content from 3 to 9 wt% and are given in Fig. 5 (a) and (b), respectively. The spectral characteristics for both emission and excitation spectra are similar for all STP Sm PiG samples except for their intensity. The excitation spectra consist of narrow bands in the near-UV and visible regions, arising from the 4f-4f absorption transitions of the Sm^{3+} ions while monitoring the 595 nm emission. The peaks at 361, 376, 401, 416, 439 and 473 nm can be ascribed to the excitation transitions of Sm^{3+} ions from the ground state $^6H_{5/2}$ to the higher energy levels $^4D_{3/2}$, $^6P_{7/2}$, $^4F_{7/2}$, $^6P_{5/2}$, $^4G_{9/2}$+ $^4I_{15/2}$, $^4I_{11/2}$+ $^4M_{15/2}$, respectively [23]. The intensity of excitation spectra is increases with the increasing content of phosphor incorporated in the PiG matrix. The most intense excitation peak is observed at 401 nm, arising from the $^6H_{5/2}$ to $^4F_{7/2}$ transition of Sm^{3+} ions [23, 24]. So, to maximise emission intensity, 401 nm is selected as the excitation wavelength to record emission spectra. The 401 nm excitation wavelength of the prepared samples aligns well with the emission of commercial blue LEDs, making them suitable for inorganic encapsulates and colour converters in pc-LEDs [25].

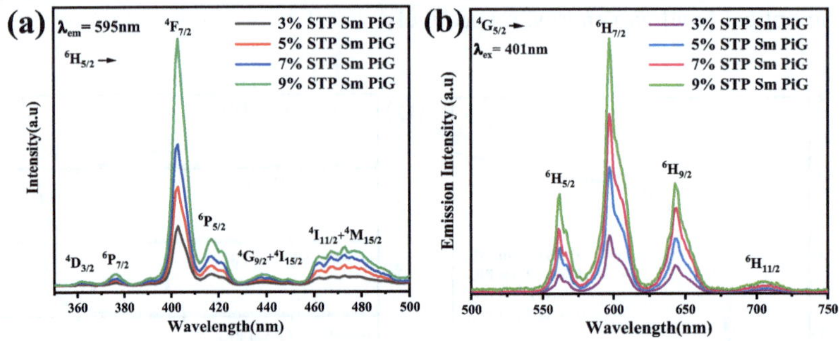

Fig. 5. (a) PL excitation spectra with 595 nm as emission wavelength and (b) emission spectra under 401 nm excitation of STP Sm PiGs.

The emission spectra under 401 nm excitation consist of the characteristic emission peaks of Sm^{3+} ions spanning the orange-red spectral region. The intense peaks at 561, 595, 642 and 704 nm arise from the intraconfigurational emission transitions of Sm^{3+} ions from the higher energy state, $^4G_{5/2}$, to the lower energy levels, $^6H_{5/2}$, $^6H_{7/2}$, $^6H_{9/2}$ and $^6H_{11/2}$, respectively. Among these four emission transitions, the peak centred at 595 nm exhibits the highest intensity, contributing significantly to the emission in the orange-red spectral region. The emission intensity of the prepared STP Sm PiG increases with increasing concentration of the phosphor. This monotonous enhancement of emission intensity without any quenching, further confirms the existence and favourable spectral performance of the phosphor, within the tellurite glass matrix.

In order to investigate the time correlated emission characteristics and to determine the life time, PL decay analysis of the representative sample has been recorded. Fig. 6(a) depicts the decay plot of 9% STP Sm PiG for the most intense transition ($^4G_{5/2}$ to $^6H_{7/2}$) by fixing excitation and emission wavelengths at 401 and 595 nm, respectively. The decay curve follows single exponential decay, which can be represented as;

$$I(t) = I_0 + A\, exp\left(\frac{-t}{\tau}\right) \tag{2}$$

Here, I(t) represents the time-dependent emission intensity, A is the fitting constant, and τ denotes the first-order lifetime. The estimated lifetime for 9% STP Sm PiG is 1.31 ms.

The millisecond-scale lifetime of the prepared PiG indicates their potential for display, imaging, and sensing applications.

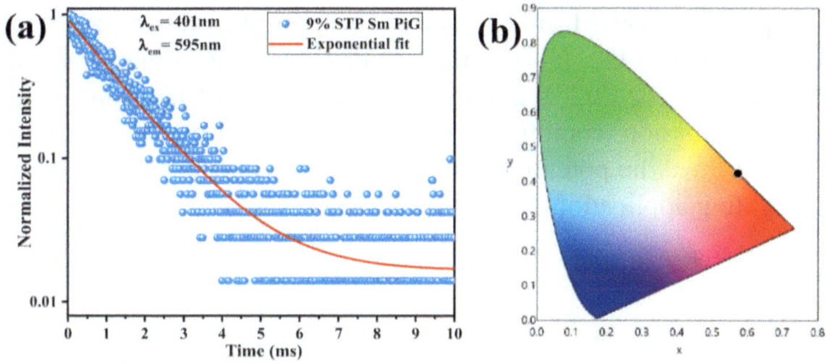

Fig. 6. (a) PL decay plot and (b) CIE chromaticity diagram of 9% STP Sm PiG.

The CIE (Commission Internationale de l'Éclairage) 1931 colour space is a widely accepted method to describe the true emission colour and to evaluate chromaticity properties of luminescent materials. The emission colour of the prepared PiG falls in the orange-red region, and the estimated colour coordinates for 9% STP Sm PiG is given by (0.571, 0.422), depicted in Fig. 6(b).

Moreover, the colour purity of the prepared PiG is calculated using the equation [26];

$$Colour\ Purity\ (\%) = \sqrt{\frac{(x_s - x_i)^2 + (y_s - y_i)^2}{(x_d - x_i)^2 + (y_d - y_i)^2}} \times 100 \tag{3}$$

Where (x_s, y_s), (x_i, y_i) and (x_d, y_d) represent the colour coordinates of the PiG, white point and dominant wavelength, respectively. With an estimated colour purity of 98.2%, the STP Sm PiG demonstrates excellent potential as a highly pure orange-red emitter for various photonic applications.

Correlated colour temperature (CCT) is another important parameter that distinguishes the nature of emitted light, indicating whether the emission is cool or warm [27]. The CCT value of prepared PiG material is estimated by employing McCamy's relation given by [27];

$$CCT = -437n^3 + 3601n^2 - 6861n + 5514.31 \tag{4}$$

Where $n = \frac{(x_s - x_e)}{(y_s - y_e)}$ and (x_e, y_e) represent the chromaticity epicentre. The estimated CCT value of STP Sm PiG is 1726 K, indicating the warm nature of the emission colour.

4 Conclusions

STP Sm PiGs, based on Sm^{3+} activated $SrTi(PO_4)_2$ phosphor embedded in the sodium-zinc tellurite glass matrix, were successfully fabricated via one-step melt-quenching method. A set of PiGs were fabricated with varying phosphor content from 3 to 9 wt%.

Structural studies, including TEM and FTIR, confirmed the effective incorporation of the phosphor into the tellurite glass matrix with minimal phosphor-glass interaction. The increase in optical band gap energy with increasing phosphor content was attributed to the creation of non-bridging oxygen in the PiG matrix. The prepared PiGs exhibited four major emission peaks under 401 nm excitation, spanning the orange-red region. These emissions corresponded to the 4f–4f intraconfigurational transitions from the excited state $^4G_{5/2}$ to ^6H_j (J = 5/2, 7/2, 9/2, and 11/2), regardless of phosphor content. The intense orange-red emission, combined with excellent colour purity and microsecond-scale lifetime, confirmed the potential of the prepared PiGs as efficient inorganic color converters for lighting applications.

References

1. Wang, S., et al.: A new strategy to the phosphors for plant growth LEDs: Far red emission from the $Ca_9MY_{0.667}(PO_4)_7$ (M = Li, Na):Eu^{3+} phosphors due to the Eu^{3+}:$^5D_0 \rightarrow {}^7F_4$ transition. J. Lumin. **225**, 117404 (2020). https://doi.org/10.1016/j.jlumin.2020.117404
2. Feng, Y., Huang, J., Li, C., Hu, G., Liu, J., Yu, X.: Structure, energy transfer and tunable photoluminescence of $Sr_7Zr(PO_4)_6$:Tb^{3+}, Eu^{3+} phosphors for near-UV white LEDs. J. Alloys Compd. **706**, 478–484 (2017). https://doi.org/10.1016/j.jallcom.2017.02.199
3. Chen, H., et al.: A novel double-perovskite Gd_2ZnTiO_6:Mn^{4+} red phosphor for UV-based w-LEDs: Structure and luminescence properties. J. Mater. Chem. C. **4**, 2374–2381 (2016). https://doi.org/10.1039/C6TC00313C
4. Yang, C., et al.: Facile fabrication and luminescence characteristics of Ce:YAG phosphor glass thick films coated on a glass substrate for white LEDs. Ceram. Int. **42**(13), 14526–14532 (2016). https://doi.org/10.1016/j.ceramint.2016.06.065
5. Lin, H., Hu, T., Cheng, Y., Chen, M., Wang, Y.: Glass ceramic phosphors: Towards long-lifetime high-power white light-emitting-diode applications-A review. Laser Photonics Rev. **12**, 1700344 (2018). https://doi.org/10.1002/lpor.201700344
6. Nishiura, S., Tanabe, S., Fujioka, K., Fujimoto, Y.: Properties of transparent Ce:YAG ceramic phosphors for white LED. Opt. Mater. **33**, 688–691 (2011). https://doi.org/10.1016/j.optmat.2010.06.005
7. Secu, M., Secu, C.E., Polosan, S., Aldica, G., Ghica, C.: Crystallization and spectroscopic properties of Eu-doped CaF_2 nanocrystals in transparent oxyfluoride glass-ceramics. J. Non-Cryst. Solids. **355**(37–42), 1869–1872 (2009). https://doi.org/10.1016/j.jnoncrysol.2009.04.062
8. Liu, T., et al.: Highly efficient and thermally stable MnII-based phosphor-in-glass towards warm WLED. Ceram. Int. **49**(4), S0272884222037841 (2022). https://doi.org/10.1016/j.ceramint.2022.10.203
9. Xiang, C., et al.: Versatile phosphor-in-glass based on Sn–F–P–O matrix with ultra-high color rendering for white LEDs/LDs. J. Lumin. **255**, 119593 (2023). https://doi.org/10.1016/j.jlumin.2022.119593
10. Zhang, Y., et al.: A high quantum efficiency $CaAlSiN_3$:Eu^{2+} phosphor-in-glass with excellent optical performance for white light-emitting diodes and blue laser diodes. Chem. Eng. J. **401**, 125983 (2020). https://doi.org/10.1016/j.cej.2020.125983
11. Lee, Y.K., Lee, J.S., Heo, J., Im, W.B., Chung, W.J.: Phosphor in glasses with Pb-free silicate glass powders as robust color-converting materials for white LED applications. Opt. Lett. **37**(15), 3276–3278 (2012). https://doi.org/10.1364/OL.37.003276

12. Xu, X., et al.: High refractive index coating of phosphor-in-glass for enhanced light extraction efficiency of white LEDs. J. Mater. Sci. **53**, 1335–1345 (2018). https://doi.org/10.1007/s10853-017-1571-y
13. Zhou, H., et al.: Facile preparation and luminescence performance of transparent YAG:Ce phosphor-in-tellurate-glass inorganic color converter for white-light-emitting diodes. J. Non-Cryst. Solids. **481**, 537–542 (2018). https://doi.org/10.1016/j.jnoncrysol.2017.11.043
14. Zhang, R., Wang, B., Zhu, W., Li, C., Wang, H.: Preparation and luminescent performances of transparent screen-printed Ce^{3+}:$Y_3Al_5O_{12}$ phosphors-in-glass thick films for remote white LEDs. J. Alloys Compd. **720**, 340–344 (2017). https://doi.org/10.1016/j.jallcom.2017.05.270
15. Jose, J.R., Amogh, M.S., Oommen, S.A., Joseph, C., Biju, P.R.: Tailoring high purity orange-red emission in Sm^{3+} activated $SrTi(PO_4)_2$ phosphor for lighting and display devices. J. Mol. Struct. **1312**, 138442 (2024). https://doi.org/10.1016/j.molstruc.2024.138442
16. Krishnapriya, T., Mohan, P.R., Jose, A., Jose, J.R., Joseph, C., Biju, P.R.: Synthesis and luminescence characteristics of $Na_6Ca_{(1\text{-}X)}P_2O_9$:X Dy^{3+} phosphor for outdoor lighting applications. Mater. Chem. Phys. **280**, 125827 (2022). https://doi.org/10.1016/j.matchemphys.2022.125827
17. Jose, J.R., Gopinath, A., Oommen, S.A., Joseph, C., Biju, P.R.: Novel red emitting Phosphor-in-Glass (PiG) based on $Na_6CaP_2O_9$: Eu^{3+} phosphor as an inorganic colour converter for w-LED applications. Opt. Mater. **148**, 114809 (2024). https://doi.org/10.1016/j.optmat.2023.114809
18. Elkhoshkhany, N., Khatab, M.A., Kabary, M.A.: Thermal, FTIR and UV spectral studies on tellurite glasses doped with cerium oxide. Ceram. Int. **44**(3), 2789–2796 (2018). https://doi.org/10.1016/j.ceramint.2017.11.019
19. Badamasi, S., Tanko, Y.A.: Thermal properties of TeO_2-ZnO-Na_2O glasses: effect of Dy_2O_3 doping. Sci. World J. **13**(4), 1–7 (2018)
20. Kaur, G., Pickrell, G., Kumar, V., Pandey, O.P., Singh, K., Arya, S.K.: Mechanical, dielectric and optical assessment of glass composites prepared using milling technique. Bull. Mater. Sci. **38**, 1003–1008 (2015). https://doi.org/10.1007/s12034-015-0947-3
21. Farag, M.A., Ibrahim, A., Hassaan, M.Y., Ramadan, R.M.: Enhancement of structural and optical properties of transparent sodium zinc phosphate glass–ceramics nano composite. J. Aust. Ceram. Soc. **58**, 653–661 (2022). https://doi.org/10.1007/s41779-022-00716-3
22. Mallur, S.B., Czarnecki, T., Adhikari, A., Babu, P.K.: Compositional dependence of optical band gap and refractive index in lead and bismuth borate glasses. Mater. Res. Bull. **68**, 27–34 (2015). https://doi.org/10.1016/j.materresbull.2015.03.033
23. Carnall, W.T., Fields, P.R., Rajnak, K.: Electronic Energy Levels in the Trivalent Lanthanide Aquo Ions. I. Pr^{3+}, Nd^{3+}, Pm^{3+}, Sm^{3+}, Dy^{3+}, Ho3+, Er^{3+}, and Tm^{3+}. J. Chem. Phys. **49**(10), 4424–4442 (1968). https://doi.org/10.1063/1.1669893
24. Xu, J., Ju, Z., Gao, X., An, Y., Tang, X., Liu, W.: Na_2 $CaSn_2$ Ge_3 O_{12}: A novel host lattice for Sm^{3+}-doped long-persistent phosphorescence materials emitting reddish orange light. Inorg. Chem. **52**(24), 13875–13881 (2013). https://doi.org/10.1021/ic401262w
25. Ning, Q., Quan, B., Shi, Y.: Effect of alkali metal ions on the spectra of $CaZn_2(PO_4)_2$: Sm^{3+} phosphor analyzed by J-O theory. J. Lumin. **206**, 498–508 (2019). https://doi.org/10.1016/j.jlumin.2018.10.101
26. Chandana, M.R., et al.: Simple fabrication of novel Sm^{3+} doped $BaGd_2ZnO_5$ nanophosphors for flexible displays, improved data security applications, and solid-state lighting applications. Mater Today Sustain. **22**, 100397 (2023). https://doi.org/10.1016/j.mtsust.2023.100397
27. McCamy, C.S.: Correlated color temperature as an explicit function of chromaticity coordinates. Color. Res. Appl. **17**(2), 142–144 (1992). https://doi.org/10.1002/col.5080170211

Fabrication and Characterization of p-Cu_2O:B/n-TiO_2 Heterojunction UV Detector

Manu Shaji[1,2](✉), K. J. Saji[3], and M. K. Jayaraj[1]

[1] Department of Physics, Cochin University of Science and Technology, Kochi, Kerala, India 682022
manushaji@cusat.ac.in
[2] Department of Physics, Fatima Mata National College, Kollam, Kerala, India 691001
[3] International School of Photonics, Cochin University of Science and Technology, Kochi, Kerala, India 682022

Abstract. Rutile TiO_2 nanorods were grown on FTO-coated glass substrate via the hydrothermal method. Photoluminescence study of the TiO_2 revealed the presence of traps and defect states in the nanorods. A p-n heterojunction diode was fabricated by depositing Cu_2O: B layer on the TiO_2 through sputtering. The fabricated diode showed rectification behaviour with a rectification ratio of 44 at 2.5 V. The diode shows deviation from the ideal character, and detailed studies suggest that additional charge transport mechanisms come into play because of the traps and surface states present in the interface. A double logarithmic plot revealed three distinct regions which exhibit power law behaviour. The diode shows significant photo response on UV illumination with an increased rectification ratio of 96 at 2.5 V. Fabricated heterojunction can be developed for sensor, UV detector and photocatalytic applications.

Keywords: Heterojunction · Metal oxide · UV detector

1 Introduction

Metal oxide-based p–n junction devices have attracted significant interest as cost-effective alternatives to silicon, owing to their versatility in optoelectronic applications. Despite their simple structure, these junctions are widely employed in LEDs, solar cells, UV detectors, photocatalysts, gas sensors, and power diodes [1, 2]. Heterojunction-based detectors and sensors have been reported, where the built-in electric field aids in efficient electron-hole separation, enhancing performance. TiO_2 is a widely studied n-type semiconductor known for its wide bandgap and versatile morphologies. Its unique catalytic and electronic properties make it important among inorganic metal oxides [3, 4]. Nanostructured TiO_2 improves device performance due to its high surface-to-volume ratio and efficient charge transport in one-dimensional forms like nanorods, nanowires, and nanotubes. TiO_2-based Schottky diodes and metal-semiconductor-metal UV detectors have been demonstrated by several research groups [5, 6]. It has been reported that, modification of the band structure of cuprous oxide (Cu_2O) by introducing suitable cations, can

A. Antony et al. (Eds.): icONMAT 2025, SPPHY 328, pp. 10–15, 2025.
https://doi.org/10.1007/978-981-95-2697-0_2

tune the optical and electrical properties. One of our previous studies indicated that doping trivalent boron into the Cu_2O lattice leads to significant improvements in its optical and electrical properties [7]. In this paper, TiO_2 nanorods were grown on commercially available FTO-coated glass substrates via hydrothermal method. A Cu_2O:B layer was deposited over the TiO_2 under optimized sputtering conditions to form a heterojunction. The structural and electrical properties of the resulting diode were thoroughly analyzed, and its potential application as a UV detector was evaluated.

2 Results and Discussions

TiO_2 nanorods (NRs) were grown on FTO substrates using the hydrothermal method. Prior to the process, the FTO substrates (resistivity: 15–25 Ω/sq., Solaonix) were cleaned sequentially in acetone, isopropyl alcohol (IPA), and deionized (DI) water. A TiO_2 seed layer was deposited to promote nanorod growth by immersing the substrates in a 0.03 M solution of titanium tetra-isopropoxide (TTIP, analytical grade) in isopropanol, followed by air annealing at 400 °C for 1 h. The hydrothermal precursor solution was prepared by adding 0.6 ml of TTIP to a 1:1 volume mixture of concentrated HCl and DI water, stirred at room temperature. The hydrothermal reaction was conducted at 180 °C for 3 h. After synthesis, the samples were rinsed with DI water and dried.

To fabricate the TiO_2/Cu_2O:B heterojunction diode, a 230 nm-thick Cu_2O:B layer was deposited onto the TiO_2 nanorods via RF sputtering. High-purity (99.99%) copper and boron targets (Kurt J. Lesker) were employed, with RF powers of 40 W for copper and 3 W for boron, as optimized in our previous study [7]. The deposition was carried out at a chamber pressure of 9×10^{-3} mbar, in presence of argon and oxygen without deliberate substrate heating. A 100 nm-thick gold (Au) layer was thermally evaporated on top of the Cu_2O:B to form the top electrode, while the FTO substrate functioned as the bottom electrode.

Fig. 1 (a) shows the XRD pattern of TiO_2 nanorods (NRs) grown on seed-coated FTO substrates. The peaks correspond to the tetragonal rutile phase of TiO_2 (ICDD No. 00–001-1292), with preferred growth along the (101) plane promoted by the seed layer. Fig. 1(b) shows the Raman spectra of FTO and TiO_2 NRs ($\lambda_{excitation} = 514$ nm), showing characteristic modes at 144 cm^{-1} (B_1g), 446 cm^{-1} (Eg), 610 cm^{-1} (A_1g), and a second-order peak at 240 cm^{-1}. The FESEM image (inest: Fig. 2(a)) shows a dense array of vertically aligned NRs with square top facets, inclined to the substrate with average length and diameter are ~700 nm and ~ 50 nm, respectively. The optical bandgap was estimated to be 3.04 eV from UV–Vis absorption measurements using a Tauc plot (Fig. 2(a)).

Fig. 2(b) shows the photoluminescence (PL) emission spectrum of TiO_2 nanorods (NRs) excited at 325 nm using a 30 mW He-Cd laser. The broad emission band consists of multiple deconvoluted peaks indicating various recombination processes. The 414 nm peak corresponds to near band edge (NBE) emission from rutile TiO2 NRs, with its redshift attributed to surface states [8]. The 436 nm peak is linked to shallow trap levels near the band edge [9]. Visible emissions originate primarily from oxygen vacancy-related transitions. UV excitation generates trapped holes, and their recombination with electrons at oxygen vacancy sites results in visible light emission [10]. The 485 nm

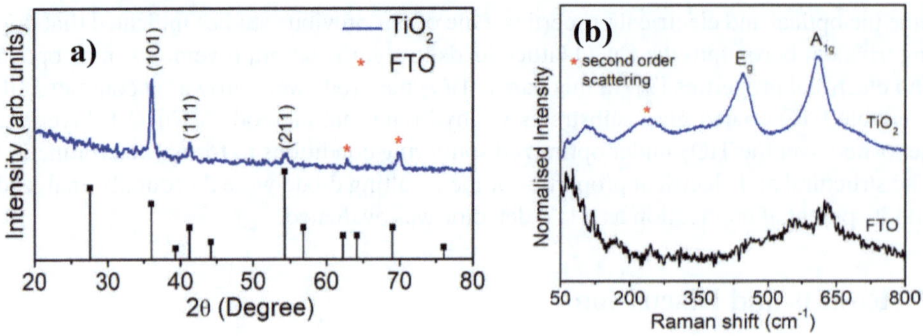

Fig. 1. a) X-ray diffraction pattern and b) Raman spectra of TiO_2 nanorods grown over FTO coated glass substrate.

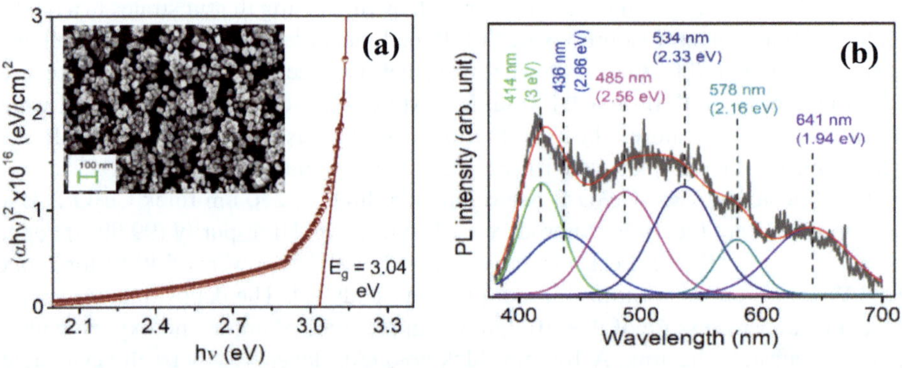

Fig. 2. a) Bandgap of TiO_2 nanorod (inset:FESEM image of TiO_2 nanorods) b) Deconvoluted PL Spectra of TiO_2 NRs showing fitting components (marked by dashed lines) and the cumulative fitted curve (red solid curve) superimposed on experimental data (black solid curve).

peak arises from electrons trapped at oxygen vacancies and Ti^{3+} to O^{2-} charge transfer within the TiO_6^{8-} complex [8, 11]. The 534 nm emission is associated with radiative recombination of self-trapped excitons at surface states [12], while the 578 nm peak is attributed to deep-level oxygen vacancies [13].

Fig. 3(a) shows the FESEM image of the cross-sectional view of the fabricated heterojunction. The I–V characteristics of the individual TiO_2/FTO and Cu_2O:B/Au contacts, shown in Fig. 3(b) and 3(c), respectively, confirm their ohmic behavior. Semi log plot of the current density-voltage (J-V) characteristics of FTO/n-TiO_2/p-Cu_2O:B heterojunction at room temperature shown in Fig. 3(d). The heterojunction exhibits non-linear J–V characteristics with rectifying behavior, confirming the formation of a p–n junction. The diode displays a turn-on voltage of 2.1 V and a rectification ratio of ~44 at 2.5 V. The diode exhibited a series resistance (R_s) of 662 kΩ and a reverse saturation current density of 1.7×10^{-9} A/cm^2. The high series resistance (R_s) likely contributes to the elevated turn-on voltage, which may originate from surface states in the TiO_2 nanorods and the resulting rough TiO_2/Cu_2O:B interface. The ideality factor, extracted from the semi-log J–V plot, was ~15, indicating strong deviation from ideal p–n junction behavior.

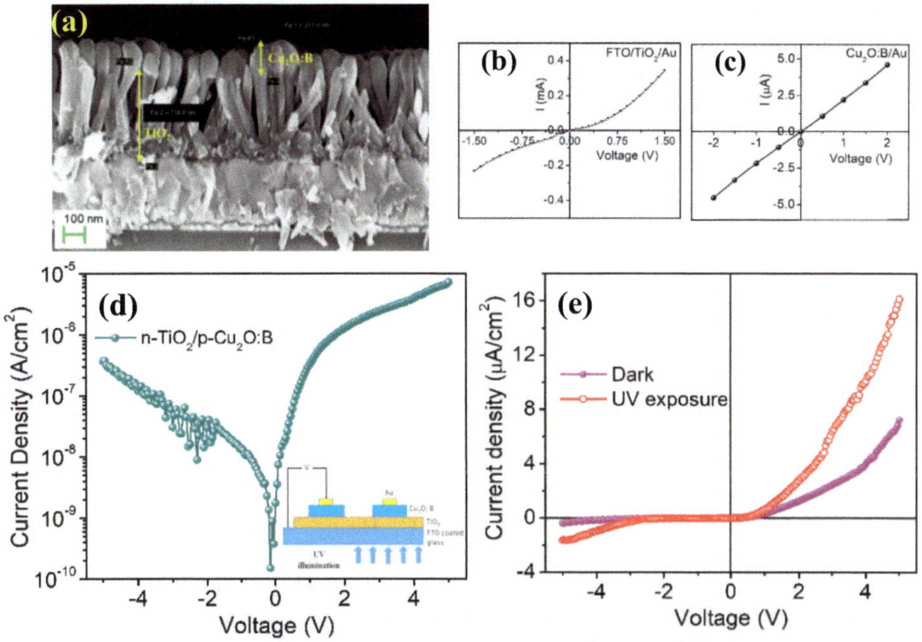

Fig. 3. a) cross section of Cu_2O:B coated TiO_2 NRs. I-V curve obtained for the contacts b) FTO/TiO_2 and c) Cu_2O:B/Au. d) Semi log plot of the J-V characteristics of the diode (inset: Schematic diagram of n-TiO_2/p-Cu_2O:B heterojunction diode). e) J-V characteristics under dark and UV illumination.

The deviation from ideal behaviour suggests the involvement of multiple transport mechanisms, including carrier recombination via surface states and trap-assisted conduction within the space-charge region. Similar high ideality factors have been reported by various groups, such as ~21 for a CuO/NiO diode [14–16]. Such non-ideal behavior is often attributed to high Rs, interfacial defects, and charge carrier recombination in the depletion region. The ~700 nm length of the TiO_2 nanorods increases defect density and contributes to higher R_S. These factors, confirmed by PL analysis, likely account for the non-ideal diode response.

Fig. 3(e) shows the J–V characteristics of the heterojunction under dark and UV (365 nm) illumination from the TiO_2 side. Both forward and reverse currents increased under UV, indicating a photo-responsive p–n junction. The enhanced reverse current confirms the response originates from the junction, not solely from the TiO_2 NRs. Under UV, the device exhibited a turn-on voltage of 1.5 V and a rectification ratio of 96 at 2.5 V. Improved performance is attributed to UV absorption and subsequent photo carrier generation. These results demonstrate the potential of the p-Cu_2O:B/n-TiO_2 heterojunction for UV detection. For practical applications, further optimization is required, including reducing TiO_2 NR length and defect density, or adopting low-defect growth methods.

3 Conclusions

Rutile-phase TiO_2 nanorods were grown on FTO-coated glass, and photoluminescence analysis confirmed the presence of trap and defect states. A Cu_2O:B layer was sputtered onto the nanorods to form a p–n heterojunction diode. The device exhibited rectifying behavior with a rectification ratio of 44 at 2.5 V. High series resistance, likely due to interface defects, resulted in an elevated turn-on voltage. The diode showed a large ideality factor (~15), indicating non-ideal behavior arising from trap-assisted transport and interfacial states. Reducing surface and deep-level defects is essential to enhance junction performance for practical use. Under UV illumination, the device demonstrated a strong photo-response, with the rectification ratio increasing to 96 at 2.5 V.

References

1. Kopidakis, N., Neale, N.R., Zhu, K., van de Lagemaat, J., Frank, A.J.: Spatial location of transport-limiting traps in TiO_2 nanoparticle films in dye-sensitized solar cells. Appl. Phys. Lett. **87**, 202106 (2005)
2. Yasomanee, J.P., Bandara, J.: Multi-electron storage of photoenergy using Cu_2O–TiO_2 thin film photocatalyst. Sol. Energy Mater. Sol. Cells. **92**(3), 348–352 (2008)
3. Chen, R.S., Chen, C.A., Tsai, H.Y., Wang, W.C., Huang, Y.S.: Ultrahigh efficient single-crystalline TiO_2 nanorod photoconductors. Appl. Phys. Lett. **100**, 123108 (2012)
4. Lin, L., Liu, L., Musselman, K., Zou, G., Duley, W.W., Zhou, Y.N.: Plasmonic-radiation-enhanced metal oxide nanowire heterojunctions for controllable multilevel memory. Adv. Funct. Mater. **26**(34), 5979–5986 (2016)
5. Xue, H., et al.: TiO_2 based metal-semiconductor-metal ultraviolet photodetectors. Appl. Phys. Lett. **90**(20), 201118 (2007)
6. Kong, X., et al.: Metal-semiconductor-metal TiO2 ultraviolet detectors with Ni electrodes. Appl. Phys. Lett. **94**(12), 123502 (2009)
7. Shaji, M., Saji, K.J., Jayaraj, M.K.: Fabrication and characterization of p-type boron doped Cu_2O thin film and Cu_2O:B/n-Si heterojunction. J. Mater. Sci. Mater. Electron. **31**, 10724–10730 (2020)
8. Rajabi, M., Shogh, S., A, I.z.: Defect study of TiO_2 nanorods grown by a hydrothermal method through photoluminescence spectroscopy. J. Lumin. **157**, 235–242 (2015)
9. Yamada, Y., Kanemitsu, Y.: Determination of electron and hole lifetimes of rutile and anatase TiO_2 single crystals. Appl. Phys. Lett. **101**(13), 133907 (2012)
10. Vishwas, M., Narasimha Rao, K., Chakradhar, R.P.S.: Influence of annealing temperature on Raman and photoluminescence spectra of electron beam evaporated TiO_2 thin films. Spectrochim. Acta Part A Mol. Biomol. Spectrosc. **99**, 33–36 (2012)
11. Saini, C.P., et al.: Impact of self-trapped excitons on blue photoluminescence in TiO_2 nanorods on chemically etched Si pyramids. J. Phys. Chem. C. **121**(21), 11448–11454 (2017)
12. Selman, A.M., Hassan, Z., Husham, M.: Structural and photoluminescence studies of rutile TiO_2 nanorods prepared by chemical bath deposition method on Si substrates at different pH values. Measurement. **56**, 155–162 (2014)
13. Chung, C.K., Liao, M.W., Lai, C.W.: Effects of oxygen flow ratios and annealing temperatures on Raman and photoluminescence of titanium oxide thin films deposited by reactive magnetron sputtering. Thin Solid Films. **518**(5), 1415–1418 (2009)
14. Prabhu, R.R., Saritha, A.C., Shijeesh, M.R., Jayaraj, M.K.: Fabrication of p-CuO/n-ZnO heterojunction diode via sol-gel spin coating technique. Mater. Sci. Eng. B. **220**, 82–90 (2017)

15. Ghosh, A., et al.: Electrochemical synthesis of p-CuO thin films and development of a p-CuO/n-ZnO heterojunction and its application as a selective gas sensor. RSC Adv. **4**, 51569–51575 (2014)
16. Mridha, S., Basak, D.: Investigation of a p-CuO/n-ZnO thin film heterojunction for H_2 gas-sensor applications. Semicond. Sci. Technol. **21**, 928–932 (2006)

Scalable Synthesis of Shaped MIL-101 (Cr) Extrudates for Efficient H_2S Capture in Humid Environments

H. Jesmitha[1], M. J. Priya[1], P. P. Subha[2], Muhammed Zahid[3], P. P. Shandev[4], and Aldrin Antony[1,5](✉)

[1] Department of Physics, Cochin University of Science and Technology, Kochi, Kerala, India
aldrin@cusat.ac.in
[2] Department of Physics, St. Paul's College, Kalamassery, Kochi, Kerala, India
[3] Aqoza Technologies Pvt. Ltd., Kochi, Kerala, India
[4] Department of Applied Chemistry, Cochin University of Science and Technology, Kochi, Kerala, India
[5] Centre of Excellence in Advanced Materials, Cochin University of Science and Technology, Kochi, Kerala, India

Abstract. The removal of hydrogen sulphide (H_2S) from industrial gas streams is a critical requirement in petrochemical, biogas, and wastewater treatment applications. Conventional adsorbents, such as activated carbon, suffer from reduced performance under humid conditions and limited regeneration potential. In this study, we present a market-oriented evaluation of shaped MIL-101(Cr) metal–organic framework (MOF) extrudates as a promising alternative for H_2S capture under conditions relevant to industrial applications. The material was synthesised *via* a hydrofluoric acid (HF)-free route using nitric acid (HNO_3) as a safer and more scalable alternative. It was subsequently shaped into mechanically stable extrudates using methyl cellulose as a binder. Adsorption experiments were carried out using a dynamic breakthrough setup at 70% relative humidity to simulate realistic operating conditions. A breakthrough time of 1 min was observed for the adsorption process. During the desorption process at 200 °C, a maximum desorption concentration of 39 ppm was recorded after 110 min, and complete desorption was achieved in 165 min. The integration of HF-free synthesis, shape engineering, and high-humidity testing represents a crucial step toward the industrial deployment of MOF-based adsorbents. This work highlights the potential of MIL-101(Cr) extrudates as a next-generation sorbent for sustainable H_2S removal in humid gas purification systems.

Keywords: H_2S adsorption · Shaping of MOFs · MIL MOF Extrudates

1 Introduction

Hydrogen sulphide (H_2S) is a highly toxic, corrosive, and foul-smelling gas encountered in various industrial operations such as petroleum refining, natural gas purification, and wastewater treatment. Its effective removal is crucial to prevent corrosion of equipment,

A. Antony et al. (Eds.): icONMAT 2025, SPPHY 328, pp. 16–27, 2025.
https://doi.org/10.1007/978-981-95-2697-0_3

protect catalysts, meet environmental regulations, and ensure workplace safety [1]. Traditional porous adsorbents, including activated carbon, zeolites, and metal oxides, are widely employed in industry due to their low cost and acceptable performance under dry conditions [2, 3]. For instance, commercial activated carbon typically exhibits H_2S adsorption capacities ranging from 10 to 50 mg/g under dry conditions [4]. However, this capacity drastically diminishes in the presence of moisture, and regeneration efficiency is often low, limiting its long-term applicability [5]. One of the other challenges with traditional materials like activated carbon is the high-temperature process needed to recycle the materials for reuse.

In recent years, metal–organic frameworks (MOFs), a class of crystalline porous materials constructed from metal ions and organic linkers, have emerged as promising alternatives due to their exceptionally high surface areas (often $>3000\ m^2/g$), tunable pore sizes, and surfaces that can be easily functionalized [6]. Among them, MIL-101(Cr) has gained considerable attention for gas adsorption owing to its large pore volume ($\sim 1.9\ cm^3/g$), high surface area ($\sim 4100\ m^2/g$), and robust thermal and chemical stability [7]. Several studies have demonstrated its superior H_2S uptake; for instance, Amvrosios et al. reported a capacity of 96.75 mg/g for modified MIL-101(Cr), significantly higher than that of most activated carbons [8]. Moreover, MOFs exhibit low desorption temperatures, making them attractive for reuse due to the ease of material regeneration.

Despite these advantages, a significant gap persists between laboratory-scale research and industrial application. Most MOF-based H_2S adsorption studies have been conducted under dry or mildly humid conditions, which do not adequately simulate real gas streams that often contain high levels of humidity [9, 10]. Moisture can competitively adsorb on active sites or degrade certain MOFs, thereby reducing adsorption efficiency. For example, the capacity of MIL-101(Cr) can drop by up to 40–60% under humid conditions if not properly optimised [11]. Therefore, it is imperative to evaluate MOF performance under realistic humidity levels, such as 60–80% relative humidity (RH), which are typical in industrial settings.

Another significant barrier to commercialisation lies in the scalability, safety, and environmental hazards of the synthesis process. Many MOFs, including MIL-101(Cr), are traditionally synthesised using hydrofluoric acid (HF), a hazardous reagent that poses serious environmental risks during handling and disposal [12]. Replacing HF with a safer alternative like nitric acid (HNO_3) is essential to facilitate industrially scalable synthesis while maintaining material performance. Hydrofluoric acid acts as a mineralising agent and enhances the crystallinity of the sample during the hydrothermal reaction; however, it is highly corrosive and should be handled with more care than other mineral acids. The use of such corrosive acids in the synthesis process hinders large-scale synthesis and limits the practical applications of MOFs. As the study focuses on industrial H_2S removal, this modification of substituting HF with HNO_3 enhances the safety and scalability of the synthesis procedure.

Therefore, in this work, we focus on bridging the gap between laboratory research and real-world applications by evaluating the adsorption and desorption behaviour of shaped MIL-101(Cr) extrudates towards H_2S, synthesised via an HF-free method, at 70% relative humidity, which closely mimics industrial conditions, especially in tropical regions. The objective is to assess the viability of MIL-101(Cr) as a practical alternative to activated carbon for H_2S capture in moisture-rich environments, and to demonstrate its robust performance, shaping compatibility, and industrial potential.

2 Experimental

2.1 Reagents and Materials

The materials used in this study for synthesis and purification of MIL-101 (Cr) are chromium nitrate nonahydrate (Sigma Aldrich), 1,4-benzene dicarboxylic acid (Sigma Aldrich), N, N-dimethyl formamide (Spectrochem Pvt. Ltd.), Nitric acid, 69% (Merck Life Sciences Pvt. Ltd.), Premium grade ethanol, 100% (Hayman group Ltd.). Methyl cellulose, 2% in water: 3000–5000 mPa.s at 20 °C (Loba Chemie Pvt. Ltd.) and chitosan, 0.5% in 0.5% Acetic Acid: 200–600 mPa.s at 20 °C (Tokyo Chemical Industry Co., Ltd.) were used in the shaping of MIL-101 (Cr) into extrudates. All these chemicals used in this study were of analytical grade without further purification.

2.2 Synthesis and Purification of MIL-101(Cr)

MIL-101(Cr) MOF was synthesised by the hydrothermal method as previously reported in the literature [13], with some modifications. Chromium nitrate nonahydrate ($Cr(NO_3)_3 \cdot 9H_2O$) and terephthalic acid served as precursors. The significant modification was the replacement of hydrofluoric acid (HF) with nitric acid (HNO_3) as the mineralising agent. The detailed synthesis procedure is as follows. Chromium nitrate nonahydrate (10 mmol) and terephthalic acid (10 mmol) were dissolved in 50 mL of deionised water under constant stirring for 1 h to ensure complete dissolution of the precursors. To this solution, 0.1 M HNO_3 was added and stirred for an additional 30 min. Here, nitric acid acts as the modulator, which promotes nucleation and facilitates crystal growth during the hydrothermal process. This solution was transferred into a Teflon-lined autoclave and heated at 220 °C for 4 h. The autoclave was allowed to cool naturally to room temperature after the hydrothermal reaction. The resulting green suspension was then filtered to remove any unreacted terephthalic acid, which appears as needle-like structures floating on the surface of the reaction mixture. The filtrate was centrifuged at 8000 rpm for 10 min to separate the green-coloured powder, which was subsequently washed multiple times with deionised water, DMF, and ethanol to remove impurities and dried at 60 °C overnight in an oven.

2.3 Preparation of MIL-101 (Cr) MOF Extrudates

MOFs are typically synthesised as fine powders; however, this form is not ideal for practical applications. Powders are difficult to handle, prone to spillage, and provide poor heat and mass transfer. They also cause significant pressure drops when packed and lack mechanical strength and space efficiency [14, 15]. To make MOFs more suitable for real-world use, it is essential to transform them into shaped forms with enhanced mechanical durability. This must be achieved without significantly compromising their inherent properties, such as surface area and pore volume. Shaping MOFs is a crucial step toward unlocking their full potential in industrial and commercial applications. One of the key challenges in shaping MOFs lies in identifying a suitable binder and optimising the MOF-to-binder ratio to ensure robust shape retention without significantly compromising surface area. Considering the focus on industrial application, environmentally friendly binders such as chitosan and methyl cellulose were selected for shaping MIL-101(Cr) into extrudates, ensuring sustainability and compatibility with large-scale processing. Initially, we explored chitosan as a potential binder across various ratios. However, the resulting extrudates exhibited poor mechanical stability, noticeably deforming upon drying. In contrast, methyl cellulose proved to be an effective binder, yielding extrudates with excellent shape retention. Notably, the addition of just 2 wt.% methyl cellulose to MIL-101 (Cr) produced mechanically strong extrudates, making them promising candidates for adsorption-based applications.

The shaping process began by mixing pristine MIL-101(Cr) powder with 2 wt.% methyl cellulose, followed by the gradual addition of equal volumes of ethanol and deionised (DI) water to form a homogenous, thick paste suitable for extrusion. The resulting paste was loaded into a syringe and extruded manually to form uniform cylindrical strands of diameter 2.4 mm (Fig. 1). These strands were then cut into smaller segments of appropriate length and subjected to a drying process at 80 °C for 12 h to ensure complete removal of moisture and solvent residues. The dried extrudates exhibited sufficient mechanical integrity for handling, packing, and downstream processing. To ensure safe and contamination-free transport for adsorption studies, the dried extrudates were neatly packed into airtight sample vials. These shaped MIL-101(Cr) samples were subjected to the H_2S adsorption studies under realistic conditions.

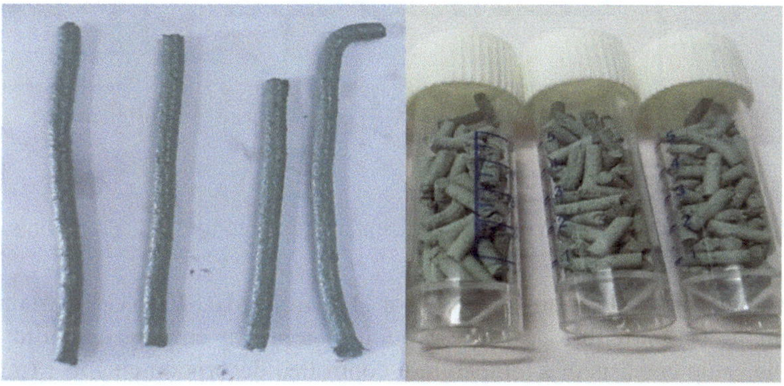

Fig. 1. Freshly extruded MIL-101 (Cr) strands; dried and segmented MIL-101 (Cr) extrudates packed in vials for the testing of H_2S adsorption capacity.

2.4 Experimental Setup for Dynamic H_2S Adsorption and Desorption Studies

The experimental setup designed to evaluate the adsorption behaviour of H_2S on MIL-101 (Cr) extrudates (Fig. 2) consists of a calibrated gas mixing and delivery arrangement integrated with an H_2S sensor to monitor the gas concentration during the adsorption process. Clean, dry air and 5% calibrated H_2S gas are introduced into the system through mass flow controllers (MFCs) to precisely regulate the flow rates. Two separate lines are utilised for passing the clean, dry air; one serves as a dilution stream, and the other is directed through a water bubbling chamber to introduce controlled humidity into the system, simulating real environmental conditions. The gases pass through a sequence of mixing vessels where thorough homogenisation occurs. These vessels ensure uniform distribution of the H_2S-air mixture before it contacts the adsorbent material. The mixed gas stream is subsequently directed through an adsorption tube packed with shaped MIL-101 (Cr) extrudates. This adsorption tube is designed to facilitate efficient interaction between the gas molecules and the porous adsorbent under controlled flow and environmental conditions. Downstream of the adsorption bed, an H_2S sensor continuously monitors the outlet gas concentration, enabling real-time tracking of H_2S breakthrough and the overall adsorption dynamics. This modular setup allows for flexible adjustment of humidity, gas composition, and flow rates, thus supporting systematic investigation of adsorption kinetics, breakthrough behaviour, and regeneration performance under varied operating conditions.

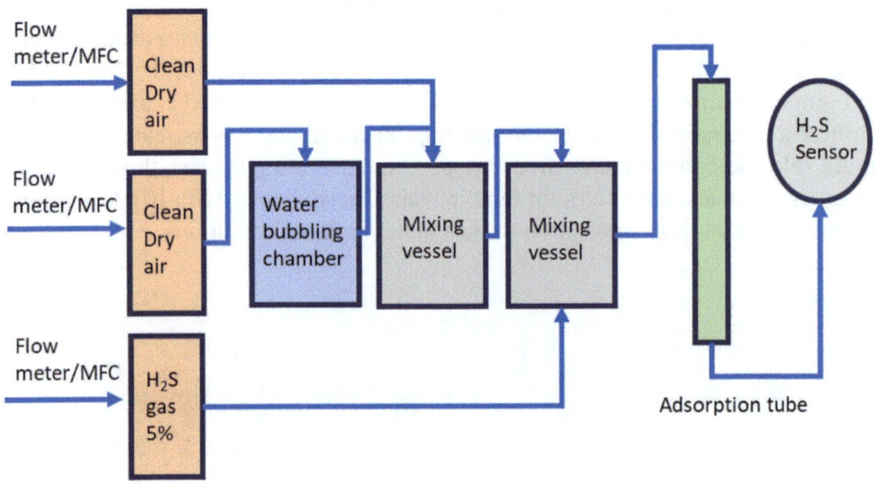

Fig. 2. Schematic illustration of the experimental setup used for H_2S adsorption studies.

To evaluate the desorption behaviour of H_2S from MIL-101 (Cr) extrudates after adsorption, a custom-built thermal desorption system was employed, as illustrated in Fig. 3. The setup consists of a temperature-controlled desorption chamber (muffle furnace), a downstream gas analysis line equipped with an H_2S sensor, and a pump to drive the flow of desorbed gases through the detection system. Following the adsorption experiments, the MIL-101 (Cr) extrudates were transferred to the desorption tube placed within a muffle furnace, which serves as the controlled heating environment. The furnace enables precise temperature programming to facilitate thermal desorption of the adsorbed H_2S. During desorption, the pump generates a steady flow through the system, drawing the evolved gas from the heated desorption tube toward the H_2S sensor. The sensor continuously monitors the concentration of H_2S in the desorbed stream, allowing for real-time evaluation of the desorption profile as a function of temperature and time.

This arrangement enables the study of desorption kinetics, regeneration efficiency, and thermal stability of the MIL-101 (Cr) adsorbent. The temperature ramp and hold protocols can be tailored depending on the thermal activation requirements and desorption conditions under investigation.

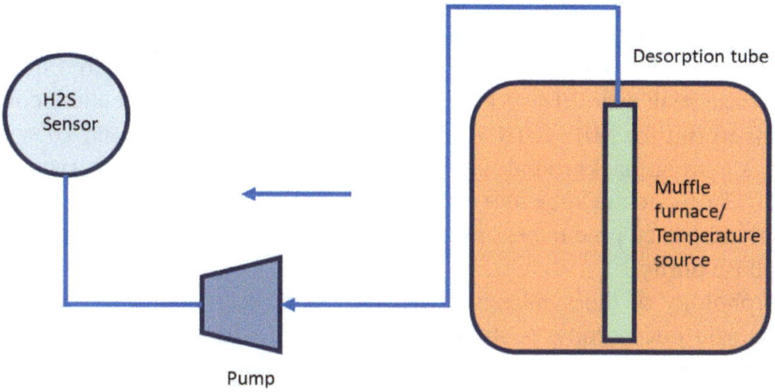

Fig. 3. Schematic illustration of the experimental setup used for H_2S desorption studies to monitor the release of adsorbed H_2S from MIL-101 (Cr) extrudates

2.5 Characterizations

The powder X-ray diffraction (PXRD) patterns of the sample were recorded using a Rigaku Mini Flex 600 Benchtop powder X-ray diffractometer with Cu Kα radiation ($\lambda = 1.54$ Å) in the 2θ range 3–20°. The morphology analysis was done using Carl Zeiss field-emission scanning electron microscope (FE-SEM). The thermal stability of the sample was measured using a Hitachi STA 7300 TGA/DTA analyser at a heating rate of 20 °C per minute under a nitrogen atmosphere, with the temperature ramped from room temperature to 750 °C. The nitrogen adsorption-desorption isotherm of the sample was measured at 77 K and up to a pressure of 101 kPa, using a Micromeritics BET analyser. Before adsorption measurements, the samples were activated by degassing at 200 °C for 4 h and at 90 °C overnight. Degassing is a crucial step before BET analysis to remove moisture, solvents, and other adsorbed impurities from the sample surface and pores. These contaminants can block pore access, leading to inaccurate surface area measurements and distorted isotherms. Proper degassing ensures the sample is clean and that the pores are fully accessible to the adsorbate gas. The specific surface area was determined using the Brunauer-Emmett-Teller (BET) equation, the pore diameter was obtained using the Barrett-Joyner-Halenda (BJH) desorption isotherm, and the pore volume was obtained using the single-point adsorption method. H_2S adsorption experiments were performed in a fixed-bed reactor under controlled conditions of temperature, pressure, and humidity.

3 Results and Discussions

The powder X-ray diffraction pattern of the synthesised MIL-101(Cr) is shown in Fig. 4. (a), which exhibits several sharp and intense peaks in the low-angle region ($2\theta = 3°$-$10°$), characteristic of MIL-101(Cr). Key reflections are observed at approximately 2θ values of 3.2°, 4.5°, 5.2°, 6.0°, 8.5°, and 9.1°, which align well with previously reported diffraction patterns for MIL-101(Cr) [16]. These values are attributed to the crystalline nature of the sample, characterised by large mesoporous cages and a well-ordered framework. The most intense peak near $\mathbf{2\theta \approx 9.1°}$ is consistent with the (531) reflection, commonly used as a fingerprint for MIL-101(Cr) phase identification [17]. Beyond $2\theta \approx 10°$, the pattern shows a broad background with weak and less intense peaks, as typically seen for MIL-101(Cr) due to the large unit cell and porous structure. No additional peaks corresponding to unreacted precursors or secondary phases (such as Cr_2O_3) were observed, indicating phase purity.

The morphology of the synthesised MIL-101 (Cr), recorded using field emission scanning electron microscopy, is shown in Fig. 4. (b), which exhibits a well-defined octahedral geometry, consistent with the typical morphology reported for MIL-101(Cr) in the literature [18]. The particles were relatively uniform in size and shape, indicating a high degree of crystallinity and controlled nucleation during the synthesis process. The observed morphology is in agreement with previous studies that have identified truncated octahedral structures for MIL-101(Cr), composed primarily of [111] facets with [100] truncations [19].

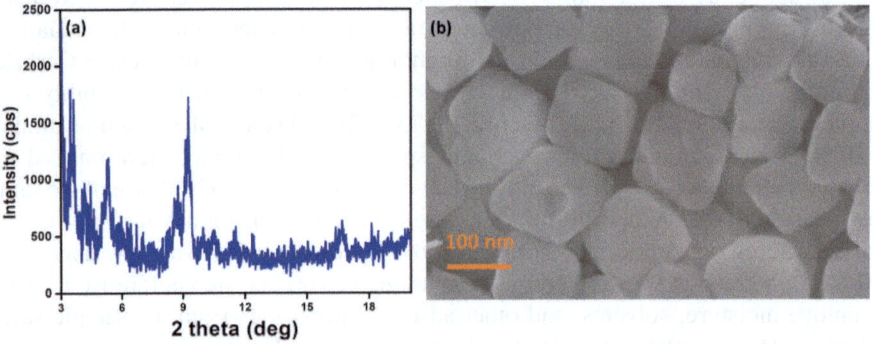

Fig. 4. (a) Powder X-ray diffraction (PXRD) pattern of MIL-101(Cr), confirming its crystalline structure; Panel (b) is an electron microscopy image displaying a cluster of nanoscale particles with well-defined octahedral morphology, with a scale bar indicating 100 nm.

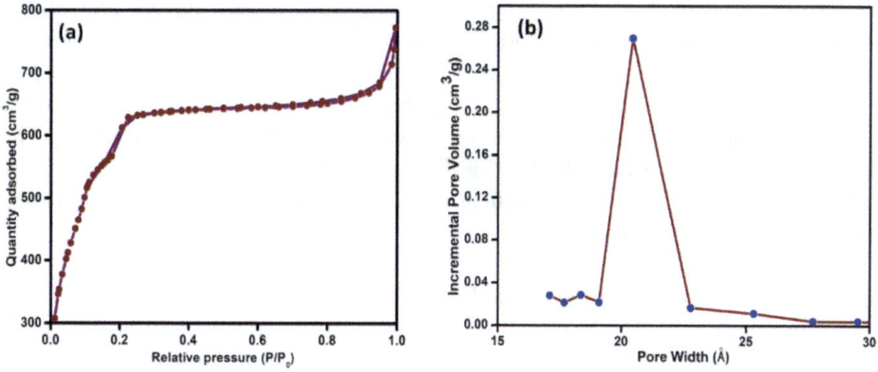

Fig. 5. (a) Nitrogen adsorption–desorption isotherm of MIL-101(Cr) measured at 77 K, indicating a typical type I/IV isotherm characteristic of micro/mesoporous materials; (b) Corresponding pore size distribution (PSD) curve derived from the adsorption branch using the BJH method.

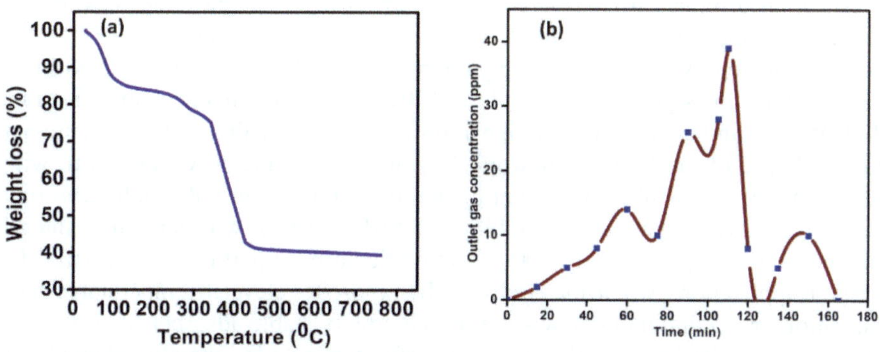

Fig. 6. (a) Thermogravimetric analysis (TGA) curve of MIL-101(Cr), showing ~30% weight loss up to 350 °C attributed to the loss of guest molecules; (b) H_2S desorption profile of MIL-101(Cr) extrudates, illustrating the thermal release behavior of adsorbed gas.

The nitrogen adsorption–desorption isotherm (Fig. 5. (a)) and corresponding pore size distribution analysis confirm the formation of a highly porous MIL-101 (Cr) framework with hierarchical porosity. The isotherm exhibits a combination of Type I and Type IV behavior, characterised by a steep uptake at low relative pressures ($P/P_0 < 0.1$), indicative of micropore filling, followed by a gradual rise up to $P/P_0 \approx 0.4$ and a sharp increase near saturation ($P/P_0 \approx 0.9$–1.0), which suggests the presence of mesopores and interparticle voids. A small hysteresis loop is evident at high relative pressures, suggesting the presence of mesopores or intercrystalline textural porosity. The shape of the loop aligns with an H4-type hysteresis, which is often associated with narrow slit-like pores or cage-like mesoporous materials. The overall isotherm confirms that MIL-101(Cr) possesses a hierarchical porosity, predominantly microporous with some mesoporosity. This dual-pore structure is advantageous for high surface area and pore volume, which enhances gas adsorption capacity. The total adsorbed quantity nearing 750 cm^3/g at $P/P_0 \approx 1.0$ suggests that the sample is of high crystallinity and purity, with minimal pore blockage or framework collapse. In this study, the synthesised MIL-101(Cr) demonstrated a high specific surface area, with a BET surface area of 2132 m^2/g and a Langmuir surface area of 3095 m^2/g. The single-point desorption total pore volume was determined to be 1.06 cm^3/g, indicating the presence of a highly porous structure suitable for gas adsorption applications.

The Barrett-Joyner-Halenda (BJH) pore size distribution (PSD) curve (Fig. 5. (b)) shows a dominant peak centred around 20 Å (2.0 nm), which is consistent with the expected mesoporous cage size of MIL-101(Cr), confirming the well-defined pore architecture of the material. The sharp and narrow pore distribution indicates uniformity in pore structure and minimal framework defects. These results collectively demonstrate the successful synthesis of MIL-101(Cr) with its characteristic microporous-mesoporous nature.

The thermal stability of MIL-101(Cr) was evaluated using thermogravimetric analysis (TGA), and the resulting curve is shown in Fig. 6(a). The TGA profile reveals an initial weight loss of approximately 7–10% below 150 °C, which can be attributed to the desorption of physically adsorbed water and residual solvent molecules from the porous framework. A subsequent moderate weight loss between 150 °C and 330 °C corresponds to the removal of more strongly bound guest molecules or coordinated solvent species. A sharp and significant weight loss (~40%) is observed in the temperature range of 330–450 °C, indicating the decomposition of the organic linker (terephthalic acid) and the collapse of the MOF framework. Beyond 450 °C, the curve levels off, with a residual mass of ~40%, likely representing the formation of thermally stable chromium oxide (Cr_2O_3). The high thermal stability of MIL-101(Cr), with structural integrity maintained up to approximately 300 °C, is particularly important for industrial H_2S adsorption applications, where materials may be exposed to elevated temperatures during regeneration or in continuous flow systems. Thermally stable adsorbents ensure structural robustness, reusability, and consistent performance under operational conditions, making MIL-101(Cr) a promising candidate for such high-demand applications.

The adsorption experiment was conducted at 70% relative humidity (RH) to closely mimic real-world atmospheric conditions, particularly where humidity can significantly impact H_2S capture performance. Breakthrough behaviour was continuously monitored using a calibrated electrochemical H_2S gas sensor placed at the outlet of the column. The breakthrough time, defined as the time at which H_2S was first detected at the outlet, was recorded as 1 min, indicating rapid saturation of the active sites under the specified test conditions. This relatively short breakthrough time could be attributed to the moderate affinity of pristine MIL-101 (Cr) toward H_2S, particularly in the presence of high humidity, which may compete for adsorption sites.

Following adsorption, the desorption behaviour of the material was studied by thermally regenerating the saturated adsorbent. The adsorption tube was transferred to a muffle furnace and heated to 200 °C, enabling the desorption of retained H_2S molecules. The desorbed gas concentration was continuously monitored using the same H_2S gas sensor, placed at the outlet of the furnace setup. The desorption profile revealed a gradual increase in H_2S concentration, indicating a slow and controlled release of the gas from the adsorbent framework. The maximum desorbed concentration observed was 39 ppm at 110 min. After reaching this peak, the concentration began to decline, and the desorption process was considered complete at 165 min. This prolonged desorption period suggests strong binding of a fraction of H_2S molecules within the MOF structure, possibly via interaction with coordinatively unsaturated metal sites or hydrogen bonding with the framework's hydroxyl groups [20].

The shaped MIL-101(Cr) extrudates demonstrated the capability for H_2S adsorption and thermal regeneration. While the breakthrough time was relatively short under humid conditions, the material showed stable desorption behaviour and reusability potential upon moderate heating. The findings emphasise the importance of post-synthetic modifications, such as metal impregnation or composite formation, to enhance the H_2S capture performance of MIL-101(Cr) in real-world applications where humidity plays a significant role.

4 Conclusion

This study presents a realistic evaluation of the H_2S adsorption and desorption performance of pristine MIL-101 (Cr) MOF extrudates synthesised via a safer, HF-free route and tested under 70% relative humidity to mimic industrial gas purification conditions. The material was shaped into mechanically robust extrudates and tested using a dynamic breakthrough setup. While the breakthrough time and adsorption capacity under high humidity were modest compared to modified industrial adsorbents, the performance of MIL-101 (Cr) remains significant considering that the material was evaluated in its unmodified, pristine form.

It is important to note that in industrial applications, activated carbon is rarely used in its pristine form; it is typically impregnated with metal oxides or other additives to enhance performance under humid conditions and improve selectivity toward H_2S [21]. Likewise, the results presented here establish a performance baseline for MIL-101 (Cr), suggesting strong potential for further improvement through similar strategies. Future efforts involving impregnation with active species, surface functionalization, or composite formation with hydrophobic or catalytic components may lead to substantial enhancement in both adsorption capacity and moisture tolerance. Overall, this work demonstrates that pristine MIL-101 (Cr) is a promising candidate for next-generation H_2S capture systems and paves the way for developing tailored MOF-based adsorbents for industrial-scale applications.

References

1. Chan, Y.H., et al.: A state-of-the-art review on capture and separation of hazardous hydrogen sulfide (H_2S): Recent advances, challenges and outlook. Environ. Pollut. **314**, 120219 (2022). https://doi.org/10.1016/j.envpol.2022.120219

2. de Oliveira, L.H., Meneguin, J.G., Pereira, M.V., do Nascimento, J.F., Arroyo, P.A.: Adsorption of hydrogen sulfide, carbon dioxide, methane, and their mixtures on activated carbon. Chem. Eng. Commun. **206**(11), 1533–1553 (2019). https://doi.org/10.1080/00986445.2019.1601627
3. Ozekmekci, M., Salkic, G., Fellah, M.F.: Use of zeolites for the removal of H2S: A mini-review. Fuel Process. Technol. **139**, 49–60 (2015). https://doi.org/10.1016/j.fuproc.2015.08.015
4. Bagreev, A., Bandosz, T.J.: H_2S adsorption/oxidation on unmodified activated carbons: Importance of prehumidification. Carbon. **39**(15), 2303–2311 (2001)
5. Lee, K.M., Lim, Y.H., Park, C.J., Jo, Y.M.: Adsorption of low-level CO_2 using modified zeolites and activated carbon. Ind. Eng. Chem. Res. **51**(3), 1355–1363 (2012). https://doi.org/10.1021/ie2013532
6. A.G. Georgiadis, N.D. Charisiou, I. V. Yentekakis, M.A. Goula, Removal of Hydrogen Sulfide (H_2S) using MOFs: A review of the latest developments Chem. Proc., in: MDPI AG, 2020 2 1 27. https://doi.org/10.3390/eccs2020-07586
7. Alivand, M.S., Shafiei-Alavijeh, M., Tehrani, N.H.M.H., Ghasemy, E., Rashidi, A., Fakhraie, S.: Facile and high-yield synthesis of improved MIL-101(Cr) metal-organic framework with exceptional CO_2 and H_2S uptake; the impact of excess ligand-cluster. Microporous Mesoporous Mater. **279**, 153–164 (2019). https://doi.org/10.1016/j.micromeso.2018.12.033
8. Georgiadis, A.G., Charisiou, N., Yentekakis, I.V., Goula, M.A.: Hydrogen Sulfide (H_2S) removal via MOFs. Materials. **13**(16), 3640 (2020). https://doi.org/10.3390/ma13163640
9. Lincke, J., et al.: A novel copper-based MOF material: Synthesis, characterization and adsorption studies. Microporous Mesoporous Mater. **142**(1), 62–69 (2011). https://doi.org/10.1016/j.micromeso.2010.11.017
10. Darabdhara, J., Ahmaruzzaman, M.: Recent developments in MOF and MOF based composite as potential adsorbents for removal of aqueous environmental contaminants. Chemosphere. **304**, 135261 (2022). https://doi.org/10.1016/j.chemosphere.2022.135261
11. Andirova, D., Lei, Y., Zhao, X., Choi, S.: Functionalization of metal–organic frameworks for enhanced stability under humid Carbon Dioxide capture conditions. ChemSusChem. **8**(20), 3405–3409 (2015). https://doi.org/10.1002/cssc.201500580
12. Zou, M., Dong, M., Zhao, T.: Advances in metal-organic frameworks MIL-101(Cr). Int. J. Mol. Sci. **23**(16), 9396 (2022). https://doi.org/10.3390/ijms23169396
13. Férey, G., et al.: A Chromium Terephthalate-based solid with unusually large pore volumes and surface area. Science. **309**(5743), 2040–2042 (2005). https://doi.org/10.1126/science.1116275
14. Liu, X.-M., Xie, L.-H., Wu, Y.: Recent advances in the shaping of metal–organic frameworks. Inorg. Chem. Front. **7**, 2840–2866 (2020). https://doi.org/10.1039/C9QI01564G
15. Wang, Z., et al.: Shaping of metal-organic frameworks: A review. Energy Fuels. **36**(6), 2927–2944 (2022). https://doi.org/10.1021/acs.energyfuels.1c03426
16. Zorainy, M.Y., Gar Alalm, M., Kaliaguine, S., Boffito, D.C.: Revisiting the MIL-101 metal–organic framework: Design, synthesis, modifications, advances, and recent applications. J Mater. Chem. A Mater. **9**, 22159–22217 (2021). https://doi.org/10.1039/D1TA06238G
17. Tripathi, J., et al.: Metal-organic framework MIL-101(Cr): An efficient catalyst for the synthesis of dihydro-pyrimidinones. Inorganica Chim Acta. **565**, 121989 (2024). https://doi.org/10.1016/j.ica.2024.121989
18. Zhang, L.J., Li, F.Q., Ren, J.X., Ma, L.B., Li, M.Q.: Preparation of metal organic frameworks MIL-101 (Cr) with acetic acid as mineralizer. IOP Conf. Ser. Earth Environ. Sci., Institute of Physics Publishing. **199**(4), 042038 (2018). https://doi.org/10.1088/1755-1315/199/4/042038
19. Lebedev, O.I., Millange, F., Serre, C., Van Tendeloo, G., Férey, G.: First direct imaging of giant pores of the metal-organic framework MIL-101. Chem. Mater. **17**(26), 6525–6527 (2005). https://doi.org/10.1021/cm051870o

20. Peluso, A., Gargiulo, N., Aprea, P., Pepe, F., Caputo, D.: Modeling Hydrogen Sulfide adsorption on Chromium-based MIL-101 metal organic framework. Sci. Adv. Mater. **6**(1), 164–170 (2014). https://doi.org/10.1166/sam.2014.1696
21. Zulkefli, N.N., Masdar, S., Jahim, J., Majlan, E.H.: Overview of H_2S Removal technologies from biogas production. Int. J. Appl. Eng. Res. **11**(20), 10060–10066 (2016) http://www.ripublication.com

Fabrication of Self-Biased Thermally Stable, ZnO/P3HT-Based Hybrid Photodetector

Akash Chaurasiya[1], Jogendra Singh Rana[2], and Abhishek Kumar Singh[1(✉)]

[1] Department of Electrical and Electronics Engineering, Rajiv Gandhi Institute of Petroleum Technology Jais, Amethi, Uttar Pradesh 229304, India
{akashc22ee,aks}@rgipt.ac.in

[2] Department of Electronics and Communication Engineering, Indian Institute of Information Technology, Bhopal, India
jogendra.rana@iiitbhopal.ac.in

Abstract. The influence of ZnO nanostructured morphology, combined with the active layer of P3HT, was analyzed to assess its impact on the enhanced photo response of the ZnO/Poly(3-hexylthiophene-2,5-diyl) (P3HT) hybrid. The unique properties of this hybrid material stem from the synergy between the organic semiconductor P3HT, known for its excellent hole transport and light absorption, and ZnO nanorods, recognized for their high electron mobility and large surface area. This study examines the synthesis, structural characterization, and electrical and optoelectronic properties of P3HT/ZnO nanorod composites. The I-V characteristics demonstrated a significant improvement in current under light illumination. The external quantum efficiency (EQE) of the P3HT/ZnO nanorod was measured as 21% at 375 nm, while its photoresponsivity was recorded as 65 mA/W, showing a gradual decline over time. Additionally, thermal stability tests conducted at 30 °C, 50 °C, and 70 °C indicated consistent photoresponsivity, highlighting the material's stability under varying temperatures.

Keywords: Nanorod · Nanostructure · Photoresponsivity · Photodetector

1 Introduction

Developing efficient optoelectronic devices, such as solar cells, photodetectors, and light-emitting diodes (LEDs), is essential for advancing renewable energy technologies and improving modern electronic systems. Organic-inorganic hybrid materials have gained significant attention due to their potential to combine the advantageous properties of both material classes, offering enhanced performance and new functionalities [1, 2]. Among these hybrid materials, the combination of poly(3-hexylthiophene) (P3HT), an organic semiconductor, with zinc oxide (ZnO) nanorods, an Inorganic semiconductor, has emerged as an up-and-coming candidate for optoelectronic applications. P3HT is a well-known conjugated polymer that exhibits strong absorption in the visible spectrum, high hole mobility, and good processability, making it an ideal material for organic photovoltaic (OPV) devices [3]. However, the efficiency of devices based solely on P3HT

A. Antony et al. (Eds.): icONMAT 2025, SPPHY 328, pp. 28–32, 2025.
https://doi.org/10.1007/978-981-95-2697-0_4

is often limited by issues such as low exciton dissociation efficiency and poor charge transport. To overcome these challenges, inorganic semiconductors like ZnO have been integrated with P3HT to form hybrid systems that can facilitate improved charge separation and transport [4, 5]. ZnO is a wide-band gap semiconductor with excellent electron mobility and a large surface area when fabricated in the form of nanorods. The unique one-dimensional structure of ZnO nanorods provides a direct pathway for electron transport, which can significantly enhance the charge separation efficiency when combined with P3HT. Moreover, the high aspect ratio of ZnO nanorods increases the interfacial area between P3HT and ZnO, further promoting exciton dissociation and charge collection [6]. The P3HT/ZnO nanorod hybrid structure presents a promising strategy to enhance the performance of optoelectronic devices by leveraging the complementary properties of the organic and inorganic components. Previous studies have demonstrated that P3HT/ZnO nanorod hybrids can achieve higher power conversion efficiencies in solar cells than their components. Additionally, the tunable properties of ZnO nanorods, such as their length, diameter, and density, offer further opportunities to optimize device performance [7, 8].

This research investigated the synthesis, characterization, and optoelectronic performance of P3HT/ZnO nanorod composites. The study aims to understand the underlying mechanisms of charge transport and recombination in these hybrid materials and to explore by optimizing the interface and morphology of the P3HT/ZnO nanorod hybrids. We seek to develop a more efficient material system for future applications in solar energy conversion and other optoelectronic technologies [9, 10].

2 Experimental Details

2.1 Material Synthesis and Device Fabrication

All chemicals and materials utilized to synthesize P3HT/ZnO nanorods were sourced from Merck Millipore and Sigma-Aldrich. No additional purification was performed as the chemicals were high purity (99.99%). In the process of device fabrication, ITO-coated glass samples (1 × 1 cm each) were cleansed using a wet cleaning process, sequentially involving acetone, soap solution, and isopropanol, followed by rinsing with deionized (DI) water at the end of each step. The hydrothermal method was used to synthesize the ZnO seed layer for ZnO nanorod (NR). A 0.1 M solution of zinc acetate dihydrate in 2-methoxyethanol was stirred at 60 °C for 120 min. 0.5 mL of diethanolamine (DEA) was added dropwise to stabilize the solution until the mixture turned clear and transparent. The solution was kept at room temperature for 24 h to ensure nanocluster stability. This aged solution was spin-coated onto cleaned ITO substrates at 1500 rpm for 30 s, then dried at 150 °C for 30 min. For nanorod growth, the coated sample was submerged in an equimolar (0.01 M) solution of zinc nitrate dihydrate and hexamethylenetetramine inside an autoclave chamber at 95 °C for 3 h. Afterward, the sample was thoroughly rinsed with DI water and annealed at 400 °C for 45 min to form ZnO NRs.P3HT (3 mg/mL) was disintegrated in chloroform to deposit the polymer layer and blended at 60 °C for 6 h. A 20-μL aliquot of this solution was spin-coated onto the ZnO NRs at 2000 rpm for 45 s, then dried at 110 °C for 10 min. Finally, Al electrodes (100 nm thick) in an interdigitated pattern were deposited using a shadow mask technique and thermal evaporation system

(Vacuum Coating Unit Model 12A4-D, HIND Hivac system). The electrodes featured a finger length of 10 mm and a spacing of 0.5 mm. The summarized fabrication steps outline the procedure for the proposed device structure (Fig. 1).

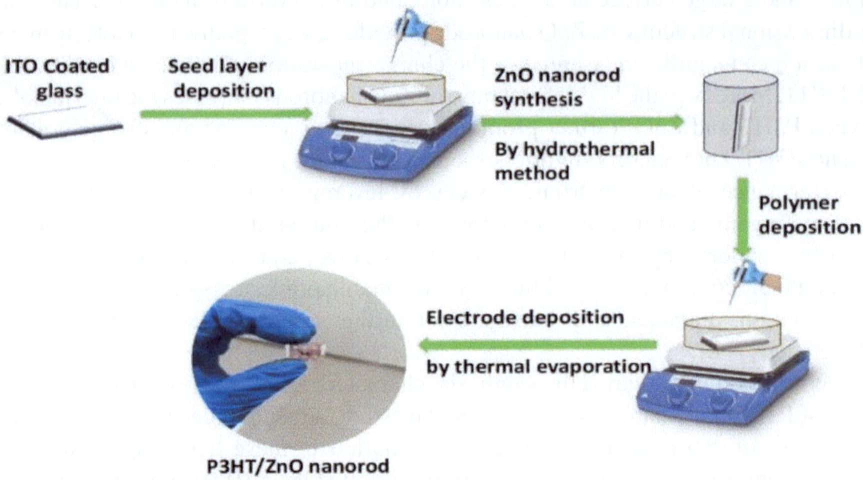

Fig. 1. Fabrication technique of P3HT/ZnO nanorod-based photodetector

3 Results and Discussion

A photodetector is a device that converts incident photons into an electrical current. When light is absorbed in the active layer of the photodetector, excitons (electron-hole pairs) are generated. These excitons subsequently dissociate into free carriers, resulting in the flow of electric current, referred to as photocurrent.

3.1 Current-Voltage (I-V) Characteristics of Fabricated Devices

The current-voltage (I-V) relationship of the P3HT/ZnO nanorod photodetector, under both dark and illuminated conditions, is presented in Fig. 2a. Measurements were conducted using the KEYSIGHT B2902B Precision Source/Measure Unit. The I-V analysis revealed a significant enhancement in performance under light illumination. In the dark, the measured current is 10.2 μA. At the same time, light illumination increased the current to 10 mA at -1 V. This pronounced improvement confirms the photosensitive behavior of the P3HT/ZnO nanorod thin film, demonstrating its potential for optoelectronic applications.

3.2 Variation of Photoresponsivity of P3HT/ZnO Nanorod with Time Intervals of 48 h

Photoresponsivity and external quantum efficiency (EQE) were calculated to evaluate the fabricated devices' performance in self-bias mode. Photoresponsivity (R) measures the electrical output per unit of optical input in a photodetector and is determined using the formula $R = (I_{ph} - I_d)/P$, where I_{ph} represents the photocurrent, I_d is the current under dark, and P is the power of incident light. The initial photoresponsivity of the synthesized nanorod was measured at 65 mA/W. Over time, a decline in photoresponsivity was observed at intervals of 2, 4, 6, 8, 10, and 12 days, with recorded values of 48 mA/W, 47 mA/W, 46 mA/W, 44 mA/W, and 41 mA/W, respectively. This gradual decrease in performance is attributed to the degradation of the polymer (P3HT) content in the P3HT/ZnO nanocomposite over time.

3.3 External Quantum Efficiency (EQE) of P3HT/ZnO Nanorod

The external quantum efficiency (EQE) is the ratio of electron-hole pairs contributing to photocurrent generation to the total number of e-h pairs formed by the absorption of incident light. It quantifies the efficiency of charge carrier collection relative to the absorbed photons. For the P3HT/ZnO nanorod photodetector, the EQE was measured at 21%.

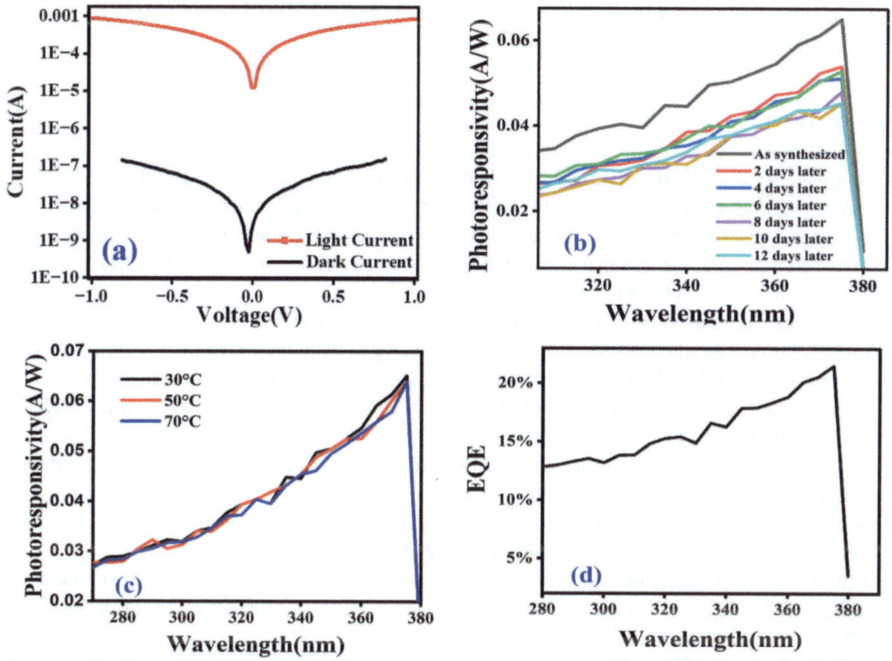

Fig. 2. (a) Current- Voltage (I-V) characteristics, (b) Photoresponsivity Vs wavelength of P3HT/ZnO nanorod, (c) Photoresponsivity Vs wavelength at 30 °C, 50 °C & 70 °C, (d) External quantum efficiency (EQE)

3.4 Thermal Stability of P3HT/ZnO Nanorod

The thermal stability of the P3HT/ZnO nanorod was evaluated by measuring its photoresponsivity at three different temperatures: 30 °C, 50 °C, and 70 °C. It was observed that the photoresponsivity remained nearly constant across this temperature range, indicating good thermal stability.

4 Conclusion

The present study fabricated completely solution-processable ZnO/P3HT hybrid self-bias, thermally stable photodetectors using Aluminum as the top electrode. This study examines the synthesis, structural characterization, and electrical and optoelectronic properties of P3HT/ZnO nanorod composites. The fabricated device is tested under dark and light conditions. The I-V characteristics significantly improved current under light illumination, confirming the sensing nature of the as-grown P3HT/ZnO nanorod composite. The external quantum efficiency (EQE) of the P3HT/ZnO nanorod was measured as 21% at 375 nm, indicating 21% electron-hole (e-h) pairs contributing to photocurrent generation to the total number of e-h pairs formed by the absorption of incident light. At the same time, its photoresponsivity was recorded as 65 mA/W, showing electrical output per unit of optical input. It decreases as time passes due to the degradation of polymer content in P3HT/ZnO nanorod. Additionally, thermal stability tests conducted at 30 °C, 50 °C, and 70 °C indicated consistent photoresponsivity, highlighting the material's stability under varying temperatures.

References

1. Shalu, C., et al.: Role of solvent used to cast P3HT thin films on the performance of ZnO/P3HT hybrid photodetector. Phys. E: Low-Dimens. Syst. Nanostruct. **115**, 113694 (2020)
2. Bilgaiyan, A., et al.: Improved photo-response of hybrid ZnO/P3HT bilayer photodetector obtained through oriented growth of ZnO nanorod arrays and the use of hole injection layer. J. Electron. Mater. **44**, 2842–2848 (2015)
3. Bilgaiyan, A., et al.: Performance improvement of ZnO/P3HT hybrid UV photo-detector by interfacial Au nanolayer. Phys. E: Low-Dimens. Syst. Nanostruct. **86**, 136–141 (2017)
4. Weng, S., Zhao, M., Jiang, D.: Use of organic polymer P3HT: PC61BM as the active layer to improve the performance of ZnO ultraviolet photodetector. J. Phys. Chem. C. **125**(37), 20639–20649 (2021)
5. Xu, S., Wang, Z.L.: One-dimensional ZnO nanostructures: solution growth and functional properties. Nano Res. **4**, 1013–1098 (2011)
6. Miikkulainen, V., et al.: Crystallinity of inorganic films grown by atomic layer deposition: Overview and general trends. J. Appl. Phys. **113**(2), 021301 (2013)
7. Liu, B., Aydil, E.S.: Growth of oriented single-crystalline rutile TiO2 nanorods on transparent conducting substrates for dye-sensitized solar cells. J. Am. Chem. Soc. **131**(11), 3985–3990 (2009)
8. Fei, X., Jiang, D., Zhao, M.: Broadening of the response spectrum and gain in the optoelectronic performance of P3HT: PC61BM/ZnO NWs "Embedded" heterojunction photodetector. Cryst. Growth Des. **23**(3), 1559–1566 (2023)
9. Lin, H., et al.: Engineering of interfacial electron transfer from donor–acceptor type organic semiconductor to ZnO nanorod for visible-light detection. J. Phys. Chem. C. **116**(41), 21657–21663 (2012)
10. Lahane, T.K., Agrawal, J., Singh, V.: Fabrication of high-performance double-sided self-biased ZnO/P3HT based hybrid photodetector. J. Mater. Sci. Mater. Electron. **35**(2), 155 (2024)

Electrochemical Performance of Polyindole in Different Electrolytes

Alvina Augusthy[1], Vishnu Binu[2], and Jayalatha Gopalakrishnan[1,2](✉)

[1] Department of Polymer Science and Rubber Technology, CUSAT, Kochi 22, India
{alvinaaugusthy,jayalatha}@cusat.ac.in
[2] Inter University Centre for Nanomaterials and Devices, CUSAT, Kochi 22, India

Abstract. Polyindole (PI) has drawn considerable interest in supercapacitor applications that show great promise in portable and wearable devices. Herein, Polyindole was synthesized using a chemical oxidative polymerization approach using two approaches, with and without dopant p-TSA. The polyindole was synthesized using the chemical oxidative polymerization method and systematically investigated using characterization studies, including FTIR, XRD, and FESEM. Further, the electrochemical capacitive behavior was examined using various electrolytes, KCl, KOH, and H_2SO_4, at 0.5 M concentration. The cyclic voltammetry studies of the samples were carried out at a scan rate of 10 mV/s-100 mV/s, and a potential window of 0 to 0.6 V. Polyindole exhibits unique redox properties, electrochemical reactivity, and better current response in 0.5 M H_2SO_4. So, the electrochemical performance of Polyindole with higher mass loading was carried out using carbon cloth as a working electrode in 1 M H_2SO_4. It exhibits a maximum discharge time and a specific capacitance of 71 F/g at 1A/g current density

Keywords: Polyindole · Supercapacitor performance · electrolyte

1 Introduction

Recent years have seen a significant increase in scientific interest in the development of ultracapacitors and supercapacitors to address the current energy storage problem. They are still governed by the same fundamental equations as conventional capacitors, but they utilize electrodes with a larger surface area and thinner dielectrics to achieve higher capacitances. The amount of energy produced by the synthesised material relies on a variety of factors that determine whether the material is suitable for use as an efficient energy source [1].

Research has focused a lot of attention on conducting polymers, polyacetylene (PA), polyaniline (PANI), polypyrrole (PPy), polyindole (PIN), poly(3,4-ethylenedioxythiophene) (PEDOT), polyfuran (PF), and other polythiophene (PTH) derivatives because of their large π-conjugated structures [2]. Conducting polymers (CPs) have become a potential electrode material for supercapacitor applications because of

A. Antony et al. (Eds.): icONMAT 2025, SPPHY 328, pp. 33–39, 2025.
https://doi.org/10.1007/978-981-95-2697-0_5

their high pseudocapacitance, remarkable electrical conductivity, and ease of manufacturing. Many energy storage applications can benefit from these materials. Polyindole (PIn), another CP, has been investigated and developed in recent years as a potential option for electrical energy storage [3]. In contrast to the other CPs, PIn has been investigated the least for energy storage [4]. This is likely because of its poor electrical and electrochemical conductivity, which is two times lower than that of PPy and PANI [5]. Despite this, PIn is a fairly important material for energy storage research due to its characteristics, which include strong redox activity [6], adjustable electrical conductivity, notable thermal stability, a moderate rate of deterioration [7], its unique structure resembling polyaniline [8] and polypyrrole, and economic feasibility [9, 10]. Moreover, polypyrrole and polyaniline are less thermally stable than polyindole. Whereas PANI degrades at about 300 °C, PPy typically degrades at temperatures above 150–200 °C. They are prone to oxidative deterioration and gradual conductivity loss, particularly in air or at elevated temperatures. Additionally, highly acidic or basic conditions might affect PANI. Their limited solubility in typical organic solvents limits their use in solution-based manufacturing techniques and makes processing them challenging [11]. To address the issues raised, make PIn technically possible, and improve its functionality, several changes are being made to the current polymer. The compounding of CP with specific additional materials (as a matrix or as a filler) might bring forth these changes. For a number of reasons, choosing an appropriate electrolyte for polyindole electrochemical analysis is essential. The performance, stability, and efficiency of polyindole during redox reactions, charge-discharge cycles, and other electrochemical processes are significantly influenced by the electrolyte. Herein, polyindole was synthesized using different synthesis approaches, and its electrochemical performance was examined in various electrolytes

2 Experimental

2.1 Materials

From SPECTROCHEM, the following chemicals were purchased: Indole monomer, Ammonium peroxodisulfate [$(NH_4)_2S_2O_8$], and p-Toluenesulfonic acid (p-TSA). Poly (Vinylidene Fluoride) (PVDF) and N-Methyl-2-pyrrolidone (NMP) were purchased from Sigma-Aldrich. KOH, KCl, and H_2SO_4 were purchased from EMPLURA. Super-P was purchased from thermoscientific. All reagents were used as received.

2.2 Synthesis of Polyindole

Polyindole was synthesized using the chemical oxidative polymerization method. In a typical synthesis, 1 g of indole monomer was dissolved in a 10 ml ethanol-water system (1:1) with nitrogen purging. Then aqueous solution of 30 ml APS(2 M) was added dropwise to the above solution to start the polymerization reaction with constant stirring. The reaction was stirred for 9 h at 15 °C. The resulting polymer was washed with IM,50 ml p-TSA solution, and dried at 60 °C for 12 h and abbreviated as PIn(P). Another sample was also synthesized using the same procedure except p-TSA washing for comparison and abbreviated as PIn [12].

2.3 Material Characterization

The phase purity and crystal structure of the samples were analyzed using a High-Resolution X-ray diffractometer (XRD; Bruker D8 Advance) supplied with Cu-Kα as radiation source ($\lambda = 1.5405$ Å) with a step width of 0.05°. Fourier Transform Infrared Spectra (FTIR) of samples were run using a PerkinElmer Spectrum Two spectrometer. The surface morphologies of the samples were characterized using Field emission scanning electron microscopy (FESEM Carl Zeiss Supra 40VP).

2.4 Electrochemical Measurements

The electrochemical properties were investigated using a Metrohm Autolab electrochemical workstation through a three-electrode system. Carbon Cloth, Ag/AgCl, and Pt-mesh acted as working electrodes, reference electrodes, and counter electrodes. The working electrode (0.5×0.5 cm^2) was modified by coating it with a homogeneous slurry of 80 wt% as-prepared material, 10 wt% conductive carbon, and 10 wt% polyvinylidene fluoride (PVDF) as the binder in N-methyl pyrrolidone (NMP) solvent and dried at 60 °C for 12 h.

2.5 Calculations

The specific capacitance of the working electrode materials was calculated using eqs. (1) or (2) from the GCD curve

$$C_m = \frac{I \times \Delta t}{m \times \Delta V} \tag{1}$$

Where C_m (Fg^{-1}) represents the mass-specific capacitance value; Cs (Fg^{-1}) represents the area-specific capacitance value; I(A) is the discharge current; Δt(s) is the discharge time; Where ΔV(V) is the potential window, m(g) mass of the active material of the electrodes; and S(cm^2) is the area of the active materials [13].

3 Results and Discussion

3.1 Structure and Morphological Analysis

The FT-IR spectrum of the prepared polyindole samples is shown in Fig. 1(a). The stretching bands near 3441 cm^{-1} correspond to the N-H stretching vibration of polyindole [14]. The bands at 2919 cm^{-1} and 2911 indicate the C-H aromatic stretching vibrations [15]. The peak observed for PIn at 1724 cm^{-1} implies the N-C=O due to the oxidation of polyindole. The hydrolytic processes during oxidation are most likely linked to these alterations [16]. The bands at 1614 and 1609 cm^{-1} show the C=C stretching. The peaks at 1575 and 1566 cm^{-1} indicate the N-H deformation of pyrrole ring [17] and peak at 1443 cm^{-1} implies C-N stretching, 1372 indicates C=N stretching [5], the band at 1116 and 1195 cm^{-1} corresponds to the C-N stretching vibration, the peak at 1032 corresponds to the decoupling of sulphonate, 817 cm^{-1} indicate S-O vibration and peak at 740 cm^{-1} implies C-H out plane vibration [5] and 685 cm^{-1} correspond to C-S coupling of the benzenoid ring. The XRD pattern demonstrates the crystal structure of the prepared samples. The broad peaks that appeared at 8.1^0, 9.6^0, and 18.9^0 correspond to the amorphous nature of polyindole. The peaks observed at 26.6^0 verify the partial crystalline nature of polyindole. FESEM images of PIn and PIn(P) were obtained to investigate the surface morphology, as shown in Fig. 1(c-d). The PIn shows an irregular morphology, and PIn(P) possesses spherical morphology with a size ranging from 0.4 to 0.8 μm, with more uniformity than PIn [5].

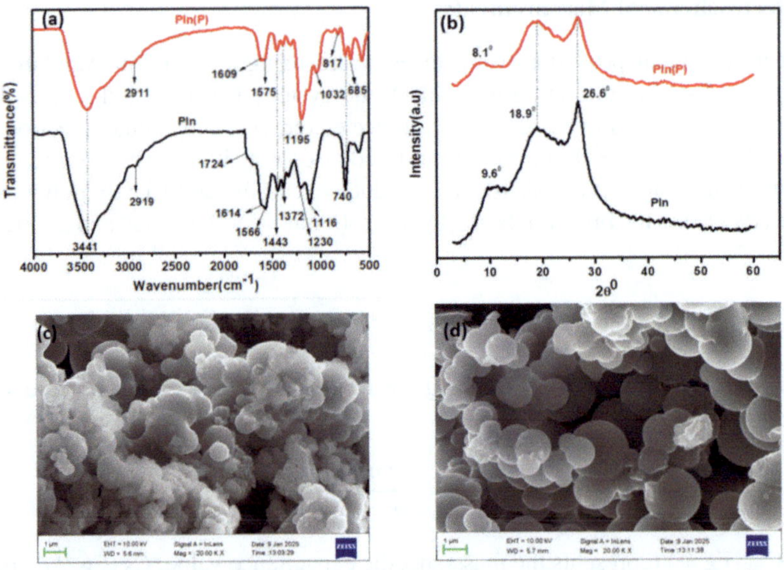

Fig. 1. (a) FTIR, (b) XRD and (c-d) FESEM analysis of PIn and PIn (P)

3.2 Electrochemical Measurements of Polyindole

The electrochemical performances, including CV, GCD, and Electrochemical Impedance Spectroscopy (EIS) of PIn and PIn(P), were investigated in a three-electrode system. Initially, the cyclic voltammetry studies of as-prepared samples were analyzed in various electrolytes, KCl, KOH, and H_2SO_4, with 0.5 M concentration at a scan rate of 100 mVs^{-1}. As shown in Fig: 2(a), the P CV curves exhibit a quasi-rectangular at 100 mVs^{-1} and a potential window from 0 to 0.6 V [18]. It is noteworthy that polyindole synthesized using p-TSA as dopant shows a better cure area and current responses compared to other polyindole synthesized without dopant, implying a higher electrochemical reactivity and capacitance. Moreover, the CV curves of PIn(P) evaluated in 0.5 M H_2SO_4 show better electrochemical performance than other electrolytes. The cyclic voltammetry study of PI(P) was performed in 0.5 M H_2SO_4 at varying scan rates from 1 mVs^{-1} to 100 mVs^{-1}. The quasi-rectangular shape was retained as the scan rate increased to 100 mVs^{-1}. Notably, these plots' symmetrical and triangular forms demonstrate strong electrochemical reversibility and remarkable capacitive behavior. The cyclic voltammetry and charge-discharge performance of PIn(P) were tested in 1 M H_2SO_4, and here, the CV curves show an EDLC charge storage mechanism. The charge-discharge performance of PIn(P) was evaluated at various current densities 1A/g,2A/g,3A/g,4A/g, and 5A/g, and it exhibited a specific capacitance of 71 F/g,63F/g,32F/g,26F/g, and 23F/g, respectively [19]. Ions have more time to diffuse into the active material's inner surface when the current density is lower, which enhances the active material's use of both its inner and outer surfaces. The specific capacitance is only influenced by the active material's outer surface at higher current densities [20]. The higher concentration of sulphuric acid results in a greater concentration of H^+ ions in the solution, which significantly enhances the electrical conductivity of the electrolyte. Higher ionic conductivity in the electrolyte allows for more efficient charge transport during the electrochemical process, improving the overall electrochemical performance of PIn(P).EIS measurement was performed on the samples to study the ions and electron transport dynamics in the frequency range of 100 kHz to 0.1 Hz, as shown in Fig. 2(f), and it shows charge transfer resistance (R_{ct}) of 20 Ω and equivalent series resistance of 3 Ω. Warburg impedance in the low-frequency region (Z_W) was close to 45 ° for PIn(P) and the longest Warburg line, revealing its smaller Z_W and faster diffusion kinetics [21].

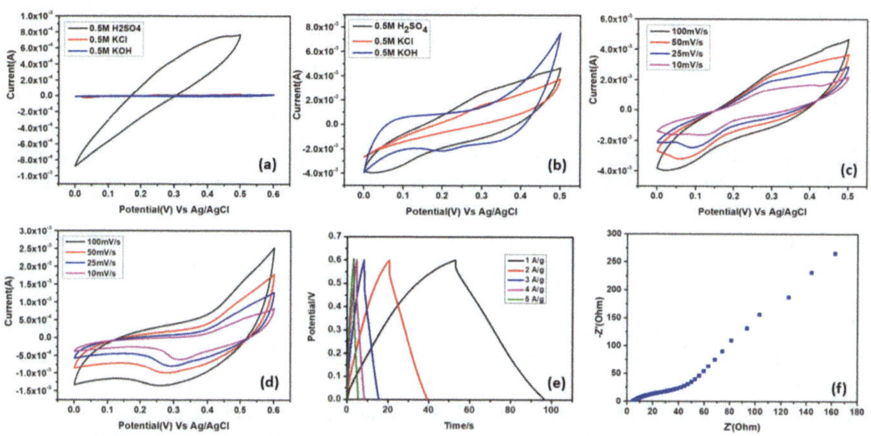

Fig. 2. **(a)** CV curve of PIn in various electrolytes at 100 mVs^{-1}, **(b)** CV curve of PIn(P) in various electrolytes at 100 mV/s **(c)** PIn(P) in 0.5 M H_2SO_4 at different scan rates and **(d)** CV curve of PIn(P) at varying scan rates in 1 M H_2SO_4 **(e)** GCD curve of PIn(P) at 1A/g, **(e)** Nyquist diagrams PIn(P) in 1 M H_2SO_4

4 Conclusion

Polyindole was successfully synthesized using the oxidation polymerization method. The electrochemical performance of as-prepared samples was studied in various electrolytes, and the sample synthesized with a dopant exhibits better current response and specific capacitance of 71 F/g at 1A/g in 1 M H_2SO_4. This strategy of polymer synthesis can be extended to other polymers and can also be explored for various electronic applications. Moreover, the composites of polyindole with other materials are expected to have wide applications in high-performance supercapacitors and other energy storage devices.

References

1. Ramya, R., Sivasubramanian, R., Sangaranarayanan, M.V.: Conducting polymers-based electrochemical supercapacitors—Progress and prospects. Electrochim. Acta. **101**, 109–129 (2013)
2. Zhao, F., Shi, Y., Pan, L., Yu, G.: Multifunctional nanostructured conductive polymer gels: synthesis, properties, and applications. Acc. Chem. Res. **50**(7), 1734–1743 (2017)
3. Mudila, H., Prasher, P., Kumar, M., Kumar, A., Zaidi, M., Kumar, A.: Critical analysis of polyindole and its composites in supercapacitor application. Mater. Renew. Sustain. Energy. **8**(9), 1–19 (2019)
4. Phasuksom, K., Sirivat, A.: Synthesis of nano-sized polyindole via emulsion polymerization and doping. Synth. Met. **219**, 142–153 (2016)
5. Mackintosh, J.G., Redpath, C.R., Jones, A.C., Langridge-Smith, P.R., Mount, A.R.: The electro polymerization and characterization of 5-cyanoindole. J. Electroanal. Chem. **388**(1–2), 179–185 (1995)
6. Billaud, D., Maarouf, E.B., Hannecart, E.: Chemical oxidation and polymerization of indole. Synthetic Metals. **69**(1–3), 571–572 (1995)
7. Dhanalakshmi, K., Saraswathi, R.: Electrochemical preparation and characterization of conducting copolymers: poly (pyrrole-co-indole). J. Mater. Sci. **36**, 4107–4115 (2001)
8. Zhijiang, C., Ruihan, Z., Xingjuan, S.: Preparation and characterization of polyindole nanofibers by electrospinning method. Synth. Met. **162**(23), 2069–2074 (2012)
9. Mudila, H., Prasher, P., Kumar, M., Kumar, A., Zaidi, M.G.H., Kumar, A.: Critical analysis of polyindole and its composites in supercapacitor application. Mater. Renew. Sustain. Energy. **8**(2), 9 (2019)
10. Cai, Z.J., Zhang, Q., Song, X.Y.: Improved electrochemical performance of polyindole/carbon nanotubes composite as electrode material for supercapacitors. Electron. Mater. Lett. **12**, 830–840 (2016)
11. Zhou, Q., Zhu, D., Ma, X., Xu, J., Zhou, W., Zhao, F.: High-performance capacitive behavior of layered reduced graphene oxide and polyindole nanocomposite materials. RSC Adv. **6**(35), 29840–29847 (2016)
12. De, S., Maity, C.K., Sahoo, S., Nayak, G.C.: Polyindole booster for Ti3C2T x MXene-based symmetric and asymmetric supercapacitor devices. ACS Appl. Energy Mater. **4**(4), 3712–3723 (2021)
13. Tang, X., Li, H., Liu, Z., Yang, Z., Wang, Z.: Preparation and capacitive property of manganese oxide nanobelt bundles with birnes site-type structure. J. Power Sources. **196**(2), 855–859 (2011)
14. Majumder, M., Choudhary, R.B., Koiry, S.P., Thakur, A.K., Kumar, U.: Gravimetric and volumetric capacitive performance of polyindole/carbon black/MoS2 hybrid electrode material for supercapacitor applications. Electrochim. Acta. **248**, 98–111 (2017)
15. Erdönmez, S., Karabul, Y., Kılıç, M., Güven Özdemir, Z., Esmer, K.: Structural characterization and dielectric parameters of polyindole/WO3 nanocomposites. Polym. Compos. **42**(3), 1347–1355 (2021)
16. Plachý, T., et al.: Electrorheology of polyindole. Polymer. **217**, 123448 (2021)
17. Konwar, G., Deka, S., Mahanta, D.: Template-free one-step synthesis of polyindole microspheres for binder-less electrochemical capacitors. J Energy Storage. **62**, 106847 (2023)
18. Zhu, J., Xu, Y., Wang, J., Lin, J., Sun, X., Mao, S.: The effect of various electrolyte cations on electrochemical performance of polypyrrole/RGO-based supercapacitors. Phys. Chem. Chem. Phys. **17**(43), 28666–28673 (2015)
19. Zaghloul, M.M.Y., Ebrahim, S., Anas, M., Soliman, M., El Nady, J.: Synthesis and characterization of nanocomposites of polyaniline and polyindole with multiwalled carbon nanotubes for high-performance supercapacitor electrodes. Electrochim. Acta. **475**, 143631 (2024)

20. Thadathil, A., Edavan Chathoth, N., Ismail, Y.A., Anjukandi, P., Periyat, P.: Three-dimensional mesoporous polyindole architectures for supercapacitor applications. J. Phys. Chem. C. **126**(40), 16965–16982 (2022)
21. Wang, W., Ren, G., Wang, M., Liu, Y., Wu, S., Shen, J.: A novel composite for energy storage devices: core–shell MnO 2/polyindole nanotubes supported on reduced graphene oxides. J. Mater. Sci. Mater. Electron. **29**, 5548–5560 (2018)

Enhanced Electrochemical Performance of Hydrothermally Synthesized SnO_2 Anodes for Li-ion Batteries Using PVP with Varied Molecular Weights

Neethu Simon, S. Jayalekshmi, and Vineeth Mohanan Parakkat(✉)

Division for Research in Advanced Materials Laboratory (DREAM), Department of Physics, Cochin University of Science and Technology, Cochin, Kerala, India 682022
vineethmp@cusat.ac.in

Abstract. Tin dioxide (SnO_2), an n-type semiconductor, is widely studied due to its applicability in solar cells, sensors, and energy storage devices. In particular, SnO_2 has garnered considerable interest as an anode material for Li-ion batteries due to its high theoretical capacity of 1494 mAh/g and affordability. Lithiation in SnO_2-based anodes entails conversion and alloying reactions. SnO_2 first irreversibly reacts with lithium ions to form Sn and Li_2O, resulting in substantial capacity loss. Thereafter, Sn reacts with Li through reversible alloying, which leads to large volume expansion, pulverization of the particles, and steady growth of the solid electrolyte interface. These issues collectively contribute to lower capacity retention and restrict cycling stability. Reducing the crystallite size is a suitable method of overcoming these limitations as it increases electronic conductivity, reduces lithium-ion diffusion length, and better handles volume variations. In this work, two various morphologies of SnO_2 samples, i.e., SnO_2-Faceted and SnO_2-Spherical particles, were synthesized. This has been done by hydrothermal synthesis through polyvinylpyrrolidone (PVP) of different molecular weights, leading to different crystallite sizes and morphologies. These structural differences improve electrochemical performance by increasing the accessible active surface area and reducing the lithium-ion diffusion distance, resulting in increased specific capacity and improved cycling stability.

Keywords: Li-ion batteries · SnO_2 · Crystallite size

1 Introduction

Lithium-ion batteries (LIBs) are well known for their efficiency as energy storage solutions due to their high specific capacity, good energy density, long cycle life, and low self-discharge. As the use of electric vehicles and portable electronic devices has become increasingly prominent, the demand for superior functionality LIBs has continued to rise [1]. The energy density of a Li-ion cell is significantly influenced by the capacity of active materials. To extend the operational life and efficiency of such batteries, enhancing their energy density is crucial. This can be done either by raising the operating voltage or by increasing the active electrode material capacity [2].

A. Antony et al. (Eds.): icONMAT 2025, SPPHY 328, pp. 40–45, 2025.
https://doi.org/10.1007/978-981-95-2697-0_6

A more practical way of enhancing energy density is by creating a high-capacity anode material [3]. Graphite, the conventional anode in commercial LIBs, boasts a theoretical capacity of 372 mAh/g [4]. Other materials like tin dioxide (SnO_2) have also been of interest because they possess a much greater theoretical capacity of 1494 mAh/g [5]. In spite of this benefit, SnO_2-based anodes present severe challenges, mostly as a result of large volume changes that take place as Li-ions are inserted into and removed from the material. Upon the first lithiation process, SnO_2 undergoes both a conversion reaction (production of Li_2O and Sn) and alloy formation that follows later (Sn for Li_xSn alloys), with extensive structural expansion and contraction. These cyclic volume variations are responsible for mechanical degradation, such as pulverization and fracture of the electrode, eventually resulting in the loss of capacity at a fast rate [6].

In order to overcome these challenges, techniques like the creation of composites with porous and conductive matrices as well as reducing the crystallite size have been investigated. By lowering the crystallite size, not only increase the specific surface area, thus improving the sites for electrochemical reactions, but it also reduces the diffusion distances for lithium ions and electrons, in turn improving overall conductivity. Moreover, smaller crystallites can better accommodate volume changes, lessening mechanical stress and enhancing structural stability during cycling [7].

SnO_2 was synthesized in the present research through a hydrothermal route with polyvinylpyrrolidone (PVP) as a surfactant of two different molecular weights, 1,300,000 and 40,000 g/mol. Changing the molecular weight of PVP affects the morphology and crystallite size of SnO_2, producing two dissimilar samples, SnO_2-S (spherical morphology) and SnO_2-F (faceted morphology). Between the two, SnO_2-S had smaller crystallites, resulting in improved initial capacity and better cycling stability than SnO_2-F. This proves the pivotal role of morphological control and size reduction in enhancing the efficiency of SnO_2-based electrodes in emerging LIB technologies.

2 Experimental Methods

2.1 SnO_2 Synthesis Using Hydrothermal Method

A total of 1.56 g of $SnCl_4$ and 1.59 g of PVP (with molecular weights of 40,000 and 1,300,000) were dissolved in a 1:1 ethanol–deionized water mixture under continuous stirring. To this solution, 3.6 mL of HCl was added, forming a clear and transparent mixture. The resulting solution was transferred into a Teflon-lined autoclave, sealed, and heated at 200 °C for 12 h under hydrothermal conditions. After completion, the resulting product was washed and centrifuged multiple times with ethanol and deionized water, and then dried to yield the final material [8, 9]. The SnO_2-S and the SnO_2-F were obtained using PVP of molecular weights 1,300,000 and 40,000 g/mol, respectively.

2.2 Electrode Fabrication and Cell Assembly

Electrodes were fabricated using the doctor blade method by preparing a slurry consisting of 70% SnO_2, 20% acetylene black, and 10% PVDF dissolved in N-Methyl-2-Pyrrolidone (NMP). The prepared slurry was uniformly coated onto a copper foil current collector and then dried at 60 °C for 12 h. The dried electrodes were then punched into circular disks with a diameter of 3/8 inch. Li-ion cells were assembled in an argon-filled glove box using SnO_2 coated on copper as the working electrode, lithium metal as both the reference and counter electrode, $LiPF_6$ dissolved in a mixture of EC and DMC as the electrolyte, and a polypropylene membrane as the separator.

2.3 Characterizations

The structural and morphological properties were analyzed with X-ray diffraction using a Rigaku MiniFlex diffractometer (Cu Kα radiation, $\lambda = 1.5418$ Å) and field-emission scanning electron microscopy with a Carl Zeiss Sigma instrument. The electrochemical behaviour of the assembled cells was assessed through electrochemical impedance spectroscopy (EIS) and galvanostatic charge–discharge (GCD) measurements. These analyses were performed using a CHI6054E potentiostat for EIS, and an MTI Corporation BST8-WA eight-channel battery analyzer for GCD measurements.

3 Result and Discussions

The XRD patterns of SnO_2-F and SnO_2-S are shown in Fig. 1, where the diffraction peaks at 26.5°, 33.9°, 37.9°, 38.9°, 51.86°, 54.6°, 57.8°, 61.8°, 64.8°, and 65.9° are associated with the lattice planes (110), (101), (200), (210), (211), (220), (002), (310), (112), and (301), respectively [10]. These peaks align well with the standard JCPDS file No. 41–1445, confirming the formation of SnO_2-S and SnO_2-F without any impurities, with crystallite sizes calculated from the high-intensity peak being 6 nm and 29 nm using Scherrer equation, respectively.

The morphology of the samples was examined using FESEM, as shown in Fig. 2. The analysis reveals that the SnO_2-F sample possesses a faceted structure, whereas the SnO_2-S sample exhibits a spherical morphology composed of clusters of smaller SnO_2-S crystallites.

Figure 3a presents the electrochemical impedance spectroscopy results for Li-ion cells assembled with SnO_2-F and SnO_2-S as working electrode. The charge transfer resistance of SnO_2-F (550 Ω) was observed to be lower than that of SnO_2-S (700 Ω), attributed to the faceted structure of SnO_2-F, which facilitates efficient charge transfer through its edges [9]. However, SnO_2-F exhibited higher diffusion resistance, likely due to its comparatively larger crystallite size. Fig. 3b and Fig. 3c illustrate the GCD analysis of cells based on SnO_2-S and SnO_2-F cycled at a current density of 0.15 A/g. The Li-ion cell utilizing SnO_2-S delivers an initial discharge capacity of 1324 mAh/g, while the SnO_2-F-based cell achieves 543 mAh/g. This difference is likely attributed to the lower diffusion resistance associated with the smaller crystallite size in the SnO_2-S-based cell. By the second cycle, the discharge capacity decreases to 812 mAh/g for the SnO_2-S-based cell and 183 mAh/g for the SnO_2-F-based cell.

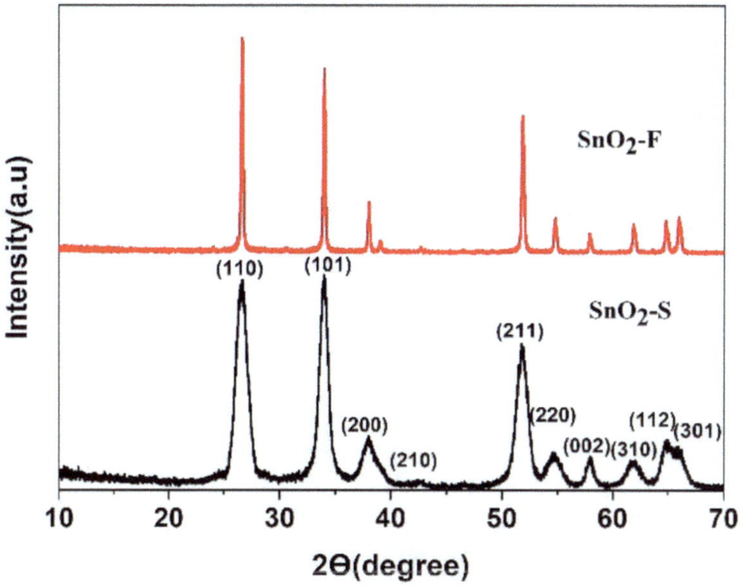

Fig. 1. XRD patterns of SnO_2-S and SnO_2-F.

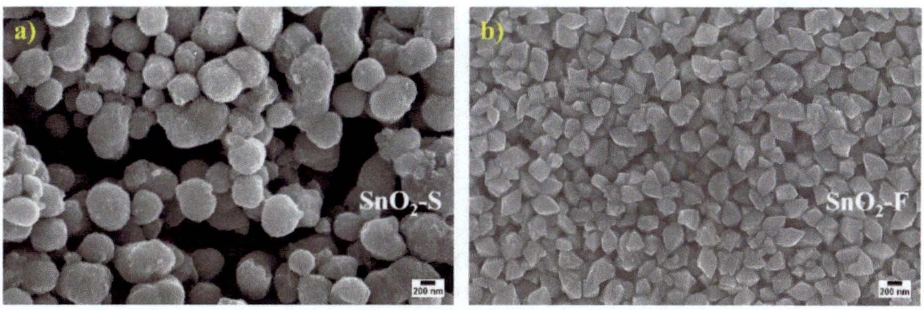

Fig. 2. FESEM images of SnO_2-S and SnO_2-F.

Figure 3d, e, and f depict the discharge capacity, charge capacity, and coulombic efficiency plots for SnO_2-F and SnO_2-S based cells. The data clearly demonstrates that the SnO_2-S-based cell exhibits superior cycling stability compared to the SnO_2-F-based cell. By the thirtieth cycle, the discharge capacity for the SnO_2-F-based cell drops to 20 mAh/g, whereas the SnO_2-S-based cell retains a discharge capacity of 260 mAh/g. This enhanced cycling stability is likely due to the smaller crystallite size of SnO_2-S, which helps to mitigate issues related to volume expansion and contraction [11].

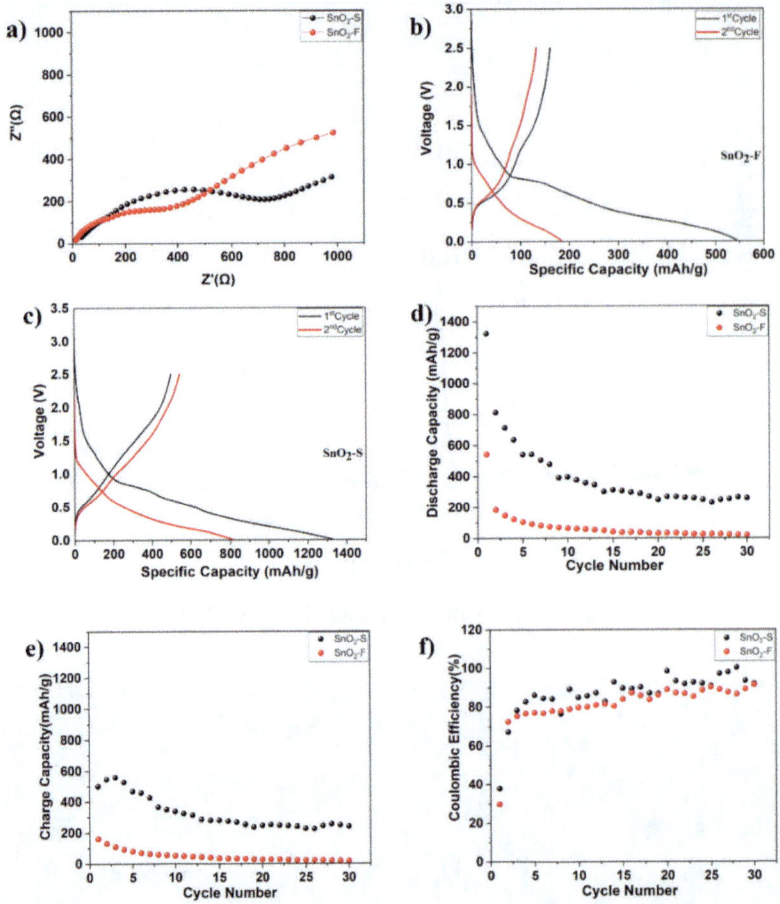

Fig. 3. Electrochemical characterizations of SnO_2-F and SnO_2-S based Li-ion cells a) EIS b) and c) first two cycles of GCD d) discharge capacity vs. cycle number e) charge capacity vs. cycle number f) coulombic efficiency vs. cycle number.

4 Conclusions

The hydrothermal synthesis of SnO_2 by PVP of different molecular weights yields two morphologies, faceted SnO_2-F and spherical SnO_2-S, having huge differences in crystallite size and electrochemical properties. The SnO_2-S sample with a smaller crystallite size and spherical shape shows better cycling stability and higher specific capacity compared to SnO_2-F. The initial discharge capacity of 1324 mAh/g and a retention of 260 mAh/g after thirty cycles highlight the enhanced performance of SnO_2-S, which can be ascribed to its lower diffusion resistance and better accommodation of volume change during cycling. These results emphasize the significance of crystallite size and morphology in maximizing the electrochemical properties of SnO_2 anodes for LIBs.

Acknowledgments. The author Neethu Simon is thankful to the Cochin University of Science and Technology (CUSAT), Government of Kerala, for the financial support in the form of a junior research fellowship. The authors would also like to acknowledge the seed money grant from CUSAT for the financial assistance.

References

1. Trana, H.H., et al.: SnO2 nanosheets/graphite oxide/g-C3N4 composite as enhanced performance anode material for lithium ion batteries. Chem. Phys. Lett. **715**, 284–292 (2019)
2. Manthiram, A.: Materials challenges and opportunities of lithium ion batteries. J. Phys. Chem. Lett. **2**(3), 176–184 (2011)
3. Shi, J.-L., et al.: High-Capacity Cathode Material with High Voltage for Li-Ion Batteries. Adv. Mater. **30**(9), 1–8 (2018)
4. Yarmolich, D., et al.: Novel binder-free carbon anode for high capacity Li-ion batteries. Nano Energy. **83**(1), 105816 (2021)
5. Kai, W., et al.: SnO2 quantum dots @ 3D sulfur-doped reduced graphene oxides as active and durable anode for lithium ion batteries. Electrochim. Acta. **291**, 24–30 (2018)
6. Huang, Z., et al.: Improved capacity and cycling stability of SnO2 nanoanode induced by amorphization during cycling for lithium ion batteries. Mater. Des. **180**, 107973 (2019)
7. Bruck, A.M., Cama, C.A., Gannett, C.N., Marschilok, A.C., Takeuchi, E.S., Takeuchi, K.J.: Nanocrystalline iron oxide-based electroactive materials in lithium ion batteries: The critical role of crystallite size, morphology, and electrode heterostructure on battery-relevant electrochemistry. Inorg. Chem. Front. **3**, 26–40 (2016)
8. Han, X., et al.: Synthesis of Tin Dioxide Octahedral nanoparticles with exposed high-energy {221} facets and enhanced gas-sensing properties. Angew. Chem. **121**(48), 9344–9347 (2009)
9. Wilson, M.K., Joseph, P.M., Antony, A., Jayaraj, M.K., Jayalekshmi, S.: Controlling the structural features of tin dioxide nanoparticles embedded reduced graphene oxide on carbon cloth to serve as high stability fast charging anode for Li-ion cells. Diam. Relat. Mater. **144**(3), 110989 (2024)
10. Narsimulu, D., Nagaraju, G., Chandra Sekhar, S., Ramulu, B., Su Yu, J.: Three-dimensional porous SnO2/carbon cloth electrodes for high-performance lithium- and sodium-ion batteries. Appl. Surf. Sci. **538**, 148033 (2021)
11. Yin, L., et al.: Ultrafine SnO2 nanoparticles as a high performance anode material for lithium ion battery. Ceram. Int. **42**(8), 9433–9437 (2016)

In-Situ $FeCl_3$ Addition Strategy to Improve the Efficiency of LiF-HCl Assisted Etching for the Synthesis of $Ti_3C_2T_x$ MXene

Varsha Vijayan[1] and Asha Arackal Sukumaran[1,2,3](✉)

[1] Nanomaterials for Emerging Solid-State Technology (NEST) Research laboratory, Department of Physics, Cochin University of Science and Technology, Kochi 682022, India
asa@cusat.ac.in

[2] Centre of Excellence in Advanced Materials, Cochin University of Science and Technology, Kochi 682022, India

[3] Inter University Centre for Nanomaterials and Devices, Cochin University of Science and Technology, Kochi 682022, India

Abstract. In this study, $Ti_3C_2T_x$ MXene was synthesized from Ti_3AlC_2 MAX phase using three distinct etching strategies: direct hydrofluoric acid (HF) etching, a milder Lithium Fluoride (LiF) – Hydrochloric Acid (HCl) route, and a modified LiF–HCl route incorporating Ferric Chloride ($FeCl_3$). X-ray diffraction (XRD) analysis revealed that the modified etching route led to the complete removal of the Al layer, evidenced by the complete disappearance of the (104) peak corresponding to the Ti–Al bond. Moreover, the modified etching method resulted in an increased interlayer spacing of 1.40 nm compared to the LiF–HCl-etched sample, indicating enhanced delamination capability and better ion intercalation. It exhibited a specific gravimetric capacitance of 31.46 F/g demonstrating that $FeCl_3$-assisted in-situ modification of the LiF–HCl etching method offers an effective and safer alternative to conventional HF etching for producing high-quality $Ti_3C_2T_x$ MXene suitable for energy storage applications such as supercapacitors.

Keywords: $Ti_3C_2T_x$ MXene · $FeCl_3$-assisted in-situ modification · Gravimetric Capacitance

1 Introduction

MXenes, a rapidly expanding family of two-dimensional (2D) transition metal carbides and nitrides, have attracted considerable interest since the first successful synthesis of $Ti_3C_2T_x$ from Ti_3AlC_2 MAX phase in 2011 [1]. Typically represented as $M_{n+1}X_nT_x$ (where M is an early transition metal, X is carbon and/or nitrogen, and T_x denotes surface terminations such as –O, –OH, and –F), MXenes exhibit exceptional electronic conductivity [2], hydrophilicity [3], and high specific surface area [4]. These characteristics make them highly promising for applications in energy storage [5], catalysis [6], electromagnetic shielding [7], and sensing technologies [8].

A. Antony et al. (Eds.): icONMAT 2025, SPPHY 328, pp. 46–53, 2025.
https://doi.org/10.1007/978-981-95-2697-0_7

Conventionally, the synthesis of $Ti_3C_2T_x$ involves selective etching of Al atoms from the Ti_3AlC_2 MAX phase using concentrated HF. While effective, HF poses significant health and environmental hazards, and often results in over etching or oxidation, compromising the structural integrity and functional performance of the resulting MXene [9]. To mitigate these drawbacks, alternative milder etching approaches have emerged, notably the in-situ generation of HF by combining LiF and HCl [10]. This method is relatively safer and provides better control over etching, producing clay-like MXenes that are more readily delaminated [10]. However, incomplete etching and residual MAX phase often persist, limiting the quality and utility of the material. Anamika et al. introduced NH_4OH into the LiF–HCl etching system, forming a buffer compound (NH_4F) that aided in removing residual Al and reducing surface oxidation, thereby enhancing the optical and morphological properties of the resulting MXene [9]. However, even in these approaches, complete elimination of the Ti–Al (104) peak and full conversion of MAX phase were not always achieved.

To further improve the etching efficiency and overcome the limitations of residual MAX phase in the LiF–HCl system, researchers have explored the use of Lewis acidic agents as alternative or supplementary etchants. Lewis acids such as $CuCl_2$ [11], and $FeCl_3$ [12] have been shown to effectively extract Al atoms from MAX phases via redox-assisted or molten-salt-mediated routes, providing a fluoride-free or reduced-fluoride pathway for MXene synthesis. However, it typically requires high temperatures and inert atmospheres, making it energy-intensive and less practical for ambient-condition processing. In a modified two-step approach, Zhang et al. employed a combination of low-concentration HF and $FeCl_3$ to etch Ti_3AlC_2 through a Fe^{3+}–Al replacement reaction [12]. Although this method reduces HF usage, it still relies on initial HF treatment which poses safety and scalability challenges. In our work, we extend this approach by integrating $FeCl_3$ into the conventional LiF–HCl etching system to achieve a safer, scalable, and more complete etching process under ambient conditions. This hybrid approach significantly enhances etching performance, leading to the complete Al extraction, increased interlayer spacing, and improved electrochemical characteristics of the resulting $Ti_3C_2T_x$ MXene, making it highly promising for energy storage applications.

2 Materials and Methods

2.1 Synthesis of $Ti_3C_2T_x$ MXene

Figure 1 shows the schematic illustration of the synthesis process. In this study, 1 g of Ti_3AlC_2 MAX phase powder ($\geq$ 99% purity, particle size 40–60 μm, Sigma Aldrich) was slowly added over the course of five minutes to an etching solution containing 5 mL of deionised (DI) water, 15 mL of concentrated HCl (37 wt.%), and 1.6 g of LiF. To this mixture, 3 M $FeCl_3$ ($FeCl_3$, anhydrous, $\geq$ 98%, Sigma Aldrich) was added dropwise to initiate an in-situ etching process. The reaction mixture was stirred continuously at 300 rpm for 24 h under ambient conditions. After the etching process, the resulting suspension was centrifuged and repeatedly washed with DI water until the pH of the supernatant reached neutral (pH $\geq$ 6). The sediment was further washed with ethanol to remove residual salts and then dried overnight. The dried multilayer $Ti_3C_2T_x$ MXene powder was collected and stored in a vacuum desiccator for further characterization.

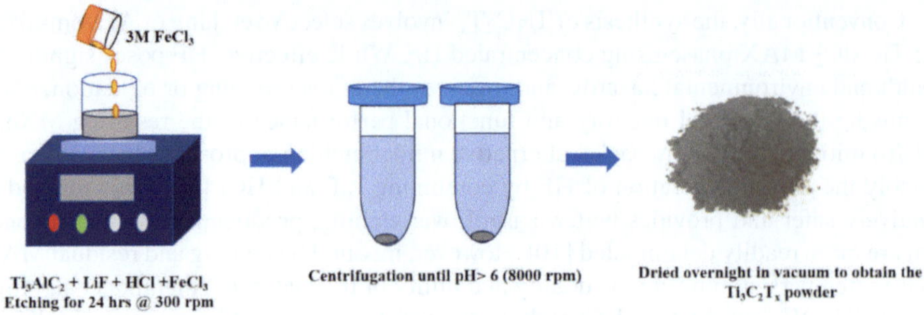

Fig. 1. Schematic illustration of the synthesis of $Ti_3C_2T_x$ MXene from Ti_3AlC_2 MAX phase

2.2 Material Characterization

Structural characterization of the synthesized samples was carried out using a Rigaku MiniFlex X-ray diffractometer with Cu Kα radiation (λ = 1.5418 Å). Surface morphology was analyzed using a ZEISS SIGMA field emission scanning electron microscope (FESEM) and HITACHI S4800 equipped with an X-ray detector. Raman spectroscopy was performed with a HORIBA JOBIN YVON LabRAM HR system. Electrochemical performance was evaluated in a three-electrode setup using a CHI-6054E electrochemical workstation by employing 2 M H_2SO_4 as the electrolyte. The gravimetric capacitance has been determined using the equation:

$$Gravimetric\ capacitance,\ C_g = \frac{I\Delta t}{m\Delta V} \tag{1}$$

where I is the current (A), ΔV is the potential window (Volt), m is the mass of the electrode (g), and Δt is the discharging time in seconds.

3 Results and Discussion

X-ray diffraction (XRD) pattern $Ti_3C_2T_x$ MXene synthesized using different etching routes is as shown in Fig. 2. a. All samples exhibited a prominent (002) peak, characteristic of the layered MXene structure. In the HF-etched sample, the (002) peak appeared at $2\theta = 8.8°$, corresponding to an interlayer spacing of 1.00 nm. While the (104) peak—associated with the Ti–Al bond—was completely absent, the significantly reduced intensity of the (002) reflection suggests possible over etching and partial degradation of the Ti layers [9]. For the sample etched using the LiF–HCl route, the (002) peak shifted to $2\theta = 6.7°$, indicating increased interlayer spacing of 1.32 nm. However, a weak remnant of the (104) peak was still observed, suggesting incomplete etching and the presence of residual Al. Notably, the $FeCl_3$-assisted LiF–HCl etched sample exhibited a further downshift of the (002) peak to $2\theta = 6.3°$, corresponding to an expanded interlayer spacing of 1.40 nm. The (104) peak was completely absent in this sample as well, confirming thorough etching of Al and complete disruption of the Ti–Al bonds. The phase purity of $Ti_3C_2T_x$ MXene can be quantitatively evaluated by analyzing the intensity ratio of the (104) diffraction peak of the residual Ti_3AlC_2 MAX phase to the (002) peak of the resulting MXene in the XRD pattern [13]. A lower $I_{(104)}/I_{(002)}$ ratio reflects more complete etching of the MAX phase. The phase purity (y) can be empirically estimated using the relation:

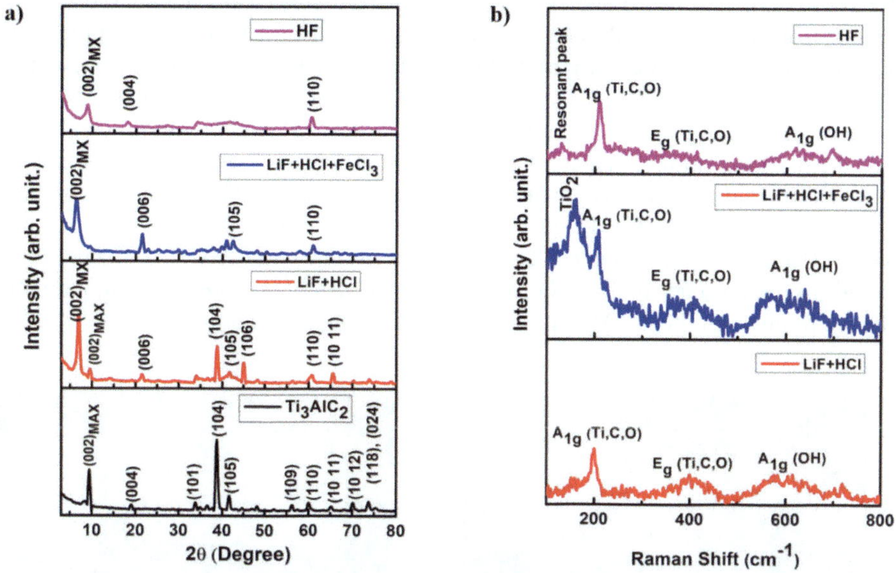

Fig. 2. a) XRD pattern b) Raman spectra taken at 532 nm laser excitation wavelength for the $Ti_3C_2T_x$ MXene prepared using the etchants LiF- HCl, $FeCl_3$ together with LiF- HCl and HF

$$y = 1 - 0.2\,x + 0.013\,x^2 \tag{2}$$

where $x = \frac{I_{(104)}}{I_{(002)}}$. x was found to be 0.17 for the $FeCl_3$-assisted LiF–HCl etched sample and 0.56 for the LiF-HCl-etched sample, demonstrating that the former yields $Ti_3C_2T_x$ MXene with significantly higher phase purity.

The Raman spectra of $Ti_3C_2T_x$ MXene synthesized via different etching routes are shown in Fig. 2. b, revealing distinct vibrational features. All samples exhibit characteristic peaks, including the A_{1g} mode (~200 cm^{-1}) and E_g mode (~380 cm^{-1}) associated with Ti, C, and surface terminations, along with a broader A_{1g}(OH) mode near 620–650 cm^{-1} [14]. The HF-etched sample shows well-defined MXene peaks with a sharp A_{1g} mode and a resonant peak at 130 cm^{-1} [15], but the reduced intensity of other modes suggests possible over etching and partial degradation of the Ti layers. While LiF–HCl etched sample exhibits clean and well-defined MXene peaks with no indication of TiO_2, the $FeCl_3$-assisted LiF–HCl etched sample have a distinct TiO_2 peak at ~145 cm^{-1}, indicating surface oxidation [15]. This oxidation is likely due to the oxidative nature of Fe^{3+} ions in $FeCl_3$, which were introduced after the initial etching step with LiF and HCl; the Fe^{3+} ions can interact with exposed Ti sites, leading to partial oxidation during the prolonged etching process [16].

The FESEM images in Fig. 3 reveal the morphological differences in $Ti_3C_2T_x$ MXene obtained via three etching methods. The HF-etched sample exhibits a well-defined, layered morphology with sharp, compact, and uniformly stacked flakes, indicating thorough Al removal and effective preservation of the accordion structure. This morphology reflects the high etching efficiency of HF, which facilitates clean delamination and the formation of high-quality MXene. In comparison, the LiF–HCl-etched sample shows a loosely packed structure with crumpled and irregular layers, suggesting incomplete etching and limited exfoliation. The $FeCl_3$-assisted LiF–HCl sample displays a similarly disordered morphology, with unevenly separated layers. These observations confirm that while the milder etching routes are safer, they do not achieve the same level of morphological clarity and quality as HF, which remains the most effective etchant for producing morphologically well-ordered $Ti_3C_2T_x$ MXene. The Energy Dispersive X-ray Spectroscopy (EDX) analysis of the $FeCl_3$-assisted LiF–HCl etched $Ti_3C_2T_x$ MXene (Fig. 4. d) confirms the presence of Ti, C, O, and F, with negligible Al and Fe content, indicating effective removal of the Ti–Al bond and minimal Fe incorporation.

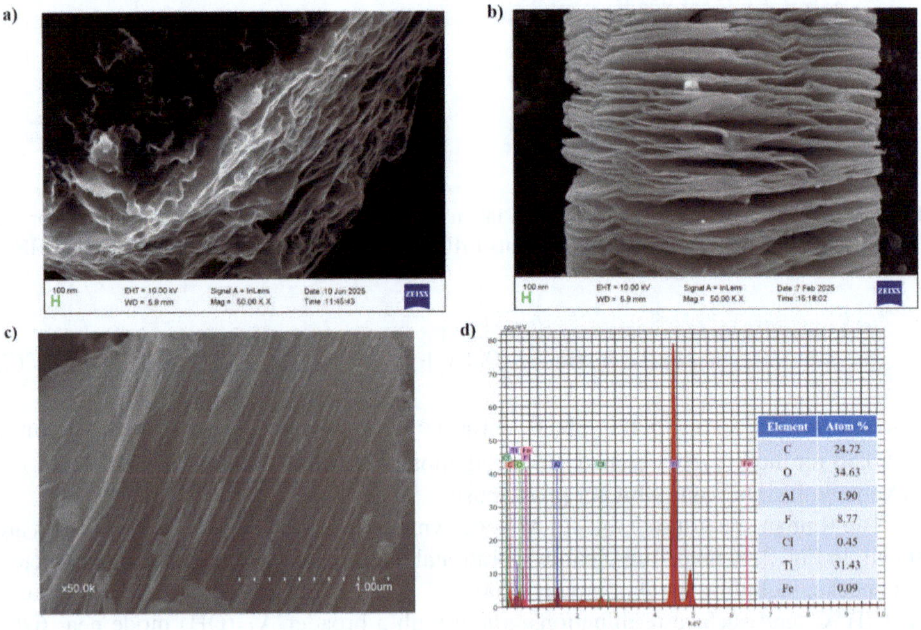

Fig. 3. FESEM image $Ti_3C_2T_x$ MXene prepared using the etchants (a) LiF- HCl, (b) HF and (c) $FeCl_3$, together with LiF- HCl d) Energy Dispersive X-ray Spectroscopy (EDX) analysis of the $FeCl_3$-assisted LiF–HCl etched $Ti_3C_2T_x$ MXene

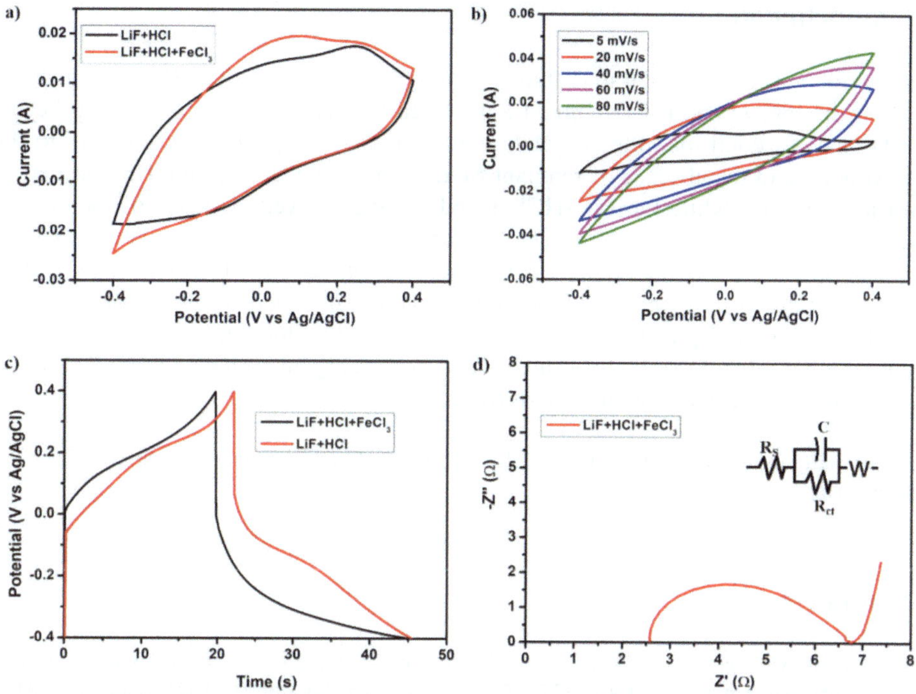

Fig. 4. Cyclic Voltammetry analysis of a) LiF- HCl etched and $FeCl_3$ assisted LiF- HCl etched $Ti_3C_2T_x$ MXene at 20 mV/s scan rate b) $FeCl_3$ assisted LiF- HCl etched $Ti_3C_2T_x$ MXene at different scan rates c) Galvanostatic Charge Discharge (GCD) analysis of LiF- HCl etched and $FeCl_3$ assisted LiF- HCl etched $Ti_3C_2T_x$ MXene at a current density of 1 A/g d) Nyquist plot of $FeCl_3$ assisted LiF- HCl etched $Ti_3C_2T_x$ MXene with equivalent circuit diagram in the inset

The electrochemical performance of $Ti_3C_2T_x$ MXene etched using LiF–HCl and $FeCl_3$-assisted LiF–HCl etched sample was assessed using cyclic voltammetry (CV), galvanostatic charge–discharge (GCD), and electrochemical impedance spectroscopy (EIS) in a three-electrode configuration with 2 M H_2SO_4 as the electrolyte. As shown in the CV curves at 20 mV/s scan rate (Fig. 4. a), both samples deliver comparable electrochemical performance. The CV curves for $FeCl_3$-assisted LiF–HCl etched MXene (Fig. 4. b) at various scan rates exhibit broad redox humps similar to LiF- HCl etched one, indicating faradaic charge storage due to surface redox transitions involving Ti^{3+}/Ti^{4+} [10]. GCD curves (Fig. 4. c) for both electrodes show deviation from ideal triangular behavior, consistent with pseudocapacitive reaction [17]. The longer discharge time for the $FeCl_3$ sample suggests improved charge retention with a slightly higher gravimetric specific capacitance of 31.46 F/g compared to 29.26 F/g for the LiF-etched sample. However, the curves also exhibit non-symmetric profiles, further confirming pseudocapacitive contributions beyond EDLC. The redox reactions are primarily driven by the reversible intercalation. EIS measurements (Fig. 4. d) reveal that the $FeCl_3$-assisted LiF–HCl etched sample exhibits a solution resistance (R_s) of 2.59 Ω and charge transfer resistance (R_{ct}) of 4.19 Ω, indicating improved ionic and electronic conductivity.

4 Conclusion

This work presents a comparative study on the synthesis of $Ti_3C_2T_x$ MXene using three etching strategies: HF, LiF–HCl, and a modified LiF–HCl route incorporating $FeCl_3$. XRD analysis confirmed the complete elimination of the peak corresponding to the Ti–Al bond and the absence of remnant MAX phase reflections in the HF and $FeCl_3$-assisted samples, while the LiF–HCl-etched sample showed evidence of incomplete etching. Raman spectroscopy revealed minor surface oxidation in the $FeCl_3$-assisted sample, attributed to the oxidative nature of Fe^{3+} ions. Despite this, the $FeCl_3$-etched MXene exhibited a gravimetric capacitance of 31.46 F/g which is comparable to that of LiF–HCl etched sample. These findings demonstrate that the $FeCl_3$-assisted LiF–HCl method is a promising and safer strategy for synthesizing high-quality $Ti_3C_2T_x$ MXene with competitive electrochemical performance.

Acknowledgments. V.V. acknowledges CUSAT for USRF. The authors acknowledge RUSA 2.0 for financial assistance.

References

1. Gogotsi, Y., Huang, Q.: MXenes: Two-dimensional building blocks for future materials and devices. Am. Chem. Soc. **15**(4), 5775–5780 (2021). https://doi.org/10.1021/acsnano.1c03161
2. Shayesteh, A., Mirkhani, S., Sun, P., Naguib, M., Braun, P., Sundararaj, U.: Improved synthesis of Ti3C2Tx MXene resulting in exceptional electrical conductivity, high synthesis yield, and enhanced capacitance. Nanoscale. **13**, 3572–3580 (2021). https://doi.org/10.1039/D0NR06671K
3. Pandey, L., Liang, W., VahidMohammadi, A., Zhang, T., Gogotsi, Y., Wanunu, M.: Hydrophilicity and surface charge modulation of Ti3C2Tx MXene based membranes for water desalination. RSC Adv. **14**(30), 21635–21643 (2024). https://doi.org/10.1039/d4ra02678k
4. Kumar, S., Kumari, N., Seo, Y.: MXenes: Versatile 2D materials with tailored surface chemistry and diverse applications. J. Energy Chem. **90**, 253–293 (2024). https://doi.org/10.1016/j.jechem.2023.11.031
5. Akhter, R., Maktedar, S.S.: MXenes: A comprehensive review of synthesis, properties, and progress in supercapacitor applications. J. Mater. **9**(6), 1196–1241 (2023). https://doi.org/10.1016/j.jmat.2023.08.011
6. Peera, S.G., Koutavarapu, R., Chao, L., Singh, L., Murugadoss, G., Rajeshkhanna, G.: 2D MXene nanomaterials as electrocatalysts for hydrogen evolution reaction (HER): A review. Micromachines. **13**(9), 1499. Sep. 01 MDPI (2022). https://doi.org/10.3390/mi13091499
7. Li, Y., et al.: Scalable manufacturing of flexible, durable Ti3C2Tx MXene/Polyvinylidene fluoride film for multifunctional electromagnetic interference shielding and electro/photo-thermal conversion applications. Compos. Part B Eng. **217**, 108902 (2021). https://doi.org/10.1016/j.compositesb.2021.108902
8. Wu, X., Ma, P., Sun, Y., Du, F., Song, D., Xu, G.: Application of MXene in electrochemical sensors: A review. Electroanalysis. **33**(8), 1827–1851 (2021). https://doi.org/10.1002/elan.202100192

9. Ashok, A., Saseendran, S.B., Asha, A.S.: Synthesis of Ti3C2Tx MXene from the Ti3AlC2 MAX phase with enhanced optical and morphological properties by using ammonia solution with the in-situ HF forming method. Phys. Scr. **97**(2), 25807 (2022). https://doi.org/10.1088/1402-4896/ac4c53
10. Vijayan, V., et al.: Fabrication of flexible supercapacitor using spray-coated $Ti_3C_2T_x$ MXene thin film. Energ. Technol. (2025). https://doi.org/10.1002/ente.202500317
11. Li, Y., et al.: A general Lewis acidic etching route for preparing MXenes with enhanced electrochemical performance in non-aqueous electrolyte. Nat. Mater. **19**(8), 894–899 (2020). https://doi.org/10.1038/s41563-020-0657-0
12. Zhang, P., et al.: Modified two-step etching method for the synthesis of Ti3C2 MXene with enhanced electrochemical performance. Electrochem. Commun. **157**, 107620 (2023). https://doi.org/10.1016/j.elecom.2023.107620
13. Ashok, A., Saseendran, S.B., Arackal Sukumaran, A.: Influence of in situ and ex situ incorporated cations on the supercapacitor performance of multilayer $Ti_3C_2T_x$ MXene. ACS Appl. Energy Mater. **7**(11), 5050–5063 (2024). https://doi.org/10.1021/acsaem.4c00931
14. Sarycheva, A., Gogotsi, Y.: Raman spectroscopy analysis of the structure and surface chemistry of Ti3C2T xMXene. Chem. Mater. **32**(8), 3480–3488 (2020). https://doi.org/10.1021/acs.chemmater.0c00359
15. Adomaviciute-Grabusove, S., et al.: Monitoring Ti3C2Tx MXene degradation pathways using raman spectroscopy. ACS Nano. **18**(20), 13184–13195 (2024). https://doi.org/10.1021/acsnano.4c02150
16. Li, W., et al.: Synchronization of Ti3C2 MXene/Fe3+ with sodium persulfate for the degradation of reactive dyes. NPJ Clean Water. **7**(1), 34 (2024). https://doi.org/10.1038/s41545-024-00330-9
17. Gogotsi, Y., Penner, R.M.: Energy storage in nanomaterials - capacitive, pseudocapacitive, or battery-like? ACS Nano. **12**(3), 2081–2083 (2018). https://doi.org/10.1021/acsnano.8b01914

Electrochemical Study of Polyaniline/Ferrite Nanocomposites for Supercapacitor Applications

Reshma Varghese[1], Alvina Augusthy[1], Theertha Thilakan[3], and Jayalatha Gopalakrishnan[1,2](✉)

[1] Department of Polymer Science and Rubber Technology, CUSAT, Kochi 682022, India
{reshmavarghesev,jayalatha}@cusat.ac.in
[2] Interuniversity Centre for Nanomaterials and Devices, CUSAT, Kochi 682022, India
[3] Sree Kerala Varma College, Thrissur, India

Abstract. The growing demand for efficient and sustainable energy storage systems has driven significant interest in the development of high-performance electrode materials for supercapacitors. In this study, polyaniline (P)/nickel ferrite (N) composites were synthesized to overcome the limitations of individual components and enhance electrochemical performance. While polyaniline offers excellent pseudocapacitive behaviour, its poor cyclic stability remains a challenge. Conversely, N, a spinel-type transition metal oxide, provides superior structural stability and redox activity. Nanocomposites were prepared by varying aniline: N ratios (1,0.05, 1,0.10, 1,0.15) via insitu polymerization. Structural and morphological characterizations using FTIR, and XRD confirmed the composite formation, while FESEM analysis revealed uniform distribution of N nanoparticles on branched P rods. Electrochemical measurements conducted in 1 M H_2SO_4 within a 0.0–0.8 V potential window demonstrated an impressive specific capacitance of 513 F/g at 1 A/g for $PN_{0.10}$ and also obtained cyclic stability (600 cycles, at 4 A/g) with good capacitance retention (99%). This signifies the composite's potential as a high-performance electrode material, paving the way for next-generation energy storage devices.

Keywords: Polyaniline · Nickel ferrite · CV · GCD · Specific Capacitance

1 Introduction

Rapid industrialization and population growth have increased the demand for energy stressing current power infrastructure. While fossil fuels remain the primary energy source, the need for efficient solutions is rising. Renewable energy sources like wind, solar, and tidal are gaining popularity but are intermittent. By 2030, energy reserve requirements are expected to triple, necessitating specialized systems. Hybrid and electric vehicles are helping reduce carbon emissions, and electrochemical energy storage, particularly supercapacitors, offers high energy and power densities. Supercapacitors, utilizes high surface area electrodes to achieve greater capacitance and power densities compared to batteries. This makes them ideal for applications like regenerative braking, biosensors, and e-textiles [1].

A. Antony et al. (Eds.): icONMAT 2025, SPPHY 328, pp. 54–59, 2025.
https://doi.org/10.1007/978-981-95-2697-0_8

The prime materials that have been used as electrodes in supercapacitors are carbon-based compounds, transition metal oxides, and conducting polymers. Conducting polymers like polyaniline, polythiophene, and polypyrrole act as pseudocapacitors through rapid redox reactions, offering higher power densities and rate capabilities than inorganic battery electrodes. However, they suffer from low cyclic stability and degrade after a few hundred charge-discharge cycles due to expansion and contraction during cycling. To improve performance, conducting polymers are often enhanced with fillers like metal oxides, carbon-based nanomaterials, and hydroxides. Incorporating spinel-type transition metal oxides, particularly ferrites, in polyaniline (P) electrodes improves cyclic stability and overall supercapacitor performance. Nickel ferrite ($NiFe_2O_4$) (N), with its inverse spinel structure, strengthens interparticle interactions and stabilizes the P chain network, preventing degradation [2]. In this study, we synthesized Polyaniline/Nickel ferrite composites with different weight percentages of $NiFe_2O_4$ (0.15, 0.1, and 0.05) relative to P and characterized using FTIR, XRD, and FESEM. Electrochemical characterization was conducted in a 1 M H_2SO_4 electrolyte using cyclic voltammetry (CV), constant current charging/discharging (GCD), and electrochemical impedance spectroscopy (EIS).

2 Experimental

2.1 Materials

Aniline (AR grade 99.5%), ammonium persulphate (APS, AR grade 98%), para- Toluene sulphonic acid (p-TSA),) was provided by Spectrochem Pvt. Ltd., Mumbai, India; Ferric chloride nonahydrate ($Fe(NO_3)_3.9H_2O$), nickel nitrate ($Ni(NO_3)_2$), and polyvinyl pyrrolidine (PVP) are purchased from Alfa-Aesar, Scientific & Trading Company, India.

2.2 Synthesis of Polyaniline (P)

The aniline monomer (1 g) was incorporated in 1 M pTSA solution and stirred continuously for 30 min. Ammonium persulfate solution (oxidizing agent) was then added dropwise and the reaction was continued at room temperature for 6 h. The precipitate was repeatedly washed with deionized water and later dried at 60 °C for 12 h.

2.3 Synthesis of Nickel Ferrite (N)

$NiFe_2O_4$ was synthesized via hydrothermal method. Fe $(NO)_3.9H_2O$ and Ni $(NO_3)_2$ solutions were introduced in the RB flask in the mole ratio 2:1 and stirred. The surfactant polyvinyl pyrrolidine (PVP) was then added followed by 3 M NaOH solution until pH = 11. The mixture was placed in the autoclave and kept at 140 °C for 10 h. The final precipitate was washed thoroughly until neutral to pH. Finally, the precipitate was dried at 80 °C for 12 h in an oven.

2.4 Synthesis of Polyaniline/Nickel Ferrite Nanocomposite (PN)

The PN nanocomposite was synthesized in different weight ratio of P:N viz., 1:0.05, 1:0.1 and 1:0.15, respectively, via in-situ polymerization. Aniline and $NiFe_2O_4$ are introduced in 1 M PTSA solution in an RB flask and the procedure as that for synthesizing polyaniline is followed. The composite is designated as $PN_{0.05}$, $PN_{0.1}$, $PN_{0.15}$.

3 Characterization and Eelectrochemical Measurement

The morphology of the samples were examined using Field emission scanning electron microscopy (FE-SEM), Zeiss Sigma HV & PV series instrument. The crystalline structures were analysed using Bruker AXS D8 Advance X-ray diffractometer equipped with a Cu Kα source ($\lambda = 1.54$ Å). Fourier Transform Infrared (FTIR) spectra of the samples were recorded with a Thermo Nicolet Avatar instrument, covering a wave scan range from 400 to 4000 cm^{-1}. Electrochemical analysis was performed in Metrohm Autolab instrument. The active material: acetylene black: PVDF (polyvinylidene fluoride) [80,10:10] in NMP was deposited as slurry on carbon cloth, which was used as the electrode in 1 M H_2SO_4 electrolyte.

4 Results and Discussion

4.1 Characterization

The FT-IR spectra of P, (Fig. 1(a)) shows characteristic peaks at 1561 cm^{-1} and 1456 cm^{-1} corresponding to C=C stretching in the quinoid (Q) and benzenoid (B) rings, respectively. Additional peaks at 1298 cm^{-1}, 1228 cm^{-1}, 1137 cm^{-1}, and 802 cm^{-1} correspond to protonation of N atoms of Q ring, C–N stretching in the B ring, N–Q–N merged with O=S=O stretching of pTSA, and C-H out-of-plane bending merged with S-O stretching, while the peak at 566 cm^{-1} corresponds to C-S stretching [3]. In the case of N, absorption bands at 3434 cm^{-1} and 1638 cm^{-1} are attributed to O-H stretching and bending vibrations of adsorbed water, while peaks at 598 cm^{-1} and 410 cm^{-1} correspond to Fe–O and Ni–O stretching, respectively. The PN composite exhibits characteristic peaks of both PANI and nickel ferrite.

In the XRD patterns (Fig. 1(b)), pure P shows peaks at $2\theta = 20°$, 25°, and 15.1°, corresponding to the (020), (200), and (011) planes, respectively. N exhibits peaks at 30.4^0, 35.76^0, 37.41^0, 43.5^0, 54.1^0, 57.5^0, 63.14^0 related to the (220), (311), (222), (400), (422), (511), and (440) lattice planes [4]. The FE-SEM image [Fig. 1(c)] of P reveals a branched rod-like morphology with an average diameter of 69 nm. In contrast, N exhibits an agglomerated spherical shape with an average size of 22 nm. The $PN_{0.1}$ composite shows that ferrite particles are uniformly distributed on polyaniline branched nanorods.

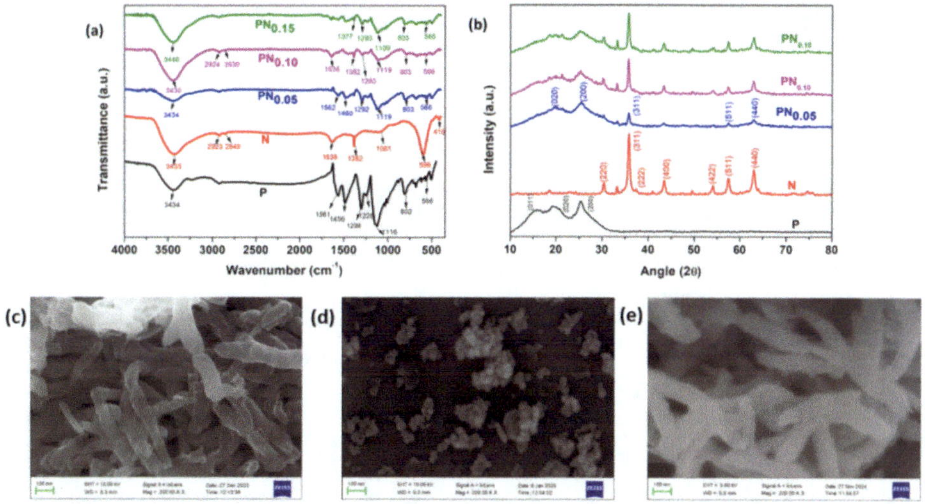

Fig. 1. (a) FTIR and (b) XRD of P, N and their composites and FE- SEM images of (c) P (d) N (e) $PN_{0.10}$

4.2 Electrochemical Performance

To evaluate the electrochemical behaviour and current response, cyclic voltametry studies of the prepared electrode materials P, N, PN composites were conducted at the scan rate 10 mV/s. The CV curve (Fig. 2(a)) shows that while P and N shows a potential window 0.5 V, PN composites, exhibited a window of 0.8 V. The CV curve of N reveal a quasi-rectangular shape [5] while P and PN composites showed a pseudocapacitive behaviour [6]. The specific capacitance from CV was calculated using (Eq. 1) and was found to be 56 F/g, 2 F/g, 214 F/g, 231 F/g, 157 F/g at 10 mV/s for P, N, $PN_{0.05}$, $PN_{0.1}$, $PN_{0.15}$ respectively.

$$Cs = \frac{A}{2m \times v x \Delta V} \tag{1}$$

where Cs, m, v, ΔV, and A are the specific capacitance (F/g), active mass (g), scan rate (mV/s), potential window (V), and the area under the curve, respectively.

Galvanostatic charge discharge (GCD) performance of the materials were performed at 1 A/g current density. The specific capacitance is calculated from the GCD profile using Eq. (2).

$$C = \frac{I \times \Delta t}{m \times \Delta V} \tag{2}$$

where C is the specific capacitance, I is the current response, Δt is the discharge time and m are the active mass. In agreement with CV curves, GCD curves also shows a quasi-triangular shape indicating pseudo capacitive charge storage mechanism. The specific capacitance was observed as 180 F/g, 33 F/g, 284 F/g, 513 F/g, 245 F/g for P, N, $PN_{0.05}$, $PN_{0.1}$, $PN_{0.15}$ respectively. To understand the charge transfer kinetics of

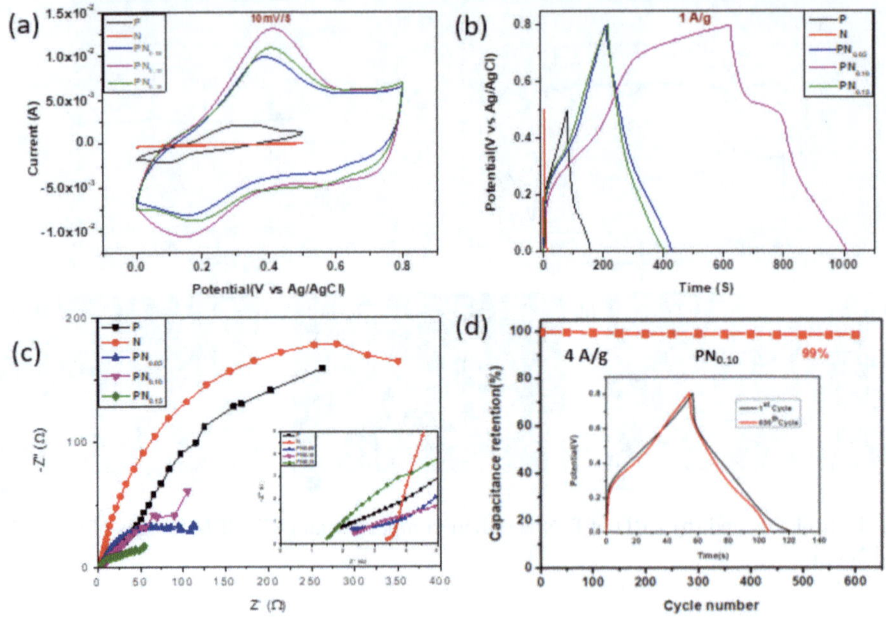

Fig. 2. (a) CV curves (b) GCD (c) EIS of P, N and PN composites (d) cyclic stability curve of $PN_{0.10}$

electrode materials EIS test (Fig. 2(c)) was performed. From Nyquistic plot the X-axis intercepts in the high-frequency range represents the equivalent series resistance (R_s). The active electrode's intrinsic resistance, the electrode-electrolyte contact resistance, and the electrolyte's bulk resistance all contribute to resistance (R_s). The diameter of the semicircle which indicates the charge transfer resistance (R_{ct}). The R_s were obtained as 1.90 Ω, 3.35 Ω, 2.76 Ω, 2.13 Ω, 1.48 Ω and charge transfer resistance values are 4.29 Ω, 4.01 Ω, 4.04 Ω, 3.61 Ω, 3.92 Ω for P, N, $PN_{0.05}$, $PN_{0.1}$, $PN_{0.15}$ respectively. A lower resistance would promote the electrolyte ions to diffuse towards the electrode surface. $PN_{0.1}$ composites exhibits lower charge transfer resistance compared to pure P. Consequently, the presence of N nanoparticles on P rods positively influences the specific capacitance, leading to a synergistic effect on the specific capacitance. To assess the cyclic durability of the $PN_{0.10}$ nanocomposite, galvanostatic charge–discharge (GCD) measurements were conducted at a current density of 4 A/g over 600 consecutive cycles. The corresponding performance is illustrated in Fig. 2(d). The results show that after 600 cycles, the $PN_{0.10}$ maintains 99% of its specific capacitance, suggesting that the composite can remain very stable over an extended period and could potentially be employed as a stable electrode material for the energy storage device.

5 Conclusion

In this work, a binary nanocomposite (P/N) was prepared by in-situ polymerization method. XRD, FTIR and FESEM confirmed the formation of composite. Among all the nanocomposites, $PN_{0.10}$ exhibits maximum capacitance 513 F/g at 1 A/g with the lowest charge resistance (3.61 Ω) and obtained cyclic stability (600 cycles, at 4 A/g) with good capacitance retention (99%). Moreover, the voltage window was extended to 0.3 V (from 0 to 0.8 V), compared to that of PANI and nickel ferrite.

References

1. Chatterjee, D.P., Nandi, A.K.: A review on the recent advances in hybrid supercapacitors. J Mater Chem A. **9**(29), 15880–15918 (2021)
2. Singh, G., Kumar, Y., Husain, S.: Nickel ferrite doped polyaniline composites: Synthesis and analysis for a high-performance symmetric pseudo-supercapacitor. Energy Storage. **6**(1), e525 (2024)
3. AM, R., Gopalakrishnan, J.: Improved dielectric and dynamic mechanical properties of epoxy/polyaniline nanorod/in situ reduced graphene oxide hybrid nanocomposites. Polym. Compos. **41**(8), 2998–3013 (2020)
4. Rana, G., et al.: Recent advances on nickel nano-ferrite: A review on processing techniques, properties and diverse applications. Chem. Eng. Res. Des. **175**, 182–208 (2021)
5. Ghasemi, A.K., Ghorbani, M., Lashkenari, M.S., Nasiri, N.: Controllable synthesis of zinc ferrite nanostructure with tunable morphology on polyaniline nanocomposite for supercapacitor application. J Energy Storage. **51**, 104579 (2022)
6. Mandal, S., Dasmahapatra, A.K.: Hierarchical polyaniline/copper cobalt ferrite nanocomposites for high performance supercapacitor electrode. J Energy Storage. **74**, 109402 (2023)

All Solid-State Li-S Cells with PEO/PVP Based Solid Polymer Electrolyte

Sreeja Ezhumayil Muraleedharan, Neethu Simon, Vineeth Mohanan Parakkat, and S. Jayalekshmi(✉)

Department of Physics, Cochin University of Science and Technology, Cochin 682022, India
jayalekshmi@cusat.ac.in

Abstract. Lithium-sulfur (Li-S) batteries are considered as promising candidates for next-generation energy storage systems due to the high theoretical capacity of sulfur (1675 mAh g^{-1}). However, their practical application is significantly hindered by the inherent low electrical conductivity of sulfur and the shuttle effect caused by soluble polysulfides. To address these challenges, this study reports the development of a lithium-enriched solid polymer electrolyte (SPE) based on polyethylene oxide (PEO)/polyvinylpyrrolidone (PVP) composite, which functions as both the electrolyte and separator in the assembled All-Solid-State Li-S cell (ASSLi-S). The incorporation of PVP enhances ionic conductivity by disrupting the crystallinity of PEO and promoting a more amorphous and conductive polymer matrix. Electrochemical characterizations were done on the ASSLi -S cell assembled using the modified sulfur cathode and the SPE. The PEO/PVP-based SPE acts as an effective physical barrier against polysulfide migration, thereby mitigating shuttle-related losses. The assembled ASSLi-S cells demonstrate enhanced specific capacity and excellent cycling stability, highlighting the potential of this composite SPE for high-performance lithium-sulfur battery applications. Assembled half cells with modified sulfur cathode and SPE deliver an initial discharge capacity of 805.19 mAh g^{-1} at 0.1 C rate.

Keywords: Solid Polymer Electrolyte · Polysulfides · Ionic Conductivity · Li-S cells

1 Introduction

Li–S batteries are gaining increasing attention as the next-generation energy storage solution due to their exceptionally high theoretical specific capacity of 1675 mAh g^{-1} and energy density reaching up to 2600 Wh kg^{-1} significantly exceeding those of traditional lithium-ion batteries. Beyond these impressive performance characteristics, sulfur's natural abundance, low cost, and environmentally friendly nature make Li–S systems especially appealing for large-scale uses, including electric vehicles and grid-level energy storage. However, several major obstacles still stand in the way of their commercial adoption. Key factors among them are the inherently low electrical conductivity of sulfur and its reduction products (Li_2S_2/Li_2S), the dissolution and migration of intermediate

A. Antony et al. (Eds.): icONMAT 2025, SPPHY 328, pp. 60–67, 2025.
https://doi.org/10.1007/978-981-95-2697-0_9

lithium polysulfides (Li_2S_x where $4 \leq \times \leq 8$), and the substantial volume expansion around 80% that occurs when sulfur is converted into lithium sulfide during charge–discharge cycles. These issues collectively contribute to reduced Coulombic efficiency, rapid capacity loss, and a shortened cycle lifespan[1, 2, 3].

Researchers have focused heavily on advancing Li–S battery technology by designing better sulfur host materials, engineered interlayer systems, and optimized electrolytes. SPEs have emerged as promising candidates, influential in multiple ways that improve the overall cell behaviour. They can significantly inhibit the migration of soluble polysulfides, replace flammable liquid electrolytes with safe, polymer-based systems, and preserve mechanical stability through many charge–discharge cycles [4].

SPEs are particularly attractive because of their advantageous properties, including ease of processing, non-flammability, mechanical flexibility, lightweight nature, and the ability to conform to various cell geometries. Their inherent elastic and plastic deformability also enable them to accommodate the volumetric changes of electrodes during charge–discharge cycles. In solid-state lithium batteries, SPEs serve dual functions as both the electrolyte and the separator, offering additional benefits of (i) structural flexibility, (ii) compact and laminated configurations, (iii) absence of electrolyte leakage, (iv) reduced self-discharge rates, and (v) mechanical stress relaxation [5, 6].

Despite these advantages, a primary limitation of SPEs is their relatively low ionic conductivity, which often falls below the practical benchmark of ~ 10^{-3} S cm^{-1}. This limitation is largely attributed to the semi-crystalline nature of commonly used polymer matrices, which restricts ion mobility. To address this issue, substantial research has been focused on enhancing the ionic conductivity of SPEs through molecular design and compositional optimization. Strategies include the development of novel polymer backbones, increasing the proportion of the amorphous phase, lowering the polymer's glass transition temperature (T_g), and optimizing salt concentration to facilitate efficient ion transport [7, 8]. These approaches aim to achieve an optimal balance between mechanical strength and ionic conductivity, both essential for the realization of high-performance solid-state Li–S batteries.

Typically, SPEs are prepared by dissolving inorganic lithium salts into a polymer matrix, where the polymer acts as the host medium for ion conduction via segmental motion of polymer chains. The interaction between metal cations and polar groups in the polymer is governed by electrostatic forces. Several factors influence the polymer–ion interactions, including the chemical nature of the functional groups on the polymer backbone, their spatial arrangement and composition, molecular weight, degree of branching, and the chemical characteristics and charge density of the metal cation and its counter ion. Understanding and manipulating these parameters are critical for tailoring the electrochemical performance of SPEs [9, 10].

In this study, we report the development and electrochemical evaluation of ASSLi–S cells, incorporating a solid polymer electrolyte with 15 wt% $LiNO_3$ and a modified sulfur cathode. The effects of the optimized electrolyte composition on ionic conductivity, interfacial stability, and long-term cycling performance are systematically investigated, demonstrating the potential of the proposed system for high-performance, solid-state Li–S batteries.

2 Experimental Methods

2.1 Preparation of SPE Film

PEs were grown using a straightforward solution casting technique. PEO (90%) and PVP (10%) were dissolved in methanol under continuous stirring to form a homogeneous PEO–PVP blend solution (10 mL). Various concentrations of $LiNO_3$, ranging from 5 wt% to 15 wt%, were then added to the solution and stirred continuously for 12 h [11]. The resulting mixture was poured into a Teflon petri dish and dried at 60 °C for 12 h to obtain transparent, free-standing SPE films. A schematic illustration of the film preparation process is shown in Fig. 1.

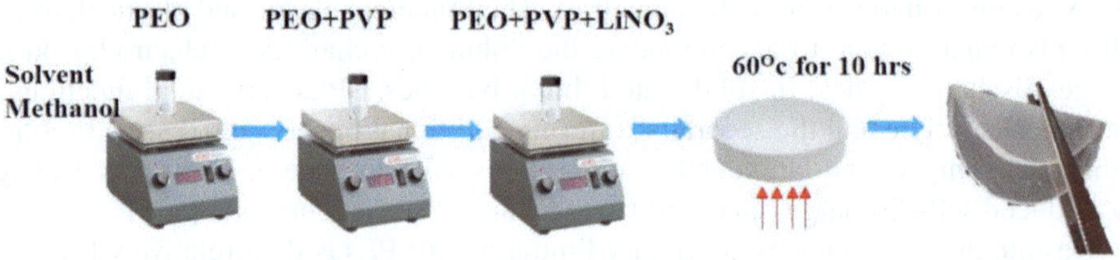

Fig. 1. Schematic representation for the preparation of SPE film

2.2 Cathode Modification and Cell Assembly

ASSLi-S cells were assembled in a' half-cell configuration in an argon filled glove box using Swagelok-type cells. The sulfur cathode was modified using polyaniline (PANI) and activated carbon (AC), as described in our previous report [12], resulting in a composite referred to as PACS. The modified cathode was made by preparing a slurry composed of 80 wt% active material, 10 wt% carbon black as the conductive agent, and 10 wt% polyvinylidene fluoride (PVDF) serving as the binder, all dispersed in suitable amount of N-methyl-2-pyrrolidone (NMP). This mixture was then evenly spray-coated onto aluminum foil and dried under vacuum at 60 °C for 12 h. The SPE film based on PEO/PVP and infused with $LiNO_3$ was employed as the electrolyte in the cell assembly.

2.3 Characterizations

The morphological characterization was with field-emission scanning electron microscopy with a Carl Zeiss Sigma instrument. The electrochemical behaviour of the assembled cells was assessed through electrochemical impedance spectroscopy (EIS) and galvanostatic charge–discharge (GCD) measurements. These analyses were performed using a HP4192A impedance analyzer in the frequency range from 5 Hz to 10 MHz for EIS, and an MTI Corporation BST8-WA eight-channel battery analyzer for GCD measurements.

3 Results and Discussion

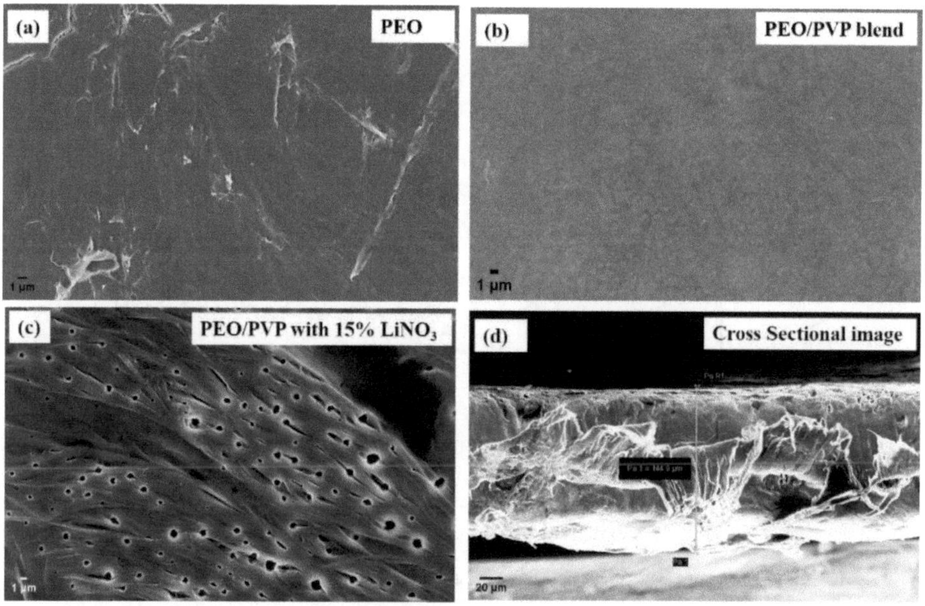

Fig. 2. FESEM images of (a) PEO, (b) PEO/PVP blend (c) PEO/PVP blend with 15% $LiNO_3$ (d) Cross sectional image of the prepared SPE with 15% $LiNO_3$

Figure 2 presents the FESEM images of (a) pure PEO, (b) the PEO/PVP polymer blend, and (c) the PEO/PVP blend incorporating 15 wt% $LiNO_3$. The surface morphology of pure PEO is characterized by a rough and irregular texture, indicating limited film uniformity. Blending PVP with the PEO matrix leads to a smoother and more uniform surface, indicating better compatibility and improved film-forming properties. When 15 wt% $LiNO_3$ is added, distinct and evenly distributed pores begin to appear in the polymer structure, as seen in Fig. 2(c). This rise in porosity is likely to promote better ion transport, which in turn contributes to enhanced ionic conductivity in the solid polymer electrolyte films.[11]. Figure 2(d) presents the cross-sectional SEM image of the SPE film having 15% $LiNO_3$.

Ionic conductivity measurements of the prepared SPE films were carried out using Electrochemical Impedance Spectroscopy (EIS). The films were sandwiched between two stainless steel (SS) electrodes, which served as blocking electrodes to suppress faradaic currents and ensure that only ionic conduction was dominant. An AC voltage was applied over the frequency range of 5 Hz to 10 MHz. From the resulting Nyquist plots, the bulk resistance (R_b) was determined from the high-frequency intercept on the real axis. The ionic conductivity (σ) was then calculated using the relation:

$$\sigma = \frac{t}{R_b A} \tag{1}$$

where t is the thickness of the SPE film, R_b is the bulk resistance of the SPE film obtained from Nyquist plot and A is the area of the SPE film. The Nyquist plots for the SPE films containing different wt% of $LiNO_3$ are given in Fig. 3.

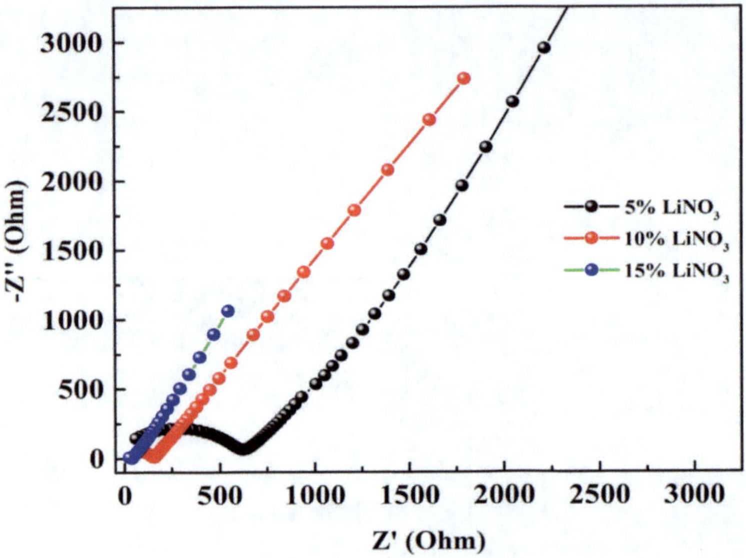

Fig. 3. Nyquist plots of the prepared SPE's for different concentrations of $LiNO_3$

The ionic conductivity was calculated using Eq. (1) for different wt% of $LiNO_3$ by obtaining the values of R_b, A and t of the SPE films. The calculated ionic conductivities are summarized in Table 1.

Table 1. Ionic conductivities of SPE films for different wt% of $LiNO_3$

$LiNO_3$ wt%	SPE film thickness (t)(μm)	R_b (Ω)	Ionic Conductivity (S/cm)
5	124.6	629.41	6.98×10^{-6}
10	131.7	161.34	2.88×10^{-5}
15	144.9	38.53	1.32×10^{-4}

A progressive increase in $LiNO_3$ concentration from 5 wt% to 15 wt% results in a marked enhancement in ionic conductivity. Specifically, the R_b decreases from 629.41 Ω to 38.53 Ω, while the ionic conductivity increases from 6.98×10^{-6} S cm^{-1} to 1.32×10^{-4} S cm^{-1}. This improvement is attributed to the higher concentration of mobile Li^+ ions and enhanced segmental mobility of the polymer chains, facilitating more efficient ion transport [5]. Among the compositions investigated, the SPE film containing 15 wt% of $LiNO_3$ demonstrates the highest ionic conductivity and is therefore selected for cell fabrication indicated in Fig. 4(c). Half-cells were assembled using the modified cathode and the optimized SPE film in a glove box under argon atmosphere to ensure an inert, moisture-free environment.

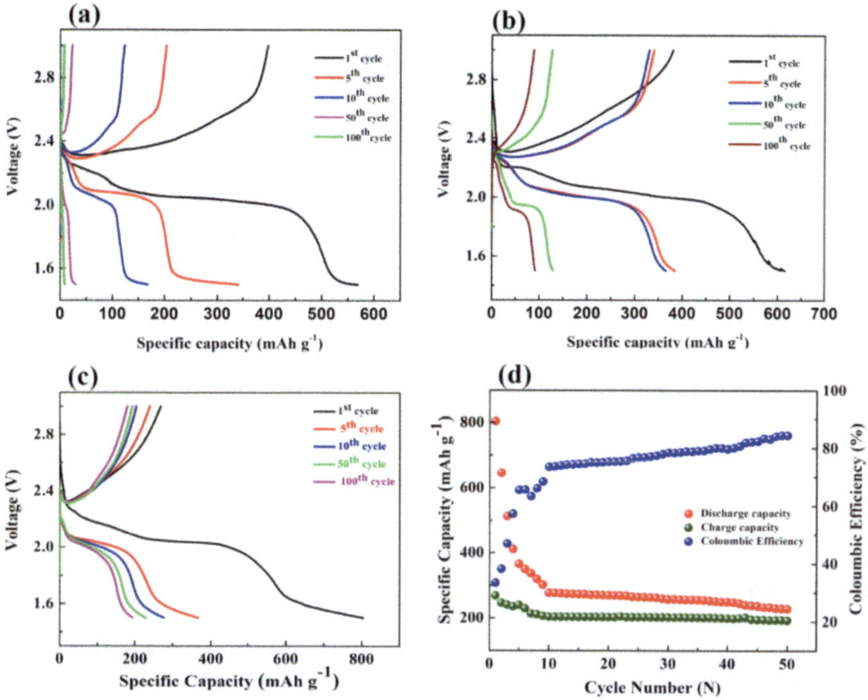

Fig. 4. Galvanostatic charge discharge studies of cells assembled with (a) 5 wt% $LiNO_3$ (b) 10 wt% $LiNO_3$ (c) 15 wt% $LiNO_3$ (d) cycling stability curve for cells assembled with 15% $LiNO_3$

The electrochemical performance of the ASSLi–S cells, assembled with SPE containing different wt% of $LiNO_3$ and the modified sulfur cathode, was evaluated at a current rate of 0.1C, and the galvanostatic charge-discharge plots are shown in Fig. 4. The cells deliver an initial discharge capacity of 569.02 mAh g^{-1}, 614.77 mAh g^{-1} and 805.19 mAh g^{-1} respectively with 5 wt%, 10 wt% and 15 wt% $LiNO_3$.The increasing trend in capacity indicates that higher $LiNO_3$ content enhances the ionic conductivity of the solid polymer electrolyte. The optimized SPE containing 15 wt% of $LiNO_3$ gradually stabilizes at approximately 200 mAh g^{-1} after 100 cycles, corresponding to a capacity retention of ~ 25% as shown in Fig. 4(d). The sharp decline in capacity observed within the first five cycles is likely attributed to the formation of a solid electrolyte interphase (SEI) layer and initial electrolyte degradation. However, beyond this initial activation phase, the cell exhibits relatively stable cycling behavior, suggesting the formation of a passivating and stable SEI. The sustained capacity over prolonged cycling demonstrates the effective suppression of polysulfide migration by the SPE and highlights good utilization of the active material. These results confirm the potential of the SPE system in enhancing the long-term performance of all-solid-state Li–S batteries.

4 Conclusions

In this work, ASSLi–S cells were developed by employing the modified sulfur cathode and the PEO/PVP-based SPE film incorporated with 15 wt% of $LiNO_3$. The SPE film was grown through a simple and scalable solution casting method, making it suitable for practical applications. The incorporation of PACS into the sulfur cathode significantly enhances its electrical conductivity and electrochemical performance. The assembled cells exhibit an initial discharge capacity of 805.19 mAh g^{-1} and sustain a capacity of 200 mAh g^{-1} after 100 cycles at 0.1 C, corresponding to a capacity retention of approximately 25%. These findings highlight the synergistic effects of $LiNO_3$ additive and cathode modification in improving ionic conductivity, interfacial stability, and long-term cyclability. Overall, this study underscores the potential of $LiNO_3$ enriched SPEs and conductive composite cathodes as effective strategies for advancing the performance and durability of ASSLi–S batteries.

Acknowledgments. The author Sreeja E M is thankful to Dr. Sasidevan V for the timeless support and guidance and Cochin university of Science and Technology (CUSAT), Government of Kerala for the financial support in the form of senior research fellowship. The authors would also like to acknowledge the seed money grant from CUSAT for the financial assistance.

References

1. Santiago, A., et al.: Salt additives for improving cyclability of polymer-based all-solid-state lithium-sulfur batteries. ACS Appl. Energy Mater. **4**(5), 4459–4464 (2021)
2. Li, Y., Shapter, J.G., Cheng, H., Xu, G., Gao, G.: Recent progress in sulfur cathodes for application to lithium–sulfur batteries. Particuology (58), 1–15 (2021).
3. Azam, S., Wei, Z., Wang, R.: Cerium oxide nanorods anchored on carbon nanofibers derived from cellulose paper as effective interlayer for lithium sulfur battery. J. Colloid Interface Sci. **615**, 417–431 (2022)
4. Machín, A., Morant, C., Márquez, F.: Advancements and challenges in solid-state battery technology: An in-depth review of solid electrolytes and anode innovations. Batteries. **10**(1), 29 (2024)
5. Li, Z., et al.: Ionic conduction in polymer-based solid electrolytes. Adv. Sci. **10**(10), 1–18 (2023)
6. Zhao, Y., et al.: Polymer electrolytes for lithium/sulfur batteries. Membranes. **2**(3), 553–564 (2012)
7. Zhang, X.L., et al.: A multifunctional nano filler for solid polymer electrolyte toward stable cycling for lithium-metal anodes in lithium–sulfur batteries. Chem. Eng. J. **444**, 136328 (2022)
8. Li, S., Li, L., Yang, H., Zhao, Y., Shan, Y.: A review of composite polymer electrolytes for solid-state lithium-sulfur batteries: Synthesis methods, optimal design, and critical challenges. Chem. Eng. J. **484**, 149433 (2024)
9. Jayathilaka, P.A.R.D., Dissanayake, M.A.K.L., Albinsson, I., Mellander, B.: Effect of nano-porous Al_2O_3 on thermal, dielectric and transport properties of the$(PEO)_9$ LiTFSI polymer electrolyte system. Electrochim. Acta. **47**, 3257–3268 (2002)
10. Tong, B., et al.: Ion transport and structural design of lithium-ion conductive solid polymer electrolytes: A perspective. Mater. Futur. **1**(4), 42103 (2022)

11. Jinisha, B., Anilkumar, K.M., Manoj, M., Pradeep, P., Jayalekshmi, J.: Development of a novel type of solid polymer electrolyte for solid state lithium battery applications based on lithium enriched poly (ethylene oxide) (PEO)/poly (vinyl pyrrolidone) (PVP) blend polymer. Electrochim. Acta. **235**, 210–222 (2017)
12. M, S.E., Thomas, A., Wilson, M.K., Abhilash, A., Jayalekshmi, S.: Self-stabilized dispersion polymerized aniline/carbon nanotubes based flexible interlayer films for boosting performance characteristics of lithium-sulfur batteries. J. Inorg. Organomet. Polym. Mater. **35**(3), 1808–1818 (2025)

Optically Transparent Conducting SnO_2 Films as Semiconductor Medium for Photoanodes in PEC Cells

Shalom Varghese[1], Reshma Joseph[1], Mahitha Madhu[1], Fazalurahman Kuttassery[2], Hiroshi Tachibana[3], Haruo Inoue[3], and Sebastian Nybin Remello[1](✉)

[1] Department of Applied Chemistry, Cochin University of Science and Technology, Kochi 682022, Kerala, India
nybinremello@cusat.ac.in

[2] Department of Chemistry, University of Calicut, Malappuram 673635, Kerala, India

[3] Department of Applied Chemistry for Environment, Graduate School of Urban Environmental Sciences, Tokyo Metropolitan University, Tokyo 1920397, Japan

Abstract. Metal oxide semiconductor nanoparticles represent a promising class of semiconductor substrates for various chemical architectures, particularly in photoelectrochemical cells. Nonetheless, their practical application in photocatalysis is frequently impeded by nanoscale particle aggregation, predominantly driven by van der Waals interactions. To overcome this limitation, the introduction of electrostatic repulsive forces among nanoparticles has been demonstrated to effectively reduce aggregation. In this study, tin oxide nanoparticles were effectively synthesized using $SnCl_4.5H_2O$ as the precursor. The resultant nanoparticles, initially in a wet suspension, were converted into an optically transparent dispersion by adding HCl and the polar protic solvent methanol. This transparent dispersion was subsequently used to fabricate transparent photoanodes by drop-casting onto FTO substrates. The visible-light-to-photocurrent efficiency of the photoanodes was assessed by sensitizing them with tin porphyrins, which served as photosensitizers. These films are anticipated to enhance photon absorption because of the absence of light scattering in the coated transparent nanoparticles.

Keywords: Transparent Photoanodes · Metalloporphyrins · PEC cells

1 Introduction

Photoelectrochemical (PEC) cells are a promising and sustainable approach for solar-to-chemical energy conversion, offering potential solutions to reduce reliance on fossil fuels and mitigate related environmental impacts [1]. Molecular catalyst-sensitized systems (MCSS) represent an emerging category of heterostructures that efficiently facilitate half-reactions in PEC devices. MCSS uses molecular catalysts that possess suitable thermodynamic potentials to facilitate the relevant half-reactions [2]. It also includes chromophore functionalities that absorb sunlight and produce one-electron oxidized or

A. Antony et al. (Eds.): icONMAT 2025, SPPHY 328, pp. 68–74, 2025.
https://doi.org/10.1007/978-981-95-2697-0_10

reduced species, which enable catalytic oxidation or reduction [3]. A critical consideration in the application of MCSS is the aggregation of semiconductor nanoparticles in aqueous environments. Aggregation leads to increased light scattering, which diminishes the effective light absorption by the molecular catalyst sensitizers (MCS) attached to the semiconductors [4]. Moreover, nanoparticle aggregation induces pH fluctuations within the nanometric interstitial spaces; as hydroxide ions are consumed and protons accumulate during water oxidation, while protons are reduced for hydrogen evolution [5]. These variations can compromise the stability of the MCS affixed to semiconductors, promoting desorption. Although buffer solutions have been proposed to stabilize pH and improve the quantum yield of redox products, these strategies do not fully resolve the issue of pH fluctuations. Recent research suggests that transparent nanoparticles, when maintained in monomeric states and devoid of aggregation, do not scatter light and effectively mitigate the pH fluctuations observed in agglomerated particles (Fig. 1) [6]. This finding presents novel opportunities for the advancement of artificial photosynthesis. Monomeric particles were synthesized via the sol-gel method under basic conditions, followed by surface protonation to induce sufficient electrostatic repulsion to prevent aggregation. To date, the integration of these transparent monomeric nanoparticles with MCSS for artificial photosynthesis remains unexplored. In this study, we fabricated transparent tin oxide-based films incorporating metalloporphyrin systems for photoelectrochemical applications. The transparent films were characterized using AFM, SEM, EIS, and the dye-sensitized film electron injection efficiency was assessed through photocurrent measurements.

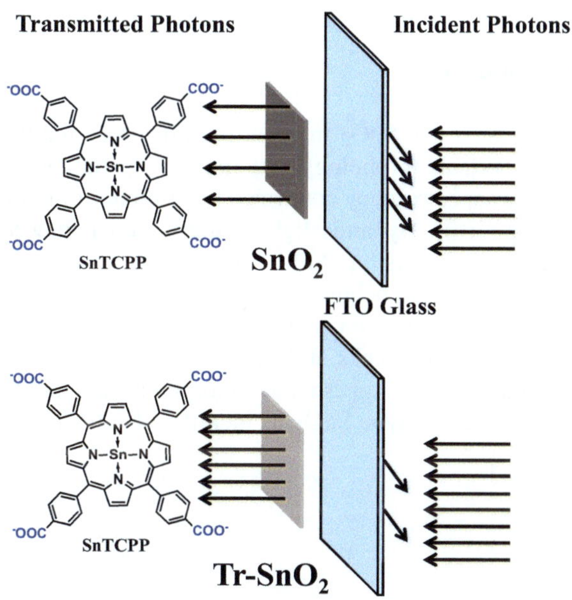

Fig. 1. Schematic illustration of enhanced photon absorption in transparent photoanodes

2 Experimental

Sodium hydroxide pellets (NaOH), tin (IV) chloride pentahydrate ($SnCl_4 \cdot 5H_2O$), hydrochloric acid (36% HCl), and methanol were procured from Spectrochem Pvt. Fluorine-doped tin oxide-coated glass (FTO) slides (100 mm × 100 mm × 2.2 mm, surface resistivity ~ 13 Ω/sq) were sourced from Merck. Acetonitrile (ACN, 99.5% for spectrophotometry) and tetrapropylammonium iodide (TPAI) were obtained from TCI. All reactions were conducted in oven-dried glassware using commercially available solvents. Diffuse reflectance spectra (DRS) were recorded using a Lambda double-beam UV-Vis spectrophotometer. Particle size was measured with a Horiba SZ-100. Atomic force microscopy (AFM) was performed using a Keysight 9500 instrument. Scanning electron microscopy (SEM) images were captured with a Zeiss Sigma FE-SEM instrument. Electrochemical impedance spectroscopy (EIS) and photoelectrochemical (PEC) experiments were conducted using a CHI600E workstation (CH Instruments, USA). Light irradiation was provided by a low-power white LED from Comtek Scientific Instruments India. Tin (IV)-tetra (4-carboxyphenyl) porphyrin (SnTCPP), the catalyst sensitizer, was also synthesized using reported methods.

The transparent SnO_2 (Tr-SnO_2) nanoparticle dispersion was prepared as described previously. The Tr-SnO_2 dispersion in H_2O/MeOH (1:9, v/v) was drop-cast onto cleaned FTO substrates and left overnight; this process was repeated eight times to form a uniform film of FTO/Tr-SnO_2. The resulting FTO/Tr-SnO_2film was soaked overnight in a 0.2 mM porphyrin solution, rinsed with ethanol, dried to get FTO/Tr-SnO_2/SnTCPP, and characterized.

3 Result and Discussion

FE-SEM analysis indicated that SnO_2 nanoparticles incorporated within the transparent film exhibited a spherical morphology with uniform spatial distribution (Fig. 2(a)). Energy-dispersive X-ray spectroscopy confirmed the presence of both tin (Sn) and oxygen (O) elements (Fig. 2(b)). Dynamic light scattering data showed an average particle size of 14 nm in the dispersion. AFM performed in tapping mode over a 4 μm area, revealed an average feature height of 69 nm and a surface roughness of 9.82 nm (Fig. 3(a)). Powder X-ray diffraction patterns (Fig. 3(b)) for the synthesized SnO_2 nanoparticles presented distinct peaks at 2θ angles of 26.6°, 33.9°, 51.9°, and 65.1°. The broad nature of these peaks suggested an amorphous structure with reduced crystallite size, aligning with rutile-type SnO_2 as identified by JPCDS reference 41–1445 [7].

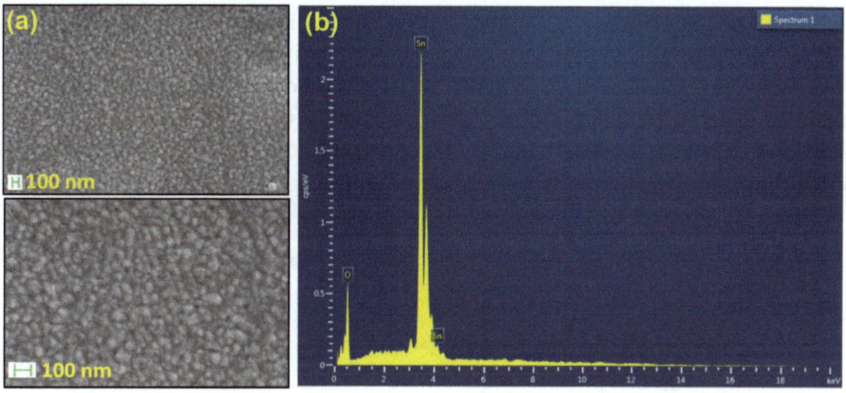

Fig. 2. (a) FE-SEM image and (b) EDAX of the SnO_2 nanoparticles

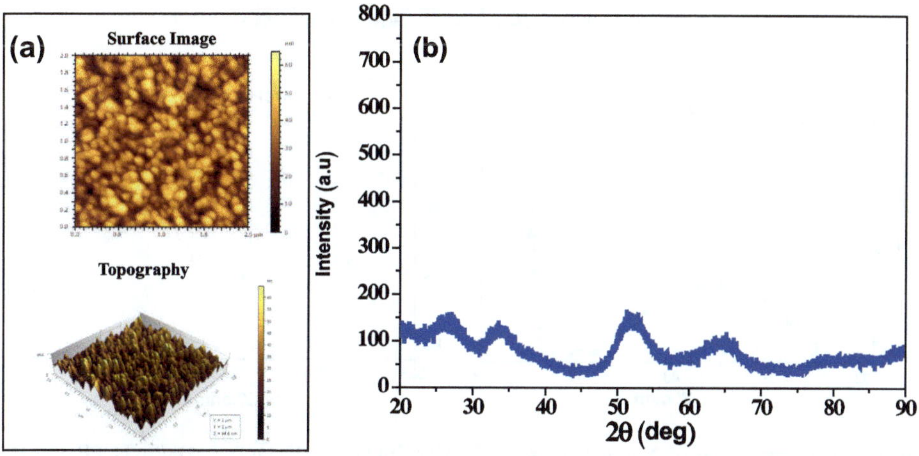

Fig. 3. (a) AFM image of the $Tr\text{-}SnO_2$ film and (b) PXRD of the SnO_2 nanoparticles

Electrochemical impedance spectroscopy (EIS) was utilized to examine charge transfer phenomena at the electrode–electrolyte interfaces of transparent SnO_2 films and FTO/Tr-SnO_2/SnTCPP composite films. During impedance measurements, a sinusoidal AC voltage is applied across a frequency range, eliciting an electrochemically induced current at the interface [8]. By measuring both real (Z') and imaginary (Z") components, Nyquist plots for FTO/Tr-SnO_2 and FTO/Tr-SnO_2/SnTCPP (Fig. 4(a)) can be generated. The transparent FTO/Tr-SnO_2 film displays a high-frequency semicircular feature in its Nyquist plot, corresponding to a charge transfer resistance ($Rct = R_b\text{-}R_s$) of 10.7 Ω (Fig. 4(b)) significantly lower than values reported for white SnO_2 powder. At lower frequencies, a linear response perpendicular to the real axis indicates predominant capacitive behavior and restricted diffusion of active species within the electrolyte. Conversely, the FTO/Tr-SnO_2/SnTCPP film demonstrates a semicircular feature associated with charge transfer resistance, together with a line exhibiting a − 45° slope, indicative of Warburg impedance and thus, diffusion of electroactive species in the electrolyte [9]. The overlapping semicircular arc and 45° line signify the simultaneous presence of slow diffusion and enhanced charge transfer processes with Rct = 0.5 Ω (Fig. 4(c)). Overall, these findings advance the understanding of electron transfer mechanisms in transparent tin oxide films functionalized with molecular catalysts.

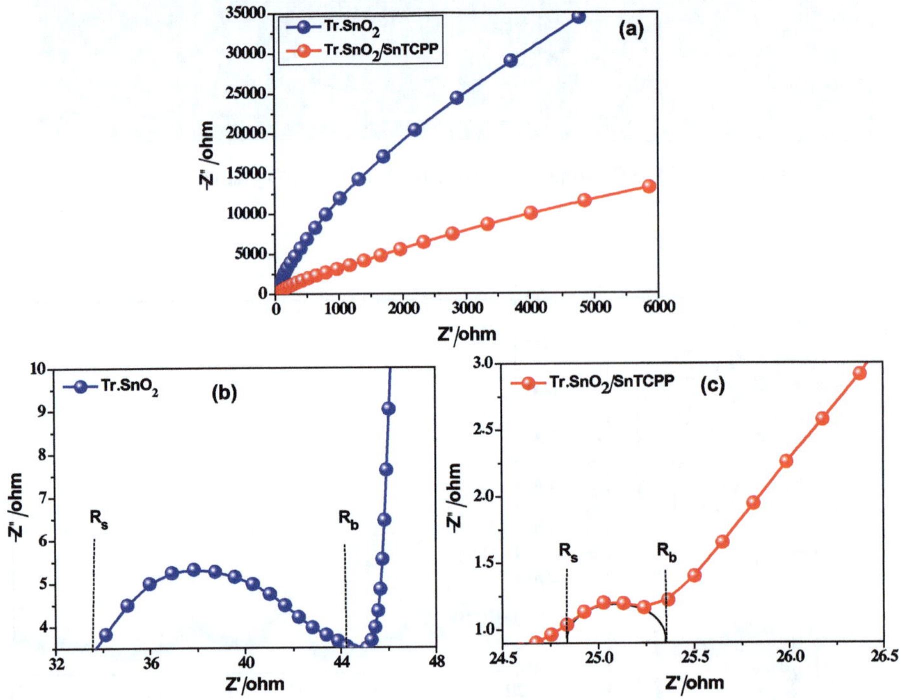

Fig. 4. (a) Comparison of Nyquist plots of FTO/Tr-SnO_2 and FTO/Tr-SnO_2/SnTCPP, the Rct determination for (b) FTO/Tr-SnO_2 and (c) FTO/Tr-SnO_2 /SnTCPP

The FTO/Tr-SnO_2 /SnTCPP transparent films exhibited broadened absorption peaks associated with porphyrin bands (Fig. 5(a)), in addition to a distinct band attributable to the SnO_2 semiconductor, thereby confirming the successful incorporation of SnTCPP. These fabricated FTO/Tr-SnO_2/SnTCPP films were evaluated as photoanodes through photocurrent measurements via linear sweep voltammetry (LSV) under chopped illumination (Fig. 5(b)). The photoelectrochemical system, comprising the FTO/Tr-SnO_2/SnTCPP photoanode and a platinum counter electrode, was investigated in an acetonitrile (ACN) solution containing 0.1 M $TBAPF_6$ and 0.1 M tetrapropylammonium iodide (TPAI) as the sacrificial electron donor. Light illumination was performed at the Soret band of tin porphyrins ($\lambda = 420$ nm) using a white LED light source. The results demonstrated a stable photocurrent density of 4 $\mu A/cm^2$ at 0 V, which marginally increased with rising potential; however, beyond 0.3 V, the photocurrent appeared to hide beneath the onset of dark current. Further analysis via LSV in ACN/H_2O enabled measurement of photocurrent generated by water splitting as the electron source. The current–potential profiles indicated an enhancement in anodic photocurrent. Notably, in ACN/H_2O system the photoanode maintained an anodic current density of 1 μA/cm^2 at 0 V, ascribed to water oxidation in the absence of sacrificial electron donors. Collectively, these findings elucidate the solar energy conversion and light-harvesting properties of the FTO/Tr-SnO_2/SnTCPP configuration utilizing transparent SnO_2, highlighting its suitability as a potential photoanode for MCSS systems. The study underscores the effective visible-light-driven photocurrent conversion efficiency achieved by the transparent SnO_2 substrate sensitized with anionic tin porphyrin (SnTCPP) on FTO glass.

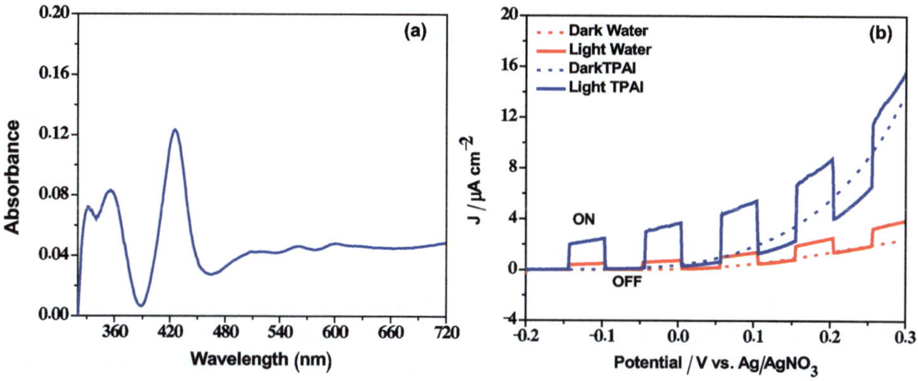

Fig. 5. (a) DRS of FTO/Tr-SnO_2/SnTCPP, and (b) LSV of FTO/Tr-SnO_2/SnTCPP under chopped white LED illumination

4 Conclusion

The integration of Tr-SnO_2 nanoparticles within the photoanodes enhances photon loss mitigation due to light scattering effects in photoelectrochemical devices. Comprehensive morphological, spectroscopic, and electrochemical analyses clearly demonstrate the structural uniformity of these films. The prepared FTO/Tr-SnO_2 exhibited improved charge transfer capabilities, and the incorporation of SnTCPP facilitated effective electron injection at the interfaces. Although the photocurrent observed for the MCSS photoanodes is moderate, they demonstrated responses even with water as the electron source. The shortfalls can be optimized by modifying parameters such as increasing the adsorption of SnTCPP through covalent linking and enhancing the thickness of the Tr-SnO_2 layer. The notable synergy between enhanced electron transfer and effective light harvesting underscores the FTO/Tr-SnO_2/SnTCPP system as a superior photoanode, particularly for a molecular catalyst sensitized system (MCSS) in PEC applications.

References

1. Jiang, C., Moniz, S.J.A., Wang, A., Zhang, T., Tang, J.: Photoelectrochemical devices for solar water splitting-materials and challenges. R. Soc. Chem. **46**(15), 4377–4810 (2017). https://doi.org/10.1039/c6cs00306k
2. Zhang, B., Sun, L.: Artificial photosynthesis: Opportunities and challenges of molecular catalysts. R. Soc. Chem. **48**, 2216–2264. Apr. 07 (2019). https://doi.org/10.1039/c8cs00897c
3. Kuttassery, F., Mathew, S., Tachibana, H., Inoue, H.: How one-photon can induce water splitting into hydrogen peroxide and hydrogen by aluminum porphyrins. Rationale of the thermodynamics. Sustain Energy Fuels. **4**(4), 1945–1953 (2020). https://doi.org/10.1039/D0SE00044B
4. Sano, K., Kuttassery, F., Shimada, T., Ishida, T., Takagi, S., Ohtani, B., et al.: Optically transparent colloidal dispersion of titania nanoparticles storable for longer than one year prepared by Sol/Gel progressive hydrolysis/condensation. ACS Appl. Mater. Interfaces **12**(40), 44743–44753 (2020). https://doi.org/10.1021/acsami.0c12951
5. Kuttassery, F., et al.: Water splitting on aluminum porphyrins to form hydrogen and hydrogen peroxide by one photon of visible light. ACS Appl. Energy Mater. **2**(11), 8045–8051 (2019). https://doi.org/10.1021/acsaem.9b01552
6. Sano, K., et al.: Photocatalytic hydrogen evolution from a transparent aqueous dispersion of quantum-sized SnO_2 nanoparticles: Effect of electron trap density within one particle. J. Phys. Chem. C. **128**(32), 13596–13605 (2024). https://doi.org/10.1021/acs.jpcc.4c03501
7. Naz, S., et al.: A simple lowcost method for synthesis of SnO_2 nanoparticles and its characterization. SN Appl. Sci. **2**(5), 975 (2020). https://doi.org/10.1007/s42452-020-2812-2
8. Mei, B.-A., Munteshari, O., Lau, J., Dunn, B., Pilon, L.: Physical interpretations of nyquist plots for EDLC electrodes and devices. J. Phys. Chem. C. **122**(1), 194–206 (2018). https://doi.org/10.1021/acs.jpcc.7b10582
9. Ariyoshi, K., et al.: Electrochemical impedance spectroscopy Part 1: Fundamentals†. Electrochemistry. **90**(10), 102007 (2022). https://doi.org/10.5796/electrochemistry.22-66071

Casein-capped Nano-curcumin: A Promising Candidate for Wound Healing Application

Dhanya P. Jacob[1], K. B. Parvathy[2], and Aldrin Antony[1,3](✉)

[1] Center of Excellence in Advanced Materials, Cochin University of Science and Technology, Kochi, Kerala, India
aldrin@cusat.ac.in
[2] Department of Physics, SNM College Malankara, MG University, Kottayam, India
[3] Department of Physics, Cochin University of Science and Technology, Kochi, Kerala, India

Abstract. Curcumin is a biphenolic active compound derived from turmeric, traditionally used for centuries in cooking, food coloring, and in traditional medicinal systems such as Ayurveda and Chinese medicine. In recent decades, its wide-ranging biological activities have been extensively studied. Curcumin exhibits numerous therapeutic properties, including anti-inflammatory, antimicrobial, antibacterial, anticancer, antioxidant, and antidiabetic effects. However, its clinical application is limited by poor water solubility and low bioavailability, necessitating high doses to achieve effective antibacterial action. These limitations can be overcome through nanoscale formulation. In this study, nanocurcumin (NC) was synthesized using an oil-in-water emulsion technique followed by probe sonication. The sonication process successfully reduced the particle size of micro-curcumin (MC), as confirmed by Field Emission Scanning Electron Microscopy (FESEM), which showed the transformation of rod-like MC structures into nanoparticles. To further improve solubility and bioavailability, NC were encapsulated in sodium caseinate, a milk-derived protein. The wound-healing efficacy of the prepared formulations was assessed using a scratch assay on L929 fibroblast cells. Both caseinate-encapsulated MC and NC exhibited significant wound closure within 18 h, highlighting their potential for therapeutic applications.

Keywords: Nanocurcimin · casein capping · wound healing

1 Introduction

Curcumin is a naturally occurring chemical compound found in the spice turmeric, derived from the rhizome of the *Curcuma longa* plant, a significant medicinal and culinary herb in Asia [1]. The hydrophobic bioactive constituent of turmeric, known as curcumin (diferuloylmethane), is a symmetric molecule composed of two aromatic ring systems with *o*-methoxy phenolic groups linked by a seven-carbon chain containing an α,β-unsaturated β-diketone moiety [2]. This unique chemical structure contributes to curcumin being more soluble in organic solvents while remaining poorly soluble in water

A. Antony et al. (Eds.): icONMAT 2025, SPPHY 328, pp. 75–82, 2025.
https://doi.org/10.1007/978-981-95-2697-0_11

at neutral and acidic pH levels. The structure of curcumin contains a variety of functional groups, including diketone groups, carbon-carbon double bonds, and phenyl rings. Curcumin is, therefore, a superior antioxidant to other powerful and distinctive antioxidant agents. Curcumin contains three reactive functional groups: a diketone moiety and two phenolic groups. Several chemical reactions are associated with its biological activity, including hydrolysis, degradation, enzymatic transformations, Michael addition, reversible and irreversible nucleophilic additions, and hydrogen donation leading to oxidation. These reactions collectively contribute to the diverse biological effects exhibited by curcumin [3].

Curcuminoids are the primary components of curcumin that have antioxidant effects. It possesses a broad spectrum of antibacterial effects against a wide range of bacteria, including antibiotic-resistant strains [3]. Due to its anti-infectious, anti-inflammatory, and antioxidant characteristics, curcumin is considered a possible wound-healing agent. Additionally, it has been discovered that curcumin accelerates the healing of wounds by encouraging the development of granulation tissue, tissue remodeling, and collagen deposition [4]. Medicinal efficacy of curcumin is constrained by some characteristics, including poor bioavailability, low water solubility, and quick metabolism. Another drawback of curcumin is that it might be toxic in high quantities when applied topically. In order to explore all the potential uses for curcumin in the field of wound healing, it is therefore interesting to incorporate various nano-delivery systems to alter the limiting parameters of this molecule [8].

Nanocurcumin (NC), a nanoscale formulation of curcumin, offers enhanced properties including reduced particle size, improved delivery to diseased tissues, better pharmacokinetic profiles, increased absorption, and reduced systemic toxicity. Nanoparticles possess a relatively large surface area, which enhances their dissolution rate and aqueous solubility, thereby increasing drug bioavailability. Additionally, the increased surface area improves the interaction of drug with specific molecular targets, enhancing its pharmacological efficacy [1]. The primary mechanism by which nanocrystals enhance skin penetration is through an increase in saturation solubility, which elevates the concentration gradient and subsequently promotes greater passive diffusion of active compounds across the skin layers [11]. The reduced particle size of NCs enables more efficient penetration through bacterial cell walls, leading to cell death. The bioavailability of nanocurcumin can be further improved by applying a protein-based coating to its surface.

In this study, we synthesized nanocurcumin (NC) using a straightforward and efficient method. To enhance its stability, bioavailability, targeting ability, and absorption, NC was encapsulated with sodium caseinate. Since phytochemicals like curcumin are typically insoluble at neutral pH but become soluble under alkaline conditions due to deprotonation, they can be co-dissolved with casein in its alkaline-dissociated form. Upon acidification, this allows for the formation of self-assembled casein nanoparticles that encapsulate the curcumin. This approach offers a low-cost, energy-efficient, and solvent-free method for nano-encapsulation, aimed at improving the anti-inflammatory efficacy of casein-based phytochemical formulations for wound healing applications [14].

2 Materials and Methods

Curcumin and Casein sodium (NaCs) from milk, were purchased from the Tokyo Chemical Industry. Sodium Hydroxide (NaOH) was obtained from Merk India. The chloroform used for the preparation was from Spectrochem Pvt. Ltd. Deionized water (DI) was used throughout the synthesis process.

The FESEM images of the synthesized particles were obtained using a Carl Zeiss field emission scanning electron microscope. UV-Vis absorption spectra were recorded using a Jasco V-570 spectrophotometer. X-ray diffraction (XRD) patterns were acquired using a PANalytical X-ray diffractometer, and Fourier-transform infrared (FTIR) spectra were recorded using a Thermo Nicolet Avatar 370 DTGS spectrometer

2.1 Synthesis of Casein-capped NC

Synthesis of NC. This follows a two-step process, wherein microcurcumin (MC) is prepared initially, which is then reduced in size to obtain NCs. MC are synthesized using a slightly modified oil-in-water emulsion method, as previously described [5] . Specifically, curcumin (0.05–0.1 wt%) was dispersed in chloroform and added dropwise into boiling water under continuous stirring, resulting in the formation of orange-colored MC particles, which were collected by centrifugation. NC was then prepared using a probe sonication method, wherein 0.1 wt% of MCs dispersed in DI water was sonicated for 5 h at 30% amplitude and a frequency of 20 kHz.

Casein-capped NC. 2 wt% sodium caseinate solution was prepared by hydrating it in DI water for over 6 h at room temperature. The pH of this solution was adjusted to 12 using 4 M NaOH, followed by a 30 min incubation. The previously prepared MC/NC were then added to the sodium caseinate solution at pH 12. After 1 h, the pH was reduced to 7 using 4 M HCl, resulting in an orange-colored solution of casein capped MC (Cs-MC) and casein capped NC (Cs-NC) particles. These solutions were lyophilized to obtain an orange-colored powder which was stored for further application.

3 Results and Discussion

The synthesized particles were further characterized to access its viability towards wound healing application. Fig. 1 (a) and (b) shows the morphology of the synthesized MC and NC particles respectively. From the image it is evident that, for the MC particles a rod like morphology with an average particle size ~2.5 μm was obtained. Fig. 1 (b) confirms the breakdown of MC particles into nano-scale to form NC with an average particle size ~192 nm. The UV-vis absorption spectra shown in Fig. 1(c) present the strong absorption peak of curcumin at 415 nm and is more predominant in its micro and nano formulations.

The structural properties of the synthesized MC particles were characterized using XRD and FTIR spectroscopy. As shown in Fig. 2 (a) and (b), there are no significant structural differences between the MC particles and bulk curcumin. Since MC serves as the precursor material for NC, it is assumed that the NC particles also retain the same structural characteristics.

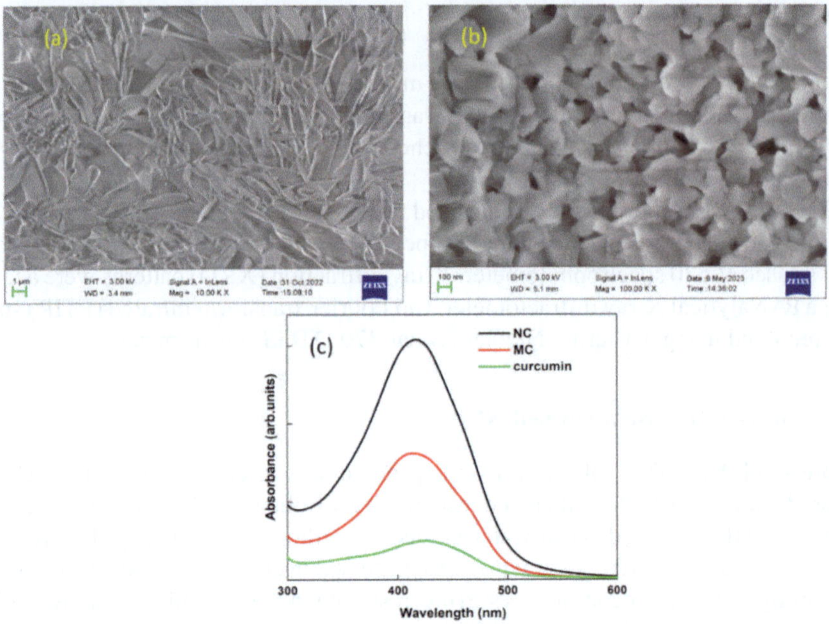

Fig. 1. FESEM images of (a) MC (b) NC (b) UV-Vis absorption spectra of Bulk curcumin, MC and NC particles

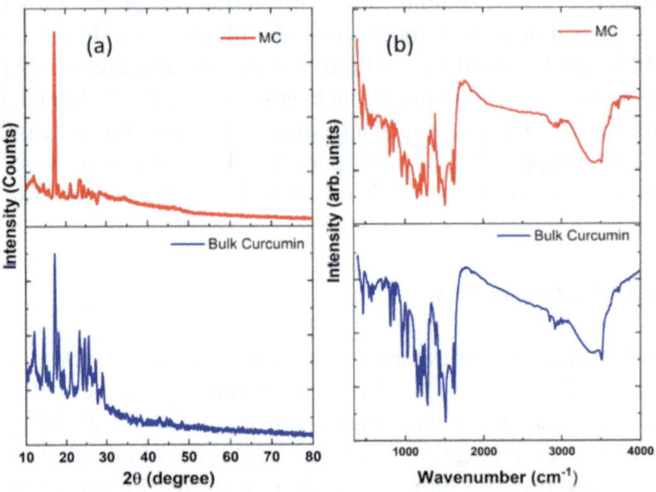

Fig. 2. XRD pattern (a) and FTIR (b) of MC and bulk curcumin

To enhance the bioavailability and water solubility of the synthesized MC and NC particles, they were coated with NaCs, a milk-derived protein. Phytochemicals like curcumin are generally insoluble at neutral pH but become soluble under alkaline conditions due to deprotonation. This property enables co-dissolution of curcumin with casein in its alkaline-dissociated form, which upon acidification, leads to the self-assembly of casein coated nanoparticles. The morphological characteristics of Cs-MC and Cs-NC were examined using FESEM, as shown in Fig. 3(a) and (b). The FESEM images clearly indicate that capping with NaCs facilitated the formation of discrete, non-agglomerated particles and significantly improved the morphology compared to their uncapped counterparts. The Casein capping has significantly improved the water solubility of the MC and NC as shown in the inset of Fig. 3 (a) and (b) respectively, and found to be stable for months. The UV-Vis absorption spectra of NaCs, Cs-MC, and Cs-NC confirm the successful incorporation of casein with the particles, as shown in Fig. 3(c). The characteristic absorption peaks of NaCs at 240 nm and 270 nm are also observed for the capped particles, with a slight blue shift. This blue shift can be attributed to the interaction and possible bond formation between the nanoparticles and NaCs. To evaluate the cytotoxic effect of capped particles, MTT assay has been conducted in L929 mouse fibroblast cells for 48 h for its application towards wound healing. The result depicted in Fig. 3 (d) clearly indicates that more than 95% cells were alive even after 48 h in the presence of NC, Cs-NC and Cs-MC compared that of bulk curcumin.

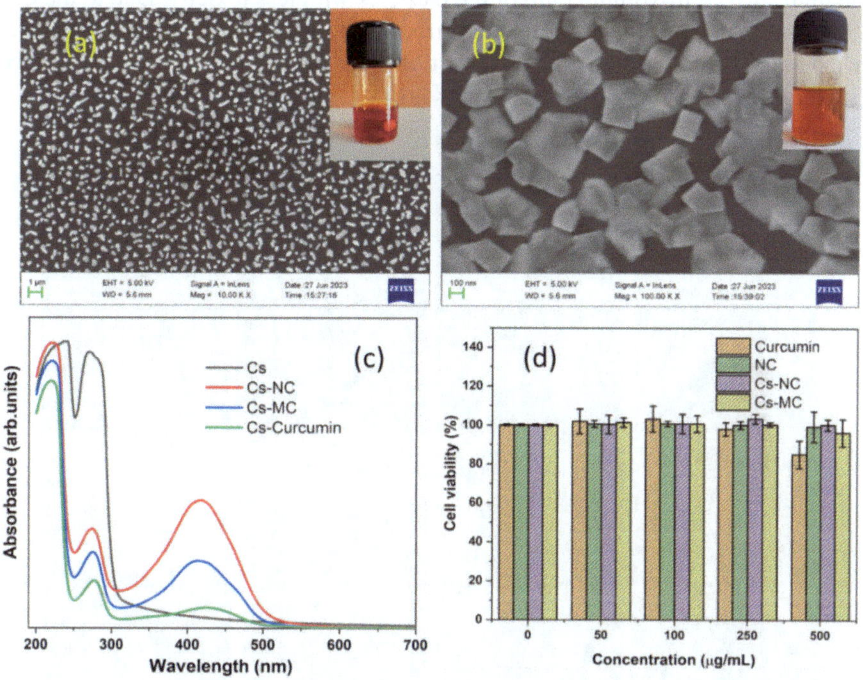

Fig. 3. FESEM images of Cs-MC (a) with an inset of aqueous solution of Cs-MC and (b) Cs-NC with an inset of aqueous solution of Cs-NC. (c) Uv-Vis spectra of the as-prepared Cs capped samples with bulk curcumin and (b) cell viability curve of the as-prepared samples

A cell migration assay/ Scratch assay is a laboratory technique used to study the movement of cells, particularly in the context of wound healing and tissue repair. It stimulates the process of cell migration that occurs naturally in the body during wound healing, tissue regeneration, and various physiological processes. One common application of this assay is in studying how cells collectively move to close a wound or gap in tissue. Cell migration assays in wound healing are crucial for understanding the underlying mechanisms of tissue repair and regeneration. They provide valuable insights into cell behavior, interactions, and response to external stimuli. Herein, we have used this assay to explore the effectiveness of synthesized liposomal particles in wound healing. In this experiment, we created a scratch using a micropipette tip in a fully confluent monolayer L929 mouse fibroblast cell in a culture dish and added 100 μg of Cs-Curcumin, Cs-MC and Cs-NC into the scratched area. By carefully observing the cell migration through an optical microscope for various intervals, we were able to infer the role of each drug in facilitating wound healing. From the assay results shown in Fig. 4, it is evident that after 6 h of injury, more cells have migrated in the presence of drugs when compared to the control. Among the different drugs, after 12 h more cells have been migrated for Cs-MC and Cs-NC than Cs-curcumin. After 18 h, the scratch has been fully confluent by cells in the presence of Cs-NC. From the results, the Cs-NC and Cs-MC showed better results than the bulk as expected. This effective and improved role of Cs-NC can be attributed to the high surface-to-volume ratio of the particles with less toxicity. Due to the Casein coating, the chance for the drug to enter directly into the cells can play a vital role not only in wound healing but also in the reduction of inflammation and swelling caused by the injury.

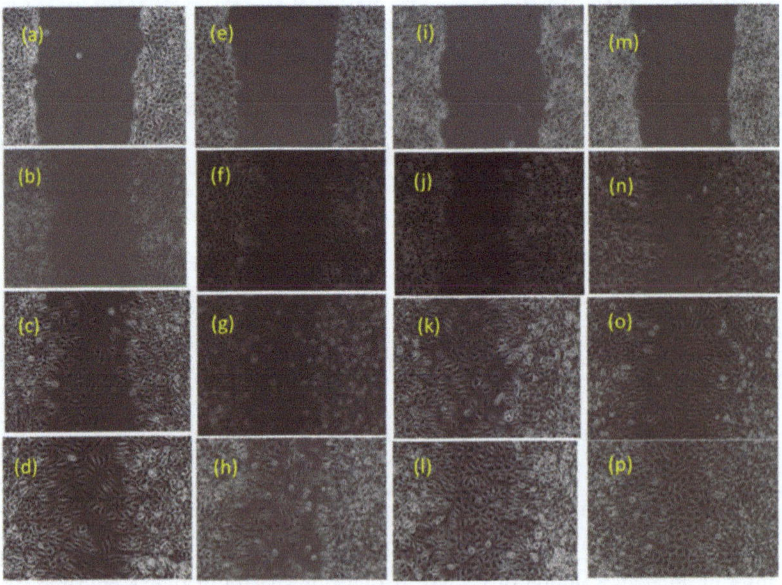

Fig. 4. Cell migration assay results in L929 cells in the presence of drug for different durations; control for 0 h (a), 6 h (b), 12 h(c), 18 h (d); Cs-Curcumin for 0 h (e), 6 h (f), 12 h(g), 18 h (h); Cs-MC for 0 h (i), 6 h (j), 12 h(k), 18 h (l); Cs-NC for 0 h (m), 6 h (n), 12 h(o), 18 h (p).

4 Conclusion

Curcumin, a naturally occurring bioactive compound found in the spice turmeric, is known for its wide range of biological properties. However, its application is limited due to poor water solubility, low bioavailability, and rapid degradation. In this study, instead of using conventional bulk curcumin, we have explored both micro- and nano-formulations synthesized through a simple two-step process. To address the limitations of solubility and bioavailability, the synthesized particles were capped with sodium caseinate, a milk-derived protein. This coating not only enhanced the water solubility but also significantly improved the colloidal stability of both micro- and nano-curcumin formulations, maintaining their stability for several months. These casein-capped curcumin formulations were further evaluated for their wound healing potential and demonstrated effective wound closure in L929 cells within 18 h, as confirmed by a scratch assay. This approach offers a practical and effective strategy to overcome inherent limitations of curcumin, paving the way for its broader application in therapeutic formulations.

Acknowledgement. Author acknowledges Cochin University of Science and Technology for fellowship Mr. Nashath K, Research scholar, Central University of Tamil Nadu for wound healing studies

References

1. Karthikeyan, A., Senthil, N., Min, T.: Nanocurcumin: A promising candidate for therapeutic applications. Front. Pharmacol. **11**, 487 (May 01, 2020). https://doi.org/10.3389/fphar.2020.00487
2. Padwal, V., Kothawade, S., Gardare, M., Waghmare, S., Kamble, H., Padwal, H., et al.: Nano-curcumin: a useful candidate in therapeutic applications. Certif. J. World J. Pharm. Res. **11**(1), 294–305 (2015). https://doi.org/10.20959/wjpr20221-22542
3. Hussain, Y., et al.: Antimicrobial potential of curcumin: therapeutic potential and challenges to clinical applications. Antibiotics MDPI. **11**(3), 322 (Mar. 01, 2022). https://doi.org/10.3390/antibiotics11030322
4. Fereydouni, N., et al.: Curcumin nanofibers for the purpose of wound healing. J. Cell. Physiol.. Wiley-Liss Inc. **234**(5), 5537–5554 (May 01, 2019). https://doi.org/10.1002/jcp.27362
5. J. Annaraj, R. Dhivya, M. Vigneshwar, K. Dharaniyambigai, G. Kumaresan, and M. Rajasekaran, "Studies on The Enhanced Biological Applications of PVA Loaded Nanocurcumin," 2014. [Online]. Available: www.indiasciencetech.com
6. Kulac, M., et al.: The effects of topical treatment with curcumin on burn wound healing in rats. J. Mol. Histol. **44**(1), 83–90 (Feb. 2013). https://doi.org/10.1007/s10735-012-9452-9
7. Emiroglu, G., et al.: The effects of curcumin on wound healing in a rat model of nasal mucosal trauma. Evid. Based Complement. Alternat. Med. **2017** (2017). https://doi.org/10.1155/2017/9452392
8. Kumari, A., et al.: Wound-healing effects of curcumin and its nanoformulations: A comprehensive review. Pharmaceutics MDPI. (Nov. 01, 2022). https://doi.org/10.3390/pharmaceutics14112288
9. Shome, S., Das Talukdar, A., Choudhury, M.D., Bhattacharya, M.K., Upadhyaya, H.: Curcumin as potential therapeutic natural product: a nanobiotechnological perspective. J. Pharm. Pharmacol. **68**(12), 1481–1500. Blackwell Publishing Ltd (Dec. 01, 2016). https://doi.org/10.1111/jphp.12611

10. Bhawana, Basniwal, R.K., Buttar, H.S., Jain, V.K., Jain, N.: Curcumin nanoparticles: Preparation, characterization, and antimicrobial study. J. Agric. Food Chem. **59**(5), 2056–2061 (2011). https://doi.org/10.1021/jf104402t
11. Ghasemiyeh, P., Mohammadi-Samani, S.: Potential of nanoparticles as permeation enhancers and targeted delivery options for skin: Advantages and disadvantages. Drug Des. Devel. Ther. **14**, 3271–3289 (2020). https://doi.org/10.2147/DDDT.S264648
12. Raghavendra, G.M., Jayaramudu, T., Varaprasad, K., Ramesh, S., Raju, K.M.: Microbial resistant nanocurcumin-gelatin-cellulose fibers for advanced medical applications. RSC Adv. **4**(7), 3494–3501 (2014). https://doi.org/10.1039/c3ra46429f
13. Pillai, R.R., Sreelekshmi, P.B., Meera, A.P., Thomas, S.: Eco-Friendly Fabrication of Nanocurcumin/Nano Iron Oxide Composite: Enhanced In Vitro Anticancer Activity Against DLD-1 Cell Lines. J. Inorg. Organomet. Polym. Mater. **33**, 3805–3814 (2023). https://doi.org/10.1007/s10904-023-02735-4
14. Pan, K., Luo, Y., Gan, Y., Baek, S.J., Zhong, Q.: PH-driven encapsulation of curcumin in self-assembled casein nanoparticles for enhanced dispersibility and bioactivity. Soft Matter. **10**(35), 6820–6830 (2014). https://doi.org/10.1039/c4sm00239c

α-MnO_2 Nanostructures as Efficient Platforms for Glutathione Sensing

J. S. Anju, P. K. Shibla, and P. M. Aneesh(✉)

Department of Physics, Central University of Kerala, Periye, Kasaragod, Kerala 671325, India
aneeshpm@cukerala.ac.in

Abstract. Glutathione (GSH) is an intracellular antioxidant that plays vital role in maintaining redox homeostasis. As a result, GSH serves as an important biomarker for understanding various physiological disorders, including cancer as well as multiple cardiovascular and liver-related diseases. Therefore, the sensitive detection of GSH is crucial in fields such as drug delivery and biological responses. In recent years, the use of transition metal oxides for GSH monitoring has gained considerable attention. In this work, we investigate the potential of hydrothermally synthesized α-MnO_2 nanostructures for GSH sensing. X-ray diffraction, FTIR, and Raman spectroscopy were employed to gain insights into the structural characteristics of the material. FESEM images revealed that the MnO_2 nanostructures exhibit an elongated nanowire morphology. Optical analysis using UV-visible spectroscopy confirmed the semiconducting nature of the sample, with a bandgap of 2.75 eV. The MnO_2 solution was prepared in citrate buffer under acidic pH conditions and upon the addition of various concentrations of GSH, a noticeable change in the color intensity of the MnO_2-citrate buffer solution was observed. This shift, as evident from the UV-visible absorption spectra, enables the colorimetric detection of GSH. These findings clearly establish α-MnO_2 nanorods as a highly effective and promising material for the sensitive detection of GSH.

Keywords: Glutathione Detection · Manganese Dioxide · Colorimetric Sensing

1 Introduction

Glutathione (GSH) is a tripeptide antioxidant produced in cells that protects them from oxidative stress and toxins [1]. Elevated levels of GSH are indicative of cellular health and the imbalance between oxidative and antioxidative processes. Hence, various physiological conditions such as neurodegenerative disorders, cancer, and cardiovascular diseases can be identified at an early stage by monitoring GSH levels in the body [2, 3]. Researchers are seeking methods that can provide a better understanding and sensitive detection of GSH. There are several methods for the detection of GSH, such as high-performance liquid chromatography, fluorescent detection, colorimetric detection, mass spectrometry, etc. [4, 5]. Among them, colorimetric assays are cost-effective and rely on specific chemical reactions that cause observable color changes and are typically measured using spectroscopy. Several sensing media are available for GSH detection,

A. Antony et al. (Eds.): icONMAT 2025, SPPHY 328, pp. 83–88, 2025.
https://doi.org/10.1007/978-981-95-2697-0_12

and transition metal oxides (TMOs) act as an effective sensing platform. Alpha–manganese dioxide (α-MnO_2) is a TMO that exhibits exceptional catalytic properties and redox behaviour due to its multiple oxidation states [6]. The structure of α-MnO_2 is characterized by a double-chain arrangement, which is formed through the sharing of edges and corners among MnO_6 octahedra [7]. Due to these characteristics, α-MnO_2 is widely used as catalyst, adsorbent, energy conversion, and storage material [8]. Hence, α-MnO_2 is a compelling candidate for designing GSH detection devices with enhanced sensitivity and reliability.

In colorimetric sensing, the redox reaction between GSH and α-MnO_2 may trigger color changes that can be observed via UV-visible spectroscopy. In this work, we aim to study the sensitivity of α-MnO_2 solution by varying GSH concentrations, and it is expected to demonstrate remarkable potential for the colorimetric sensing of GSH.

2 Experimental Section

2.1 Synthesis

The α-MnO_2 nanostructures were synthesized using a one-pot hydrothermal method. To prepare α-MnO_2 nanostructures, potassium permanganate ($KMnO_4$) and manganese sulfate monohydrate ($MnSO_4.H_2O$) were mixed with 70 ml of distilled water. After stirring for one hour, the solution was transferred to a sealed Teflon-lined stainless-steel autoclave and heated at 200 °C for 24 h. After cooling, the brown-black precipitate was washed, centrifuged, and dried at 180 °C for one hour to obtain α-MnO_2 nanostructures.

2.2 Material Characterization

The crystal structure of the nanostructures was investigated using the Rigaku MiniFlex 600 X-ray diffractometer. Fourier transform infrared (FTIR) spectroscopy was conducted with a PerkinElmer Spectrum 2 FTIR spectrometer. The Raman spectroscopy measurements were obtained utilizing a 532 nm wavelength diode-pumped solid-state laser within a Horiba Scientific Xplora Plus Raman spectrometer. JEOL JSM-IT800 Field Emission Scanning Electron Microscope (FESEM) and Agilent Cary 5000 UV-visible spectrophotometer were used for obtaining SEM images and reflectance spectra of the nanostructures, respectively.

2.3 GSH Sensing

The citrate buffer was prepared by mixing citric acid and sodium citrate solutions in an appropriate ratio to maintain a pH of 3.6. As-synthesized α-MnO_2 nanostructures were dissolved in 2 mL of the citrate buffer at a concentration of 0.05 mM and 50 μL of GSH (glutathione reduced, purchased from SRL), at varying concentrations, was added. The mixture was then shaken well for 3 min before being transferred to a cuvette for UV-visible absorption measurements.

3 Results and Discussion

X-ray diffraction (XRD) analysis was conducted to investigate the crystallographic structure of the synthesized nanostructures. Fig. 1(a) presents the XRD pattern of the as-prepared nanomaterial. The diffraction peaks observed at 12.74°, 18.07°, 28.68°, 37.6°, 49.79°, 56.23°, 60.23°, and 69.56° correspond to the (110), (200), (310), (211), (301), (411), (600), (521), and (541) planes, respectively of tetragonal α-MnO_2 as referenced by ICDD card no: 00–044-0141. The crystallite size was calculated to be approximately 17 nm, with lattice parameters a = b = 9.86 Å and c = 2.85 Å.

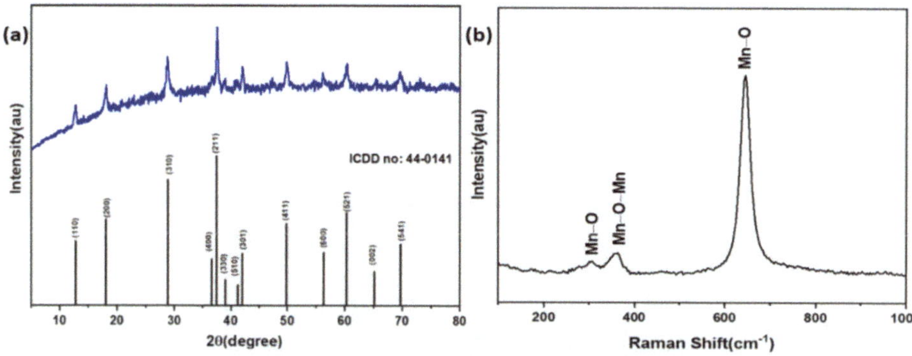

Fig. 1. (a) XRD pattern and (b) Raman spectrum of α-MnO_2 nanostructures

Raman spectrum provide information about the molecular vibrations and the symmetry. The Raman spectrum of MnO_2 nanostructures are shown in Fig. 1(b). The peak at 307 cm^{-1} corresponds to Mn—O lattice vibrations, and bending vibrations of Mn—O—Mn cause the formation of a peak at 363 cm^{-1}. The main peak at 646 cm^{-1} is due to the symmetric stretching of Mn—O bond [9].

FTIR spectroscopy is widely used to identify the functional groups and chemical bonding present in a material by analyzing its vibrational modes. The FTIR spectrum of the synthesized α-MnO_2 nanostructures, as shown in Fig. 2, provides insight into the molecular structure and bonding environment of the material. The characteristic absorption bands of α-MnO_2 are observed below 1000 cm^{-1}, which are typically associated with various Mn–O vibrational modes. These bands confirm the formation of MnO_2, as they correspond to stretching and bending vibrations within the MnO_6 octahedra [6].

FESEM analysis is primarily effective for surface morphology characterization. FESEM images of α-MnO_2 nanostructures, which highlight the unique morphological characteristics of α-MnO_2, are given in Fig. 3. The α-MnO_2 nanostructures show a nanorod-like morphology, exhibiting a slender, needle-like profile. The nanorods are scattered in random orientations and the length of the nanostructure is considerably larger than the thickness, meaning high surface area to volume ratio.

The optical properties of a material can be understood using UV-visible spectroscopy. The reflectance spectrum of the nanomaterial is given in Fig. 4(a) and α-MnO_2 is found to have reflectance in the wavelength range of 500–700 nm. The Kubelka-Munk plot shown

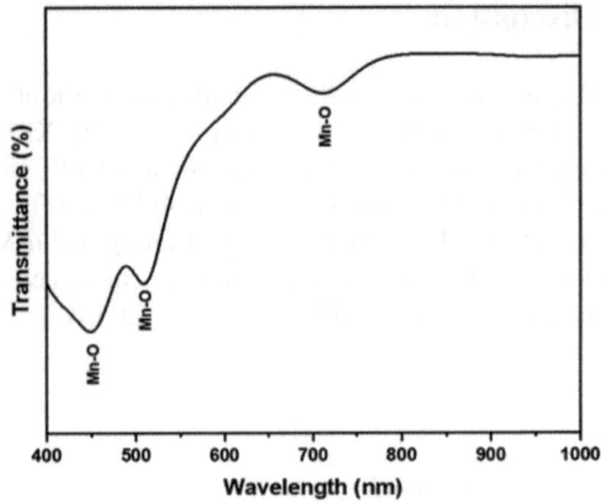

Fig. 2. FTIR spectrum of α-MnO_2 nanostructures

Fig. 3. FESEM image of α-MnO_2 nanostructures

in Fig. 4(b) confirms that the α-synthesized MnO_2 nanostructures have a direct bandgap with a value of 2.75 eV, indicating the semiconductive nature of MnO_2 nanostructures.

To assess the sensing performance of the synthesized α-MnO_2 nanostructures, we conducted glutathione (GSH) detection by monitoring changes in the UV-visible absorption spectra at varying GSH concentrations. The α-MnO_2-citrate buffer solution exhibited an absorption peak at 435 nm. We allowed different concentrations of GSH, ranging from 0.01 to 0.2 M, to react with the α-MnO_2 solution for 3 min before taking absorption measurements. With increasing GSH concentration, the color of α-MnO_2 solution gradually diminished, corresponding to a continuous decline in the absorption intensity of the characteristic α-MnO_2 peak, leading to the disappearance of the absorbance peak for a GSH concentration of 0.2 M as depicted in Fig. 5. This response is attributed to the reduction of α-MnO_2 by GSH, which alters the oxidation state and affects the optical properties of MnO_2 [10]. The correlation between the change in absorbance and

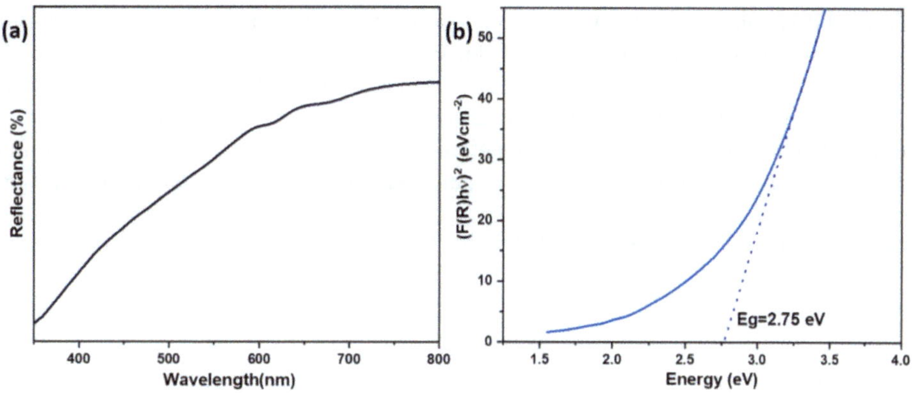

Fig. 4. (a) Reflectance spectrum (b) Kubelka-Munk (KM) plot of α-MnO_2 nanostructures

GSH concentration indicates that this system can be effectively used for quantitative colorimetric detection of GSH.

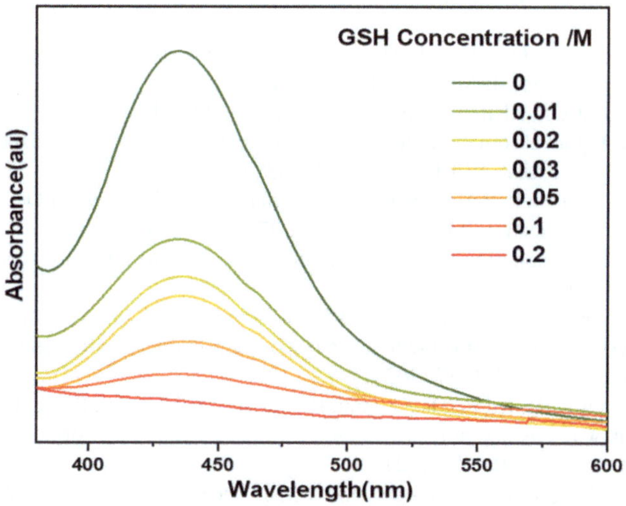

Fig. 5. UV-visible absorbance spectra of α-MnO_2 solution with the addition of different concentrations of GSH

4 Conclusion

The α-MnO_2 nanostructures with a nanorod-like morphology were synthesized using a one-step hydrothermal method. The tetragonal crystal structure was confirmed through X-ray diffraction patterns. The vibrational modes and chemical bonding were characterized through Raman and FTIR spectroscopic analysis. The surface morphology was characterized using FESEM images. The optical properties were analyzed using UV-visible spectroscopy, and the bandgap was estimated to be approximately 2.75 eV. The effect of GSH concentration on the absorbance spectra of α-MnO_2 in citrate buffer solution was monitored. The sensitivity of detection is evaluated by tracking changes in the absorption spectra with varying GSH concentrations, highlighting the potential of α-MnO_2 nanostructures for the sensitive detection of GSH and a complete reduction in the absorbance was observed at a concentration of 0.2 M. These findings indicate that α-MnO_2 nanorods exhibit effective colorimetric sensing capabilities for GSH detection.

Acknowledgments. Anju J S acknowledges the Council of Scientific and Industrial Research (CSIR), Government of India for Senior Research Fellowship.

References

1. Lushchak, V.I.: Glutathione homeostasis and functions: Potential targets for medical interventions. J. Amino Acids. **2012**, 1–26 (2012)
2. Vairetti, M., Di Pasqua, L.G., Cagna, M., Richelmi, P., Ferrigno, A., Berardo, C.: Changes in glutathione content in liver diseases: An update. Antioxidants. **10**, 364–403 (2021)
3. Matuz-Mares, D., Riveros-Rosas, H., Vilchis-Landeros, M.M., Vázquez-Meza, H.: Glutathione participation in the prevention of cardiovascular diseases. Antioxidants. **10**, 1220–1239 (2021)
4. Sun, X., Guo, F., Ye, Q., Zhou, J., Han, J., Guo, R.: Fluorescent sensing of glutathione and related bio-applications. Biosensors. **13**, 16–35 (2022)
5. Asensi, M., Sastre, J., Pallardo, F.V., Delaasuncion, J.G., Estrela, J.M., Vina, J.: A high-performance liquid chromatography method for measurement of oxidized glutathione in biological samples. Anal. Biochem. **217**, 323–328 (1994)
6. Sivakumar, S., Nelson Prabu, L.: Synthesis and characterization of α-MnO_2 nanoparticles for Supercapacitor application. Mater Today Proc. **47**, 52–55 (2021)
7. Yuan, Y., et al.: Deciphering the atomic patterns leading to MnO_2 polymorphism. Chem. **5**, 1793–1805 (2019)
8. Joseph, A., Sangeetha, C.K., Abraham, D.S., Bhagiyalakshmi, M., Aneesh, P.M.: Hydrothermally synthesised α-MnO_2 and β-MnO_2 nanorods for pseudocapacitor electrode applications. MRS Commun. **14**, 1480–1487 (2024)
9. Umar, A., Ibrahim, A., Kumar, R., Algadi, H., Albargi, H., Ahmad, F., Zeng, W., Akhtar, M.: α-MnO_2 nanowires as potential scaffolds for a high-performance formaldehyde gas sensor device. Coatings. **11**, 860–871 (2021)
10. Di, W., Zhang, X., Qin, W.: Single-layer MnO_2 nanosheets for sensitive and selective detection of glutathione by a colorimetric method. Appl. Surf. Sci. **400**, 200–205 (2017)

Investigating the Anti-cancer Potential of Silver Nanoparticles Synthesized by Chemical Reduction of $AgNO_3$ Using Trisodium Citrate and Ascorbic Acid

K. S. Dhanya[1], M. Anaswara Biju[2], U. Arun[3], and R. Sreeja[2(✉)]

[1] Department of Physics, NSS College, Nilamel, Kollam 691535, India
[2] Department of Physics, Mar Ivanios College, Thiruvananthapuram 695015, India
sreeja.r@mic.ac.in
[3] Department of Biochemistry, Government College, Kariavattom, Thiruvananthapuram 695581, India

Abstract. In order to synthesize uniform, evenly distributed silver nanoparticles (AgNPs), a straightforward chemical reduction technique was used. In this work we synthesized the particles by using ascorbic acid as a surfactant and trisodium citrate as reducing agent to reduce $AgNO_3$ to metallic silver. UV–visible spectroscopy was the primary method used to confirm the formation of AgNPs. At 422.84 nm, it displays the surface plasmon resonance. The nanoparticles were characterised by FTIR, XRD, DLS, Zeta potential, FESEM and EDAX. HR TEM and SAED are taken to measure the size and crystalline nature of the nanoparticles. It demonstrates that the produced AgNPs have an average diameter of 30 nm and were quasi spherical in shape and evenly distributed. The anticancer and cytotoxic properties of the samples were examined. MTT assay performed in mouse fibroblast L929 cells have substantiated the cytocompatibility of the silver nanoparticles in normal cells. Significant cytotoxicity was observed in SKOV-3 ovarian cancer cells. The anti-cancer potential of the silver nanoparticles was validated through AO/EtBr double staining technique which shows the apoptotic potential of the nanoparticles. Annexin V apoptotic assay established the percentage of apoptotic cells in nanoparticle treated group compared to that in control group. The results highlight the anti-microbial and anti-cancer efficacy of chemically synthesized AgNPs.

Keywords: Silver nanoparticles · chemical synthesis · ovarian cancer

1 Introduction

Many metal nanoparticles are currently receiving a lot of attention in the medical sciences because of their particular and distinctive qualities, which include high surface area, high reactivity, variable pore size, and particle shape [1]. In a number of in vitro and in vivo studies, silver nanoparticles (AgNPs) have garnered a lot of interest and seem to be one of the promising substitutes to stop the growth of tumours [2, 3]. For the manufacture

A. Antony et al. (Eds.): icONMAT 2025, SPPHY 328, pp. 89–104, 2025.
https://doi.org/10.1007/978-981-95-2697-0_13

of silver nanoparticles (AgNPs), numerous protocols have been carried out, including chemical, photochemical, and biological method [4, 5, 6]. Silver, a noble metal with numerous physical, chemical, and biological properties [7, 8, 9], has been used since ancient times for treating diseases and wound healing [10, 11]. Its diverse biomedical potential includes antiseptic, medical and surgical devices, medicines, antimicrobial potential, drug delivery, cosmetic products, biolabeling, optical properties, larvicidal applications, and food preservation [12, 13, 14]. Silver nitrate is used to treat microbial infections in hospitals. Silver ions disrupt cell membranes, inhibit DNA replication, and disrupt membrane permeability [15]. Ag NPs have antimicrobial activities at low concentrations, but silver salts have higher toxicity [16, 17].

Cancer is a disease characterized by abnormal cell growth and spread, caused by external factors like tobacco use, infectious diseases, and unhealthy diets, as well as internal factors like inherited genetic mutations and hormone dysfunction [18, 19]. It is a leading cause of death worldwide, with nine million deaths in 2022 and up to 13 million by 2030 [20]. Ovarian carcinoma is the fifth most common cancer in women and the leading cause of gynaecological cancers [21, 22]. The most frequent cause of gynaecological cancer-related mortality is epithelial ovarian carcinoma [23]. Abdominal pain and distension usually appear for a few months in postmenopausal women. The most common cause of death, intestinal blockage, is the result of these episodes, which also lead to chemoresistance [24]. Around 90% of primary malignant ovarian tumours are epithelial, likely arising from ovarian surface epithelium or inclusion cysts [25]. Treatments include chemotherapy, hormone therapy, radiation, surgery, and immune therapy [26]. Nanomedicine has revolutionized personalized cancer treatments, offering a synergistic approach to treatment. Combining drugs and nanoparticles can provide greater therapeutic effects than single-drug treatments [27]. Targeting multiple signalling pathways is a new model for cancer treatment. Nanoparticle-mediated targeted therapy is an innovative alternative, potentially more effective than conventional small-molecule chemotherapeutics [28, 29]. Nanoparticles are specifically targeted, overcoming multiple drug resistance and providing enhanced therapeutic efficacy [30]. This approach increases intracellular drug concentration and enhances the cytotoxic effect of antitumor agents.

Ag nanoparticles have been extensively studied for their potential to prevent cancer, with their effectiveness against various types of cancers. Their size, shape, and surface charge influence their anticancer activity, while dosing and exposure duration are important factors. Ag nanoparticles' anticancer effects are linked to processes such as disrupting cancer cell structure, producing reactive oxygen species, DNA damage, deactivating enzymes, and controlling signalling cascades. They also prevent angiogenesis, preventing tumor cell invasion and spread. The enhanced permeability and retention (EPR) effect in tumour cells allows for more preferential absorption, improving targeted delivery [30]. This work reports the anti-cancer potential of chemically synthesised AgNPs.

2 Materials and Methods

2.1 Materials

Silver nitrate ($AgNO_3$, extra pure 99% AR-SRL), trisodium citrate dihydrate ($Na_3C_6H_5O_7.2H_2O$ extra pure 99% AR-SRL) and ascorbic acid were used for the synthesis of the nanoparticles.

2.2 Synthesis of Silver Nanoparticles

The precursor of Ag^+ ions in this process is silver nitrate. First, 100 ml of 1 mM $AgNO_3$ dissolved in distilled water was heated to 70 °C and magnetically stirred. After that, a mixture of 5 ml, 1 mM trisodium citrate and 1 mM ascorbic acid at 70 °C was added to the above solution drop by drop with continuous stirring. The reaction was continued for 30 min at 70 °C. The solution's colour first changed to a pale yellow and then to dark brown as the reaction progresses. The colloid's yellow hue denotes the development of AgNPs. The changes in the metal oxidation state are indicated by the colour transition. Centrifugation was used to condense and purify the nanoparticles, and distilled water was used for washing. The experiment is repeated by changing the concentration of reducing agent, reaction time and reaction temperature for optimization.

2.3 Characterization of Nanoparticles

The absorption studies were carried out by using a high-performance Cary 5000 UV-Vis-NIR spectrometer operating in the 300–800 nm range to capture the UV–Vis–NIR spectra. The crystallinity, particle size and phase identification of the sample were done by X-ray diffraction analysis (XRD) using an X-ray diffractometer, Brucker, D8 Advance with DAVINCI design with CuKα radiation at 40 mA and 40 kV with a 0.02° step size in the 2θ range of 20° to 90°. The surface functionalization of AgNPs was investigated by Fourier transform infrared spectroscopy (FTIR) NICOLET iS50 (Thermo Scientific, America) utilizing KBr as a backdrop in the wavelength range of 4000–400 cm^{-1}. Dynamic light scattering (DLS) was used to analyse the size distribution and stability of the nanoparticles and Zeta sizer Ver. 7.11 was used to evaluate the zeta potential. The surface form and microstructure of Ag nanoparticles were examined using micrographs taken with a scanning electron microscope (MAIA3 TE SCAN) at an accelerating voltage of 10 kV. The size, shape and crystalline nature of the nanoparticles were analysed by HRTEM and SAED patterns.

2.4 Procedure for MTT Assay in L929 and SKOV-3 Cell Lines

Dulbecco's modified Eagles medium (DMEM) (Sigma Aldrich, USA) was used to maintain the mouse fibroblast cell line (L929) and the human ovarian cancer cell line SKOV-3, which were originally obtained from the National Centre for Cell Sciences (NCCS), Pune, India. The cell line was cultured in a 25 cm^2 tissue culture flask with DMEM, FBS, L-glutamine, sodium bicarbonate, and an antibiotic solution. The cells were kept in an incubator at 37 °C with 5% CO_2 and humidity. An inverted phase contrast microscope and the MTT assay technique [31] were utilized to assess the cells' viability. Two-day-old cells were trypsinized, then suspended in 10% growth medium and plated in 96-well tissue culture plates. Next, 100 µl of cell suspension (5×10^3 cells/well) was added. The cells were then maintained at 37 °C in a humidified incubator with 5% CO_2. The sterility of the sample solution was confirmed by passing it through a 0.22 µm Millipore syringe filter. Five serial dilutions of the extracts in 5% DMEM were performed at a weight of 150, 50, 25 g, 125 g, and 65 g. The mixture was then incubated at 37 °C in a humidified 5% CO_2 incubator after three hundred millilitres of each concentration were added to the corresponding wells in triplicate. Additionally, untreated control cells were preserved. Following a 24-h treatment, an inverted contrast microscope was used to examine the entire plate and take images. It was believed that changes in the cytoplasm of the cell, such as vacuolization and shrinkage, were indicators of cytotoxicity. After dissolving 3 ml of PBS and 15 mg of MTT (Sigma, M-5655), the liquid was filtered through a syringe with a hole size of only 0.22 µm to guarantee sterility. After the Ag samples were removed from the wells after a 24-h incubation period, around 30 µl of MTT solution was added to the test and control wells. The plates were gently shaken and then incubated for four hours at 37 °C with 5% carbon dioxide. The absorbance values at a wavelength of 540 nm were measured using a microplate reader.

The following formula was used to determine the growth inhibition percentage:

$$\%\text{of viability} = \frac{\text{Mean OD Samples}}{\text{Mean OD of control group}} \times 100 \quad (1)$$

2.5 Evaluation of Nuclear Alterations and Cell Morphology Using Acridine Orange (AO) and Ethidium Bromide (EtBr) Double Staining

Following a 24-h treatment with AgNPs at their LC_{50} concentration, the SKOV-3 cells were washed with cold PBS and stained for 10 min at room temperature using an acridine orange (AO) and ethidium bromide (EtBr) solution. The stained cells were examined using an Olympus CKX41 with an Optika Pro 5 camera and a fluorescence microscope with a blue filter after being cleaned twice with 1X PBS.

2.6 Quantitative Evaluation of Apoptosis Using Flow Cytometry

A quick and easy technique using a PI fluorescence staining kit and Annexin V-FITC was used to measure the proportion of apoptosis caused by AgNPs in SKOV 3 cells. The nanoparticles were applied to SKOV-3 cell lines at their Lethal Concentration 50 (LC50) for a duration of 24 h. For comparison, untreated control wells were also kept up to date. The cell sample was placed in a 50 ml conical flask or a 12 mm x 75 mm polystyrene tube. For fixation in a tube, a minimum of 1×10^6 cells is advised. The materials were then centrifuged at 3000 rpm for five minutes. The particles held together when the supernatant was extracted. After centrifugation, the cell pellet either shows up on the tube's bottom as a white film or as an apparent pellet. 100 µL of the Muse TM Annexin V & Dead Cell Reagent was added to each tube. By pipetting the tubes up and down or vertexing them for three to five seconds at a medium pace, the tubes were well mixed.

After that, they were left in the dark for twenty minutes at room temperature. Muse flow cytometry software was used to assess the cells after they had been examined in a flow cytometer. The Muse FCS 3.0 software was used to gate the cells versus untreated control cells and check for apoptosis.

3 Results and Discussion

3.1 Analysis of UV Visible Absorption Spectra

Figure 1 (a-c) respectively shows the UV-visible absorption spectrum of citrate-capped colloidal AgNPs taken in quartz cuvettes (1 cm optical path) at gradually increasing concentration of trisodium citrate (1 mM, 2 mM, 3 mM and 4 mM) keeping volume at 5 ml, gradually increasing temperature (30 °C, 50 °C,70 °C and 90 °C) and varying reaction time (10 min, 20 min, 30 min, 40 min, 50 min and 60 min). Strong surface plasmon resonance (SPR) peaks around 425 nm were observed indicative of the generation of AgNPs. The amount of colloid aggregation determines the position of the plasmon peak and full width at half maximum [32, 33]. A repulsive force operated along the colloidal AgNPs, prevents the particles from aggregation. The colloids exhibit broad absorption band with gradually red shifted SPR peak with increase in concentration of trisodium citrate. It is evident that as the concentration of trisodium citrate increases, the size of nanoparticles also increases. There is a blue shift in SPR peak with increase in temperature and time of reaction, indicating the formation of nanoparticles with reduced size as reaction temperature and time increased. There is no significant change in the position of SPR peak after 30 mins. By studying the position of SPR peak we optimized the best reaction parameters as 1 mM concentration of trisodium **c** citrate, 70 °C as reaction temperature and 30 min as reaction time.

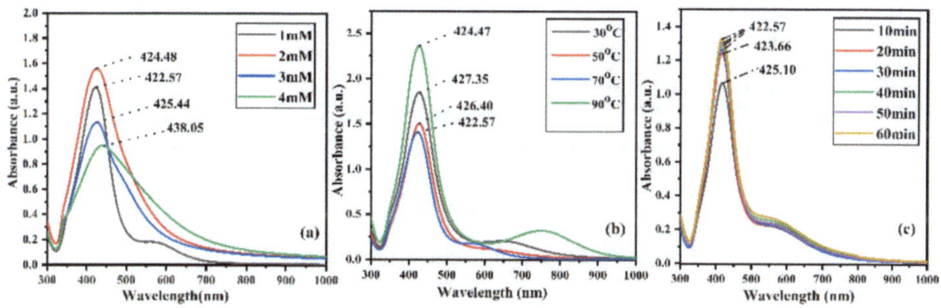

Fig. 1. UV-visible spectra of citrate-capped AgNPs synthesized at a) different concentrations of trisodium citrate, b) different reaction temperature, c) at different reaction time.

3.2 Analysis of XRD Spectrum

Using an X-ray diffractometer (XRD), citrate-capped AgNPs were investigated to ascertain the average crystallite size of the NPs. The X-ray diffraction pattern of Ag NPs is shown in Fig. 2, which shows a cubic structure with diffraction planes (111), (200),

(220), (311) and (222) at a 2θ angle of 38.326°, 44.300°, 64.569°, 77.49° and 81.15° respectively. The values are in good agreement with ICDD card number 04-0783 and earlier published literature [34, 35] space group FM 3 m with space group number 225. AgNPs' crystalline nature is confirmed by the strong and narrow peaks seen in the spectra. The AgNPs' crystallite size was determined utilizing the Scherrer equation using the values provided by Eq. (2),

$$D = \frac{k\lambda}{\beta cos\theta} \tag{2}$$

Where D, β, θ, λ and k stand for the average crystallite size, FWHM value, half the diffracting angle, X-ray wavelength (i.e., 0.154 nm), Scherrer constant (i.e., 0.9 for spherical-shaped particles). The average particle size is estimated from XRD to be 28.0106 nm.

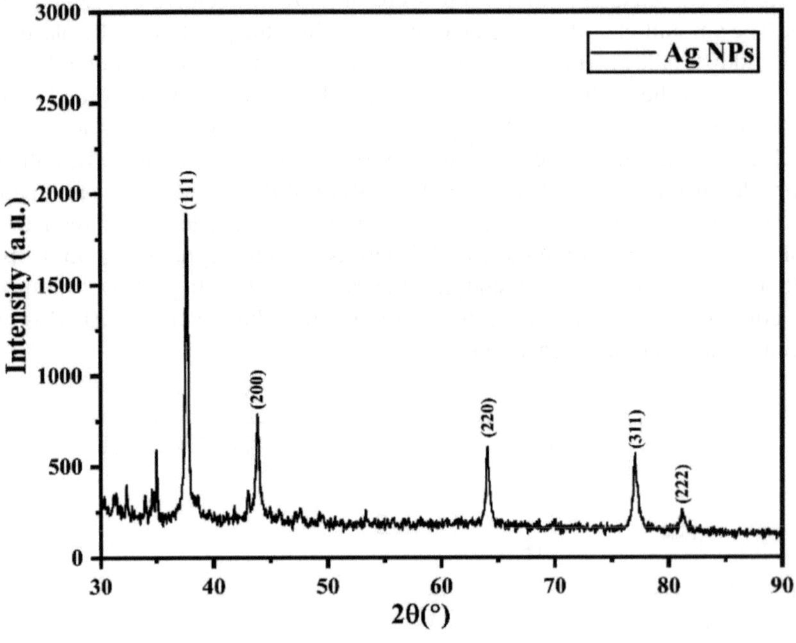

Fig. 2. XRD pattern of citrate capped AgNPs.

3.3 Analysis of FTIR Spectrum

The FTIR analysis (Fig. 3) was carried out to identify the presence of trisodium citrate on the surface of nanoparticles, which are responsible for the formation and stability of AgNPs. Characteristic C-O stretching band at 1627 cm^{-1} in trisodium citrate was present intact in the nanoparticles [36]. The peak at 2073 cm^{-1} confirms the presence of C-C stretching vibration in trisodium citrate. The intermolecular hydrogen-bonded peak at 3285 cm^{-1} of trisodium citrate is also present here. The peak at 607 cm^{-1} corresponds to stretching vibrations of metal-oxygen bond located at cubic crystal sites [37].

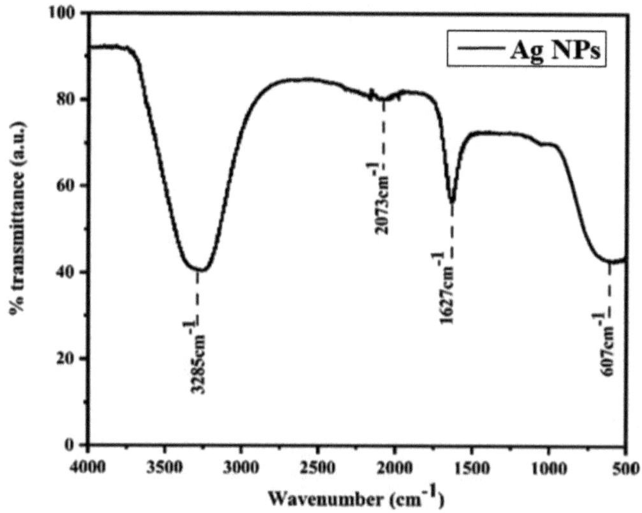

Fig. 3. FTIR of citrate capped AgNPs.

3.4 DLS and Zeta Potential Measurements

Dynamic light scattering (DLS) analysis was used to examine the size distribution of citrate-capped AgNPs, as illustrated in fig. 4(a). The measurement of particle size and dispersity in a solution is the focus of DLS. It is evident from the analysis that the majority of the particles are smaller than 50 nm. Zeta potential is the term used to describe the electro-kinetic potential of the particles in a colloidal system. It shows the degree of repulsion between neighbouring particles in solution, and the AgNPs' observed Zeta potential value of 2.71 mV, as seen in fig. 4(b), demonstrated how stable the produced nanoparticles were [38].

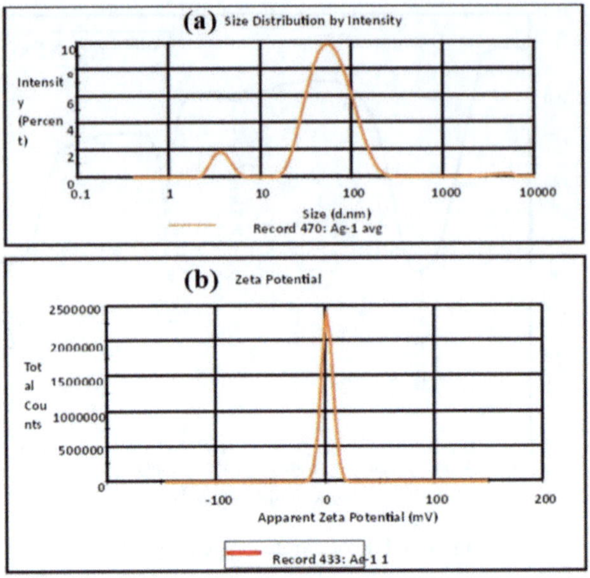

Fig. 4. (a) DLS and (b) Zeta potential of citrate capped silver nanoparticles.

3.5 FESEM and EDAX Studies

Using a FESEM image, the nanoparticles' morphological characteristics and size were determined, as seen in fig. 5. The nanoparticles' sizes range from 5 nm to 50 nm, according to the FESEM image. It can also be seen from the image that the nanoparticles had a quasi-spherical shape. With a prominent peak at 2.9 keV, the EDAX spectrum in the inset makes it evident that silver is the primary component of the sample. The spectrum also clearly shows that the sample contains components like sodium, carbon, oxygen, gold (coating), Si (glass plate) etc. According to the semi-quantitative analysis of citrate capped AgNPs, the elemental percentages of Ag, Na, O etc. are as follows (Table 1),

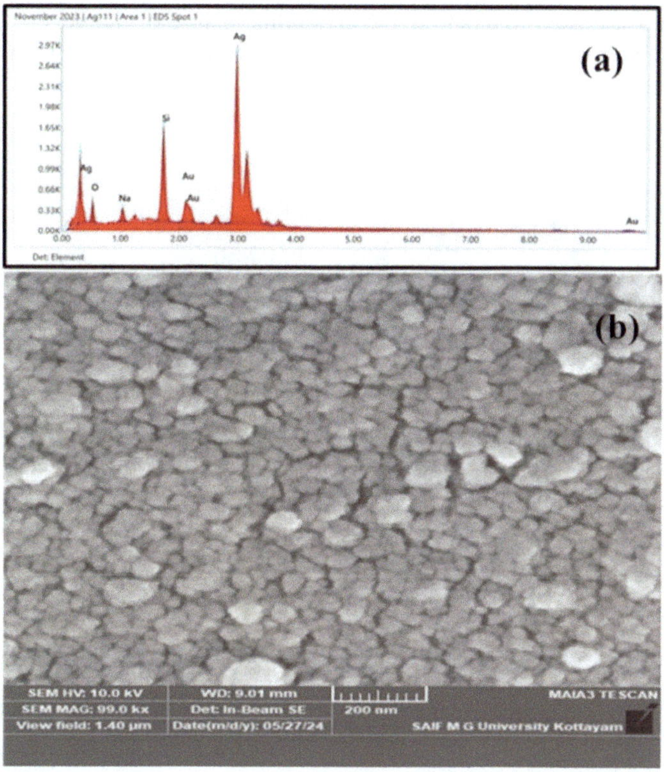

Fig. 5. (a) EDAX spectrum (b) FESEM image (200 nm) of citrate capped AgNPs

Table 1. Elemental percentage in citrate capped silver nanoparticles.

Element	Weight %	MDL	Atomic %	Net Int.	Error %
O K	10.9	1.08	37.4	75.2	14.6
Na K	3.0	0.39	7.3	61.7	14.3
Si K(glass plate)	9.2	0.19	17.9	451.4	7.0
Ag L	69.7	0.87	35.4	968.2	4.4
Au M(coating)	7.2	0.68	2.0	104.8	9.5

3.6 HR-TEM and SAED Analysis

The morphological characteristics of the produced AgNPs were examined using the HR-TEM, as shown in Fig. 6. The AgNPs were spherical in shape and varied somewhat in size, as shown by the TEM micrograph. Their size of roughly 30 nm is exactly in accordance with the value obtained from XRD using the Debye Scherrer equation.

SAED pictures are used to examine the crystalline nature of the generated nanoparticles. The SAED image's concentric rings with small dots show that the particles were polycrystalline.

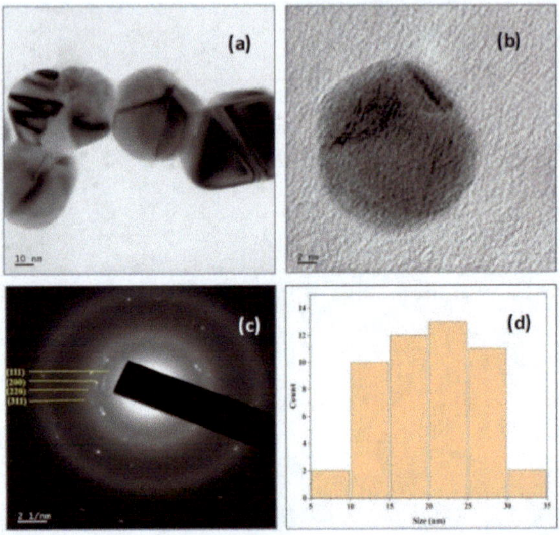

Fig. 6. HRTEM images at different resolutions (a)10 nm, (b)2 nm, (c) SAED pattern and (d) size distribution histogram of AgNPs synthesized by using trisodium citrate

3.7 Evaluation of the Cell Viability of AgNPs on L929 and SKOV-3 Cell Lines Through MTT Assay

The MTT assay examines the number of mitochondria in the cells and may be a useful tool for estimating the percentage of cell viability after treating with the nanoparticles [39]. In a dosage range of 3.125–50 μg/ml, the cell viability of citrate-capped silver nanoparticles was assessed against normal mouse fibroblast L929 and Human Ovarian cancer (SKOV-3) cells. From Fig. 7, it is observed that there is a notable decline in cell viability from 90.64% to 73.85% in cancer cells as the concentration of the nanoparticle changes from 3.125–50 μg/ml where-as maximum cell proliferation is obtained even at the highest concentration of nanoparticles in normal healthy cells. The obtained results show that as the concentration of AgNPs is increased, a steady drop in cell viability of the cancer cells is observed. The lethal concentration 50 (LC-50) value obtained for the nanoparticles in L929 is negative which means the particles doesn't show any cytotoxicity in L929 cell line and that in SKOV-3 cells lines is 91.08 μg/ml.

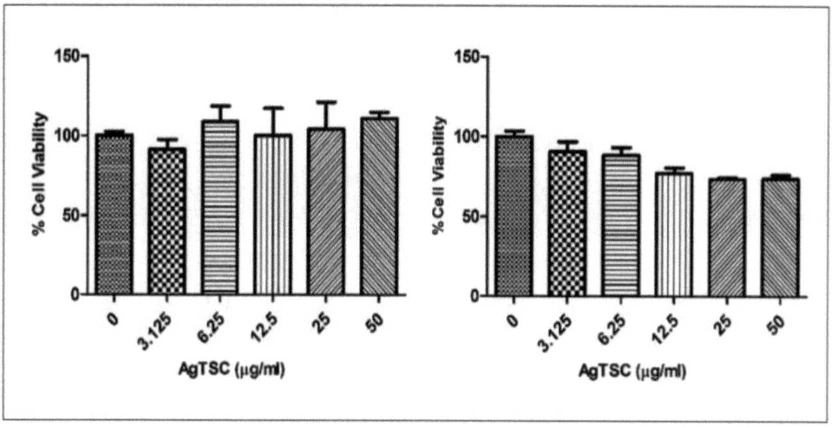

Fig. 7. Comparison of cell viability of L929 mouse fibroblasts and SKOV-3 human ovarian cancer cell line through MTT assay upon treating with AgNPs.

One of the main causes of the toxicity to cells is the generation of reactive oxygen species (ROS) after treating with the nanoparticles. Reactive oxygen species in mitochondria have the ability to affect mitochondrial respiration and death [40]. They will also cause lipids in cell membranes to oxidize. The most powerful intracellular generators of reactive oxygen species are the mitochondria and when the AgNPs come into contact with cell walls, they deform and break the membrane. Reactive oxygen species generated on the surface of the AgNPs dissolve, releasing silver ions in the process, which can lead to the production of reactive oxygen species in cells. The morphology of the cells changes upon treating with increase in concentration in both L929 normal cells and SKOV-3 cancer cells as observed using a phase contrast microscope. The membrane integrity of cells was not altered in L929 cells treated with all concentrations of the nanoparticles; cells exhibited a confluent tight monolayer (fig. 8). But the morphological changes were observed to be more significant in cancer cells (fig. 9) upon treating with the nanoparticles even at lower concentrations of the nanoparticles compared to that in normal cells.

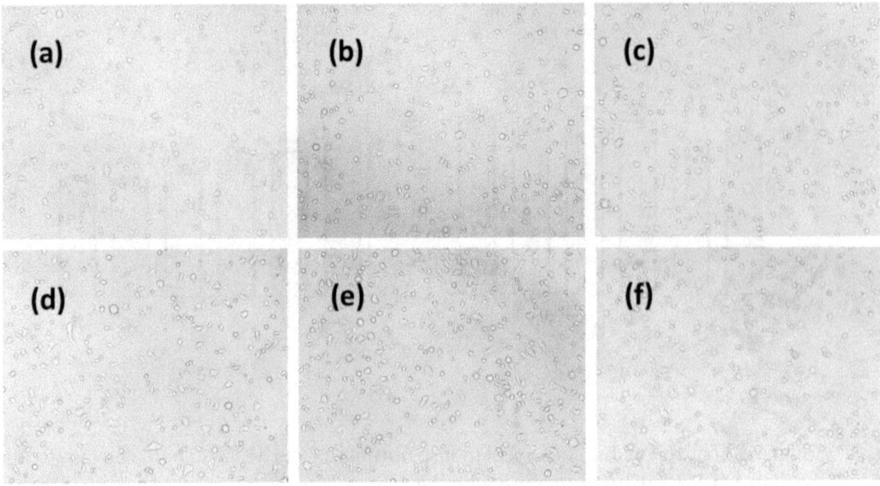

Fig. 8. Control and AgNP treated normal L929 cell lines at different concentrations of nanoparticles. (a) Control, (b) 3.125 μg/mL (c) 6.25 μg/mL, (d) 12.5 μg/mL, (e) 25 μg/mL and (f) 50 μg/mL

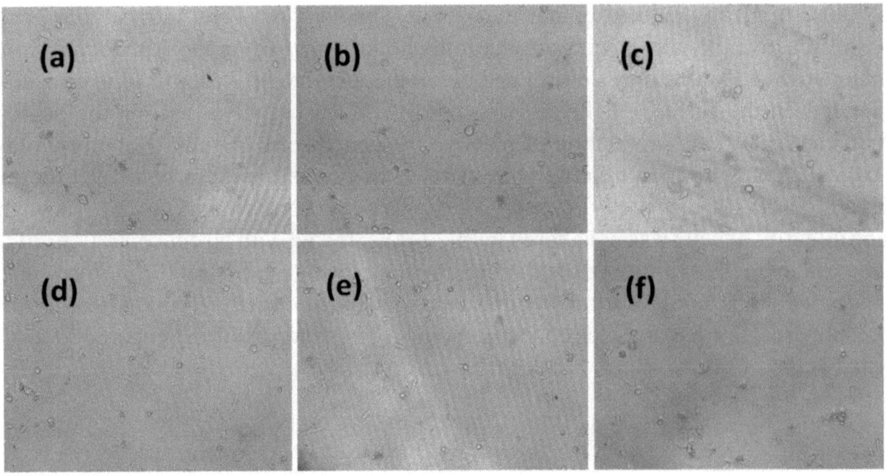

Fig. 9. Control and AgNP treated SKOV-3 cell lines at different concentrations of nanoparticles. (a) Control, (b) 3.125 μg/mL (c) 6.25 μg/mL, (d)12.5 μg/mL, (e) 25 μg/mL and (f) 50 μg/mL

3.8 Observation of the Apoptosis Using AO/EtBr Staining

Apoptosis, or programmed cell death, is an essential and dynamic regulating mechanism of cell growth and proliferation. Certain induction signals cause cells to initiate internal processes that result in discernible physiological alterations. These consist of membrane integrity loss, phosphatidylserine (PS) externalization to the cell surface, nuclear chromatin compaction and fragmentation, and particular cellular protein cleavage and degradation. It is possible to assess morphological changes in the cancer cells after treating with nanoparticles using dual AO/EtBr fluorescence labelling. This technique also makes it possible to distinguish between early and late apoptotic cells, as well as normal cells. Consequently, the method of AO/EtBr double labelling is both qualitative and quantitative in its analysis of apoptosis. In the double staining, the living cells (normal green nucleus), early apoptotic (bright green nucleus with condensed or fragmented chromatin), late apoptotic (orange-stained nuclei with chromatin condensed or fragmentation), and necrotic cells (uniformly orange-stained cell nuclei) were the four categories into which the cells were divided. AO/EtBr staining of the SKOV-3 cells is shown in fig. 10. The figure shows that the control cells show usual green colour indicative of live cells whereas in the nanoparticle treated cells we can see the orange-stained nucleus of necrotic cells, the bright green nucleus of early apoptotic cells with fragmented or condensed chromatin, the orange-stained nuclei of late apoptotic cells [41]. Thus, the double staining confirms the visualisation of the apoptotic potential of the citrate capped AgNPs.

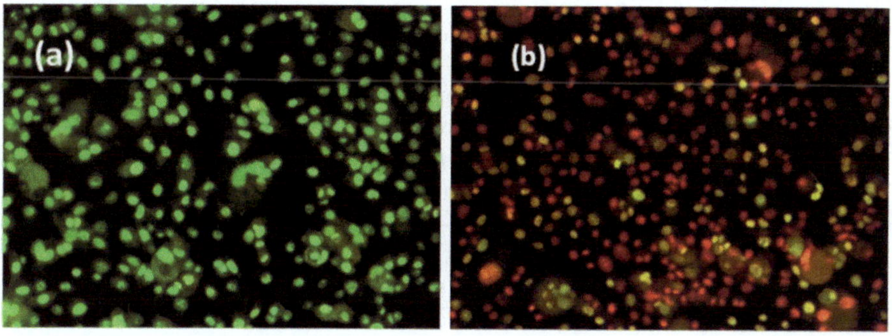

Fig. 10. Double staining of the SKOV-3 cells using Acridine Orange (AO) and Ethidium Bromide (EtBr) **(a)** Control **(b)** nanoparticle treated group.

3.9 Quantitative Assessment of Apoptosis via Flow Cytometry Technique

The analysis of apoptosis in SKOV-3 cells was also conducted using the flow cytometry methodology, which is a very thorough and quick way to assess the different toxicity criteria. The kit contains recombinant Annexin V coupled to fluorescein (FITC Annexin V), red-fluorescent propidium iodide (PI), a nucleic acid binding dye. PI cannot pass through apoptotic or live cells, but it can bind strongly to the nucleic acids in dead cells, causing a red fluorescence stain to be produced [42]. After staining a cell population with FITC Annexin V and PI in the supplied binding buffer, apoptotic cells exhibit green fluorescence, dead cells exhibit red and green fluorescence, and living cells exhibit little to no fluorescence. It is easy to differentiate between these groups using a flow cytometer and an argon-ion laser line at an excitation wavelength of 488 nm. Thus, in this experiment, four cell groups can be identified in four quadrants: dead cells, early apoptotic cells, late apoptotic cells, and non-apoptotic cells [43]. The flow cytometry analysis after treating the SKOV-3 cells with 91.08 μg/mL (LC 50) concentration of the nanoparticles is shown in fig. 11, which displays that the percentage of cells in the treated group is 22.65% where as it is only 4.45% in untreated control group.

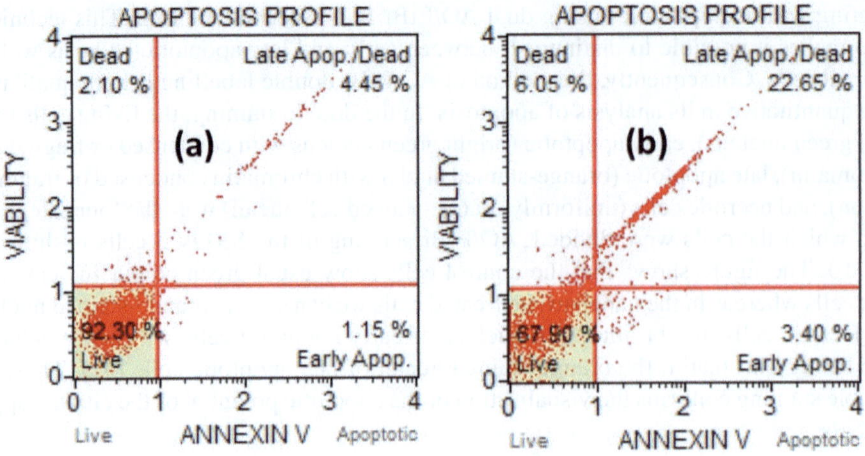

Fig. 11. Flow cytometry assay in SKOV-3 cells for the detection of early and late apoptosis using Annexin V/PI (a) Control cells (b) Cells treated with AgNPs.

However, the nanoparticle induced cancer cell toxicity and its future application necessitates additional thorough research so as to clarify the exact mechanism of action of the nanoparticles in inducing cellular apoptosis. We have to establish the effectiveness of the synthesised nanoparticles through in *vivo* studies in future in the perspective of clinical application against ovarian cancer.

4 Conclusion

Here we have successfully synthesised citrate capped AgNPs using the chemical approach. The physico-chemical characterisation of the nanoparticles confirms the suitability of the nanoparticles for biomedical applications such as smaller size, crystallinity, stability, surface functionalisation etc. It was observed from the MTT assay that the citrate capped AgNPs are less hazardous to normal cells and more toxic to ovarian cancer cells. The double staining and flow cytometry analysis revealed the apoptosis as the major mechanism of cell death. Thus, anti-cancer studies reveal that citrate capped AgNPs offer a low cost, eco-friendly therapeutic agent against the ovarian cancer.

Acknowledgements. The authors are grateful to CLIF, University of Kerala and SAIF, M.G University, Kerala for providing characterization facility. I also acknowledge PMS Dental College, Thiruvananthapuram for providing instrumental support for anticancer study.

References

1. Sreeprasad, T. S., Pradeep, T.: Noble metal nanoparticles. Springer Handbook of Nanomaterials. 1st edn. Springer Berlin, Heidelberg (2013).

2. Al-Khedhairy, A.A., Wahab, R.: Silver nanoparticles: an instantaneous solution for anticancer activity against human liver (HepG2) and breast (MCF-7) cancer Cells. Metals (Basel). **12**(1), 148 (2022)
3. Zhang, X.F., Gurunathan, S.: Combination of salinomycin and silver nanoparticles enhances apoptosis and autophagy in human ovarian cancer cells: An effective anticancer therapy. Int. J. Nanomedicine. **11**, 3655–3675 (2016)
4. Wei, L., Lu, J., Xu, H., Patel, A., Chen, Z.S., Chen, G.: Silver nanoparticles: synthesis, properties, and therapeutic applications. Drug Discov. Today. **20**(5), 595–601 (2015)
5. Pryshchepa, O., Pomastowski, P., Buszewski, B.: Silver nanoparticles: Synthesis, investigation techniques, and properties. Adv. Colloid Interf. Sci. **284**, 102246 (2020)
6. Ahmad, A., Haneef, M., Ahmad, N., Kamal, A., Jaswani, S., Khan, F.: Biological synthesis of silver nanoparticles and their medical applications (Review). World Acad Sci J. **6**(3), 22 (2024)
7. Zhang, X.F., Liu, Z.G., Shen, W., Gurunathan, S.: Silver Nanoparticles: Synthesis, Characterization, Properties, Applications, and Therapeutic Approaches. Int. J. Mol. Sci. **17**(9), 1534 (2016)
8. Duman, H., Eker, F., Akdaşçi, E., Witkowska, A.M., Bechelany, M., Karav, S.: Silver nanoparticles: A comprehensive review of synthesis methods and chemical and physical properties. Nano. **14**(18), 1527 (2024)
9. Mikhailova, E.O.: Silver nanoparticles: Mechanism of action and probable bio-application. J Funct Biomater. **11**(4), 84 (2020)
10. Wilkinson, L.J., White, R.J., Chipman, J.K.: Silver and nanoparticles of silver in wound dressings: A review of efficacy and safety. MA Healthcare Ltd. **20**(11), 543–549 (2011)
11. Chen, et al.: Three-dimensional layered nanofiber sponge with in situ grown silver- metal organic framework for enhancing wound healing. Chem. Eng. J. **443**, 136234 (2022)
12. Verma, P., Maheshwari, S.K.: Applications of Silver nanoparticles in diverse sectors. Int J Nano Dimens. **10**(1), 18–36 (2019)
13. Xu, L., Wang, Y.Y., Huang, J., Chen, C.Y., Wang, Z.X., Xie, H.: Silver nanoparticles: Synthesis, medical applications and biosafety. Theranostics. **10**(20), 8996 (2020)
14. Abass Sofi, M., Sunitha, S., Ashaq Sofi, M., Khadheer Pasha, S.K., Choi, D.: An overview of antimicrobial and anticancer potential of silver nanoparticles. J King Saud Univ Sci. **34**(2), 101791 (2022)
15. Hembram, K.C., Kumar, R., Kandha, L., Parhi, P.K., Kundu, C.N., Bindhani, B.K.: Therapeutic prospective of plant-induced silver nanoparticles: application as antimicrobial and anticancer agent. Artif Cells Nanomed Biotechnol. **46**(3), 38–51 (2018)
16. Kvitek, L., et al.: Antibacterial activity and toxicity of silver - Nanosilver versus ionic silver. J. Phys. Conf. Ser. **304**(1), 012029. IOP Publishing (2011)
17. Bruna, T., Maldonado-Bravo, F., Jara, P., Caro, N.: Silver Nanoparticles and Their Antibacterial Applications. Int. J. Mol. Sci. **22**(13), 7202 (2021)
18. Blackadar, C. B.: Historical Review of the Causes of Cancer. Baishideng Publishing Group Co. Limited 7(1), 54 (2016).
19. LNCS Homepage, https://www.cancer.gov/about-cancer/understanding/what-is-cancer.
20. Tavares, V., Marques, I.S., de Melo, I.G., Assis, J., Pereira, D., Medeiros, R.: Paradigm Shift: A comprehensive review of ovarian cancer management in an era of advancements. Int. J. Mol. Sci. **25**(3), 1845 (2024)
21. LNCS Homepage, https://www.cancer.gov/about-cancer/understanding/statistics.
22. LNCS Homepage, https://www.ncbi.nlm.nih.gov/books/NBK567760
23. Sambasivan, S.: Epithelial ovarian cancer: Review article. Cancer Treat. Res. Commun. **33**, 100629 (2022)
24. Jansen, et al.: Bowel obstruction in advanced tubo-ovarian cancer: a retrospective cohort study. Ann Med Surg. **85**(5), 1539–1545 (2023)

25. Berek, J.S., Bast Jr., R.C.: Epithelial ovarian cancer. In: Kufe, D.W., Pollock, R.E., Weichselbaum, R.R., et al. (eds.) Holland-Frei Cancer Medicine, 6th edn. BC Decker, Hamilton (ON) (2003)
26. LNCS Homepage, https://www.mayoclinic.org/diseases-conditions/ovarian-cancer/diagnosis-treatment/drc-20375946
27. Alshayma, N., et al.: Nanoparticles in cancer theragnostic and drug delivery: A comprehensive review. Life Sci. **352**, 122899 (2024)
28. Fan, D., Cao, Y., Cao, M., et al.: Nanomedicine in cancer therapy. Signal Transduct. Target. Ther. **8**(1), 293 (2023)
29. Elumalai, K., Srinivasan, S., Shanmugam, A.: Review of the efficacy of nanoparticle-based drug delivery systems for cancer treatment. Biomed. Technol. **5**, 109–122 (2024)
30. Mikhailova, E.O.: Silver nanoparticles: Mechanism of action and probable bio-application. J Funct Biomater. **11**(4), 84 (2020)
31. LNCS Homepage, https://pacificbiolabs.com/testing-services/medical-device-tests/microbiology/cytotoxicity.
32. Baset, S., Akbari, H., Zeynali, H., Shafie, M.: Size measurement of metal and semiconductor nanoparticles via UV-Vis absorption spectra. Dig. J. Nanomater. Biostruct. **6**(2) (2011)
33. Amirjani, A., Firouzi, F., Haghshenas, D.F.: Predicting the size of silver nanoparticles from their optical properties. Plasmonics. **15**(4), 1077–1082 (2020)
34. Lv, Y., et al.: Silver nanoparticle-decorated porous ceramic composite for water treatment. J. Membr. Sci. **331**, 50–56 (2009)
35. Kalpana, D., Han, J.H., Park, W.S., Lee, S.M., Wahab, R., Lee, Y.S.: Green biosynthesis of silver nanoparticles using Torreya nucifera and their antibacterial activity. Arab. J. Chem. **12**, 1722–1732 (2019)
36. Kayed, K., Issa, M., Al-ourabi, H.: The FTIR spectra of Ag/Ag2O composites doped with silver nanoparticles. J. Exp. Nanosci. **19**(1) (2024)
37. Rajkumar, M., Navamathavan, R.: Investigation of NiO/CdO/Co3O4 ternary nanocomposite for sunlight-driven degradation via photocatalyst and antimicrobial studies. Next Mater. **8**, 100684 (2025)
38. Chicea, D., Nicolae-Maranciuc, A., Doroshkevich, A.S., Chicea, L.M., Ozkendir, O.M.: Comparative synthesis of silver nanoparticles: Evaluation of chemical reduction procedures, AFM and DLS Size Analysis. Materials. **16**(15), 5244 (2023)
39. Gonnabathula, P., Mekala, N., Relangi, P.K.: Nanotoxicology: Evaluating toxic potential of drug nanoparticles. Nanomed.: Nanotechnol. Biol. Med., 235–243 (2019)
40. Fu, P.P., Xia, Q., Hwang, H.M., Ray, P.C., Yu, H.: Mechanisms of nanotoxicity: Generation of reactive oxygen species. J. Food Drug Anal. **22**(1), 64–75 (2014)
41. Kasibhatla, S., et al.: Acridine orange/ethidium bromide (AO/EB) staining to detect apoptosis. Cold Spring Harb. Protoc. **3**, 4493 (2006)
42. Jeffery, U.: Flow cytometry. In: Schalm's Veterinary Hematology, pp. 1295–1310 (2022)
43. Wlodkowic, D., Skommer, J., Darzynkiewicz, Z.: Flow cytometry-based apoptosis detection. Methods Mol. Biol. **559**, 19–32 (2009)

Sulfonic Acid Functionalized Porous Organic Polymer for Biginelli Reaction

T. P. Sreelakshmi[1], P. V. Aswathy[1], and R. Kala[1,2(✉)]

[1] Department of Applied Chemistry, Cochin University of Science and Technology, Cochin, India
kala@cusat.ac.in

[2] International and Inter University Centre for Nanomaterials and Devices, Cochin University of Science and Technology, Cochin, India

Abstract. In this study, a sulfonic acid-functionalized β-ketoenamine linked porous organic polymer (POP-SO_3H) was strategically designed and synthesized through a solvothermal Schiff-base condensation reaction between 2,4,6-trihydroxybenzene-1,3,5-tricarbaldehyde and 2,4-diaminobenzenesulfonic acid. The incorporation of sulfonic acid groups and β-ketoenamine linkages imparts high acidity, chemical robustness, and thermal stability to the polymer network. The synthesized POP-SO_3H was thoroughly characterized by Powder X-ray Diffraction (PXRD), Fourier Transform Infrared Spectroscopy (FTIR), Thermogravimetric Analysis (TGA), Scanning Electron Microscopy (SEM), and X-ray Photoelectron Spectroscopy (XPS), confirming its crystalline integrity, functional group incorporation, morphological features, and elemental composition. Owing to its high surface area, acidic functionality, and porous architecture, POP-SO_3H was explored as a heterogeneous catalyst for the Biginelli reaction under mild reaction conditions. The polymer exhibited excellent catalytic activity, affording the corresponding dihydropyrimidinone products in high yields. Furthermore, the catalyst demonstrated good recyclability and stability over multiple reaction cycles without significant loss in activity, highlighting its potential as a sustainable and reusable catalyst for multicomponent organic transformations.

Keywords: Catalyst · Porous organic polymer · β-ketoenamine · Biginelli reaction

1 Introduction

The design of environmentally benign, robust, and reusable catalytic systems has emerged as a crucial aspect of sustainable organic synthesis. In this regard, Porous Organic Polymers (POPs) represent an advanced class of functional materials characterized by their high surface area, thermal and chemical stability, and tunable pore architectures. POPs exhibit structural rigidity, ambient functionality and permanent porosity, making them ideal platforms as heterogeneous catalyst. Unlike many conventional porous materials such as zeolites or metal-organic frameworks (MOFs), POPs

A. Antony et al. (Eds.): icONMAT 2025, SPPHY 328, pp. 105–112, 2025.
https://doi.org/10.1007/978-981-95-2697-0_14

can be synthesized under mild conditions using a diverse range of aromatic aldehydes, amines, or other linkers, allowing fine-tuning of their backbone electronics and pore size distribution. The ability to introduce catalytic or binding functionalities directly into the framework combined with solvent and thermal resistance, makes POPs particularly attractive in high-temperature catalytic systems. In addition, the possibility of controlling the acid strength and spatial distribution of functional groups along the pore walls offers a unique advantage for tailoring catalytic microenvironments. Furthermore, the insoluble nature of POPs in common solvents facilitates easy catalyst recovery and reuse, aligning well with the principles of green chemistry and circular economy. The Biginelli reaction, a one-pot multicomponent reaction involving an aldehyde, a β-ketoester, and urea (or thiourea), is a prominent example of a transformation that benefits significantly from acid catalysis. First reported by Pietro Biginelli in 1893, the reaction affords dihydropyridine (DHPM) derivatives, which have attracted significant attention due to their broad pharmacological profile, including antiviral, antitumor, antibacterial, and antihypertensive properties. Notably, compounds such as Monastrol, a mitotic kinesin Eg5 inhibitor, exemplify the therapeutic potential of DHPM scaffolds. In addition, recent derivatives have shown potential in antimalarial, anti-inflammatory, and anti-HIV research, further expanding the scope and relevance of this transformation in medicinal chemistry. Despite the operational simplicity and atom economy of the Biginelli reaction, traditional protocols typically require homogeneous Bronsted Lewis acid catalysts such as HCl, H_2SO_4, $FeCl_3$, or metal triflates, which suffer from limitations including poor recyclability, catalyst contamination in products, corrosiveness, and environmental burden. These challenges have prompted the development of heterogeneous catalytic systems capable of promoting the reaction efficiently under milder and greener conditions. In this context, solid acid catalysts have emerged as a promising solution. Materials such as sulfonated carbon, acidic ion-exchange resins, heteropoly acids, and zeolites have been evaluated, but often suffer from issues such as limited surface area, pore collapse, restricted diffusion, or acid site leaching under wet conditions. Sulfonic acid functionalised POPs, by contrast, offer a superior combination of high acid site density, structural integrity, hydrophobic or hydrophilic tuning, and well-defined porous channels, facilitating enhanced catalytic turnover and recyclability [1, 2].

In this study, we introduce a sulfonic acid functionalized porous organic polymer (POP–SO_3H) as a highly effective and recyclable catalyst for the Biginelli reaction. The POP framework was synthesized using a bottom-up solvothermal method. We carried out characterization of the catalyst using techniques such as FTIR, PXRD, TGA and SEM. The catalytic performance of POP–SO_3H to the Biginelli reaction is investigated. For the effective performance, various parameters like catalytic yield, temperature, solvent and reaction time have been optimized. Importantly, it could be easily recovered and reused for multiple cycles without any significant drop in performance. This makes it a practical and eco-friendly alternative to traditional acid catalysts. Overall, this work demonstrates the potential of sulfonated POPs as next-generation heterogeneous catalysts capable of performing complex multicomponent reactions like the Biginelli condensation in a more sustainable, efficient, and scalable manner. Our findings open up new avenues for using these materials in pharmaceutical synthesis and other fine chemical transformations where clean, reusable, and as metal free catalyst without post synthetic modification.

2 Materials and Method

2,4-Dihydroxybenzenetricarbaldehyde (DHBTC) and 2,4-Diaminobenzenesulphonic acid (DAS) were purchased from Sigma Aldrich. Acetic acid, 1,4-Dioxane, Tetrahydrofuran (THF), Methanol (MeOH), Dimethylformamide (DMF), Acetone and Ethanol (EtOH) were purchased from Alpha chemicals. HPLC grade solvents were used for photophysical studies. Metal and anion salts were used as received. All chemicals and reagents were used without additional purification. The graphs were generated using Origin Pro 8.5 software, and the schemes and structures were created with Chem Draw 18.0. Various methods used for the characterizations are described as follows. The IR spectrum was achieved in Bruker, Tensor 27 model FT-IR spectrometer to recognize the functional groups. The powder XRD analysis was executed in Bruker D8 Advance instrument. The surface morphology of the polymer was scrutinized by Field Emission Scanning Electron Microscopy (FESEM) via the instrument Carl Zeiss-Sigma with SMARTSEM software. Thermo Gravimetric Analysis (TG/DTA) was carried out using Perkin Elmer diamond thermogravimetry/differential thermal analysis (TG/DTA) instrument with a heating rate of 20 °C/min in presence of continuous flow of nitrogen.

3 Synthesis of POP-SO_3H

Solvothermal method was used for the synthesis of sulphonic acid functionalized covalent organic framework with certain modification. 2,4-Diamino benzene sulphonic acid (DAS) was taken in the two neck round bottom flask containing 20 ml Tetrahydrofuran (THF) and sonicated. 2,4,6-Tris formyl phloroglucinol (TFPG) was taken in 100 ml beaker containing 20 ml THF, 2 drops of 6 M acetic acid were added and stirred for 15 min. The resultant mixture was added to the two neck round bottom flask containing DAS. Then it was refluxed for 3 day in an oil bath at 80 °C. The resultant yellow precipitate obtained was washed with acetone, THF, methanol, and then purified by soxhlet extraction in THF and methanol for 24 h. The reaction scheme is represented as Fig. 1.

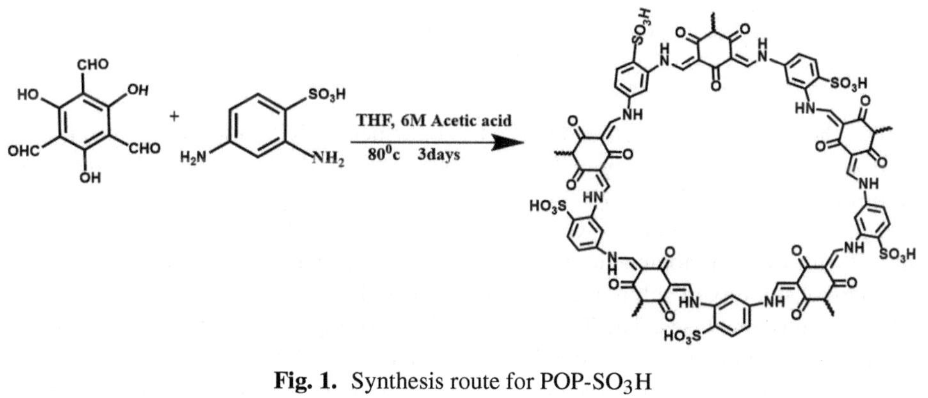

Fig. 1. Synthesis route for POP-SO_3H

4 Result and Discussion

4.1 Characterisation

The structural and morphological characteristics of the polymer were analyzed using Fourier Transform Infrared Spectroscopy (FTIR), Powder X-ray Diffraction (PXRD), Thermogravimetric Analysis (TGA), Field Emission Scanning Electron Microscopy (FESEM), High-Resolution Transmission Electron Microscopy (HRTEM), and X-ray Photoelectron Spectroscopy (XPS).

The X-ray diffraction (XRD) measurements were employed to determine the crystalline nature of the polymer using Cu $K\alpha_2$ radiation ($\lambda = 1.54060$ Å) over a 2θ range of 10° to 80°. The PXRD pattern reveals that the polymer exhibits a crystalline nature, with a crystallinity of approximately 69.2%, as confirmed by the distinct diffraction peaks observed at Bragg angles between 20° and 40° (Fig. 2a).

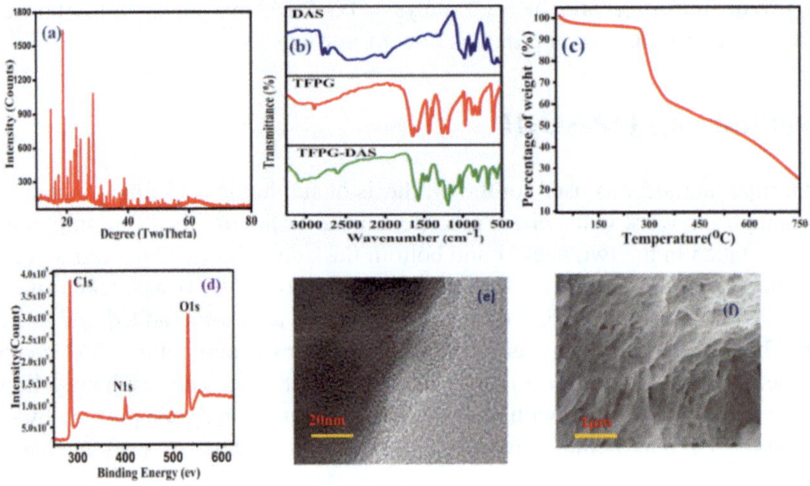

Fig. 2. (a) Powder X-ray diffraction spectrum of the sulphonic acid functionalized POP (b) FTIR spectrum of TFPG, DAS, Sulphonic acid functionalized porous organic polymer (d) XPS Survey scan (e-f) SEM and TEM image of SO_3H functionalized porous organic polymer

The chemical bonding in the POP can be confirmed by XPS Survey scan and Fourier transform infrared spectroscopy. The disappearance of aldehydic peak at 1690 cm-1 and primary amine peak at 3770 cm-1and formation of new peak at 1550 cm-1 (-C-N-), 1600 cm-1 (-C=O), 1301 cm-1 (-C=C-) and 1100 cm-1 (SO_3H), confirms β-ketoenamine linkage with free sulphonic acid present in the polymer matrix (Fig. 2(b &d)).

Thermal stability of sulphonic acid functionalized polymer was investigated by thermal gravimetric analyses. The TGA plot involve plotting percentage of weight loss as the function of temperature. The SO_3H functionalized polymer is thermally stable up to 300 °C. Weight losses observed above 300 °C could be due to degradation of SO_3H functional group, followed by the degradation of polymer framework. The degradation continues up to the complete carbonization of the organic polymer (Fig. 2c).

The structural morphology of the sulphonic acid functionalized polymer has been studied by scanning electron spectroscopy (SEM) and Transmission electron microscopy (TEM). The morphological studies confirm non uniform porous nature of the polymer with layered arrangement of the polymer matrix (Fig. 2(e-f)).

4.2 Application of Sulphonic acid Functionalised POP for Biginelli reaction

3,4-dihydropyrimidones, also known as thymine analogs are important compounds with diverse applications, particularly in medicinal chemistry and pharmaceuticals. Thymine analogs, including 3,4-dihydropyrimidones, play a crucial role in the development of pharmaceuticals. Thymine itself is one of the four nucleobases found in DNA, and its analogs can be used in the design and synthesis of nucleoside and nucleotide analogs with potential antiviral and anticancer activities. Some 3,4-dihydropyrimidone derivatives exhibit antiviral properties. These compounds can be explored in the development of antiviral drugs, targeting various viral infections. It also has been investigated for their anticancer potential. They may interfere with DNA synthesis and replication, making them potential candidates for cancer chemotherapy. It is also explored for their radioprotective properties. This could be important in situations where protection against radiation-induced damage is needed, such as in certain medical treatments or exposure scenarios. The Biginelli reaction is a multicomponent organic reaction which involves the condensation of three components: an aldehyde, a β-ketoester (or β-diketone), and urea or thiourea. The reaction results in the formation of a dihydropyrimidinone derivative as shown in Fig. 3. The sulphonic acid functionalized COFs can potentially serve as catalyst supports due to their tailored pore sizes and sulphonic acid functional group. The formation of product was confirmed by Gas chromatography, due to the existence of peak at m/z value 276.2 (Fig. 5).

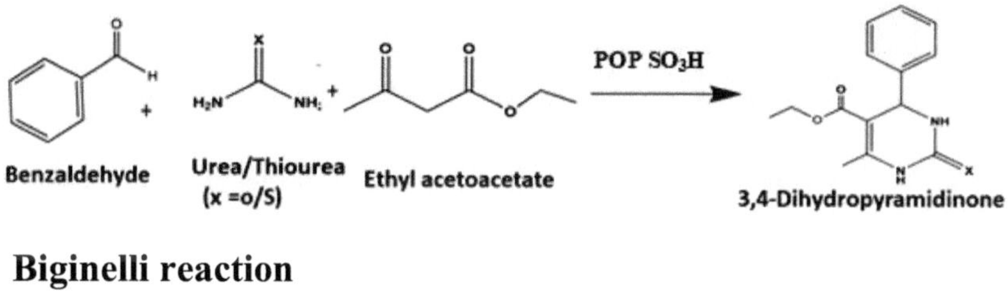

Fig. 3. Synthesis route of dihydropyramidinone

Optimization of Reaction Condition. In heterogeneous catalysis, the quantity of catalyst used plays a crucial role in determining both efficiency and the eco-friendliness of the process. A preliminary experiment was conducted using benzaldehyde, ethyl acetoacetate, and thiourea as reactants in methanol at 60 °C, but without the COF-SO_3H catalyst. After 4 h, the reaction gave a very low yield of product, emphasizing the essential role of the catalyst for achieving satisfactory yields. Increasing the catalyst load from 10 to 30 mg significantly enhanced the yield, with 30 mg as the optimum amount, resulting in a maximum yield of 96%. Under these optimized conditions, only trace amounts of by-products or impurities were detected, showcasing the catalyst's high efficiency. Beyond 30 mg, the yield did not improve significantly, while using less than 30 mg led to longer reaction times and lower yields. The increased catalyst load up to 30 mg enhanced both yield and reaction rate due to the greater availability of sulfonic acid groups, which intensified the acidic nature of the reaction environment. Further optimizations identified methanol as the most effective solvent, outperforming ethanol and acetonitrile, with yields exceeding 90%. Methanol was thus selected as the preferred solvent. The reaction time was optimized to 60 min, ensuring maximum efficiency under these conditions. Additionally, the reaction temperature was optimized, with 60 °C identified as the ideal condition for achieving maximum efficiency and yield. Optimization of various factors are depicted in Fig. 4.

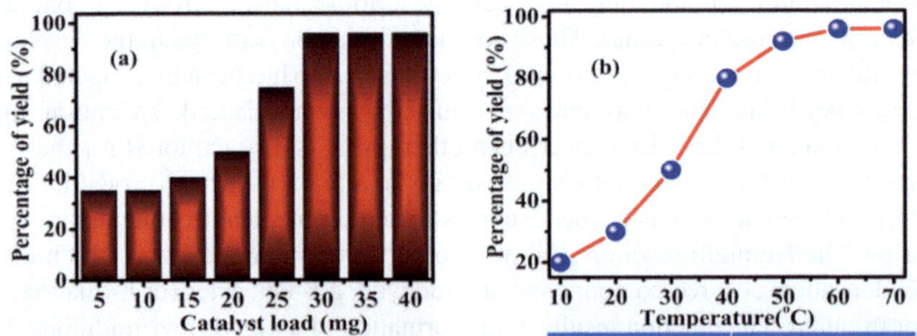

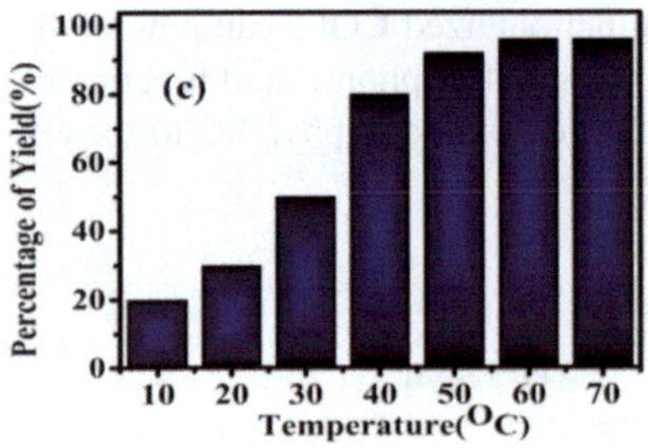

(d)

Solvent	Percentage of yield (%)
Water	------------
DMF	60
THF	65
Ethyl acetate	70
DCM	82
EtOH	90
MeOH	95

Fig. 4. Optimization of (a) Catalyst load (b) Reaction time (c) Temperature (d) Solvent

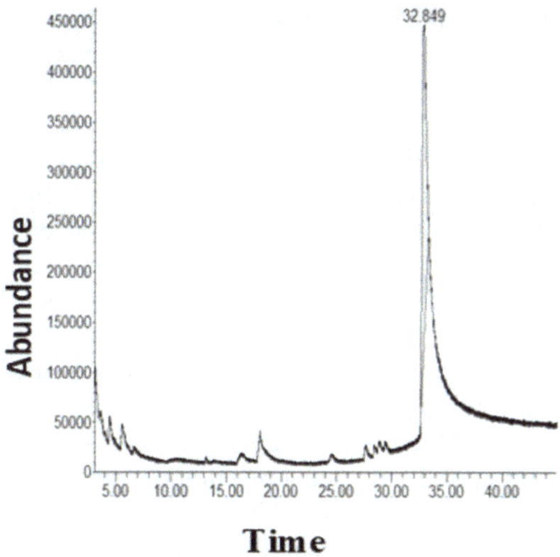

Fig. 5. GCMS profile of 3,4-Dihydropyrimidone with m/z value 276.12

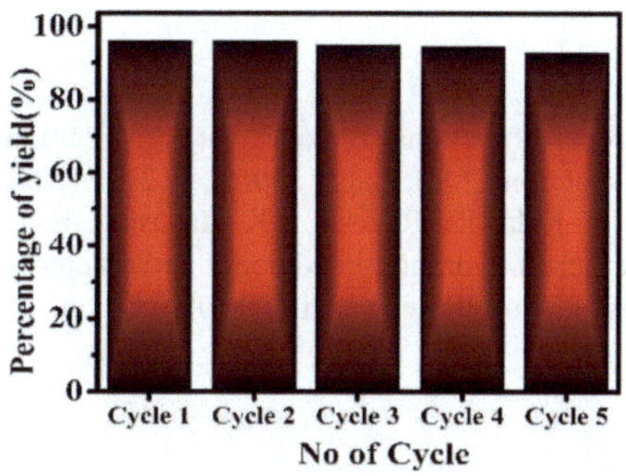

Fig. 6. Recyclability of POP SO_3H for five catalytic cycle

Reusability of Catalyst. The reusability of the POP-SO_3H catalyst was systematically investigated to evaluate its practical applicability in sustainable catalysis. The catalyst was subjected to five consecutive cycles of the Biginelli reaction under optimized conditions. After each cycle, POP-SO_3H was recovered by simple filtration, washed thoroughly with ethanol and deionized water to remove any physisorbed reactants or products, and dried under vacuum at 60 °C [2]. Impressively, the catalyst retained its high catalytic activity with only a marginal decrease in yield, suggesting minimal leaching of the active –SO_3H groups and excellent framework stability. These findings underscore the robustness, chemical stability, and reusability of POP-SO_3H, making it a highly efficient and environmentally benign catalyst for repeated use in multicomponent organic transformations. The bar graph (Fig. 6) illustrates the percentage yield of the Biginelli product using POP-SO_3H over five consecutive catalytic cycles.

A novel sulfonic acid-functionalized β-ketoenamine linked porous organic polymer (POP-SO_3H) was successfully synthesized via a solvothermal condensation reaction between 2,4,6-trihydroxybenzene-1,3,5-tricarbaldehyde and 2,4-diaminobenzenesulfonic acid. The resulting polymer was thoroughly characterized using Powder X-ray Diffraction (PXRD), Fourier Transform Infrared Spectroscopy (FTIR), Thermogravimetric Analysis (TGA) and Scanning Electron Microscopy (SEM). confirming the successful formation of a stable, highly cross-linked porous network. The incorporation of sulfonic acid groups introduced strong Brønsted acidity, while the β-ketoenamine backbone imparted excellent chemical and thermal robustness to the polymer framework. The catalytic efficiency of POP-SO_3H was evaluated using the Biginelli reaction as a model multicomponent reaction. Various reaction parameters including solvent type, catalyst loading, reaction temperature, and time were systematically optimized to achieve the highest yield. Methanol was identified as the most effective solvent, with an optimal catalyst loading of 30 mg and a reaction temperature of 60 °C, resulting in up to 96% product yield within 60 min. In addition to high catalytic activity, POP-SO_3H exhibited excellent reusability and structural integrity over five consecutive reaction cycles, with negligible loss in yield or performance. These results establish POP-SO_3H as a promising, green, and reusable heterogeneous catalyst for multicomponent organic transformations in sustainable synthetic chemistry.

References

1. Gong, K., et al.: Sulfonic acid functionalized covalent organic frameworks as efficient catalyst for the one-pot tandem reactions. Mol. Catal. **519**(6), 112139 (2022)
2. Aswathy, P.V., Athira, P., Kala, R., Rakesh, S.K., Jayasree, E.G.: A triazine based porous organic polymer as an efficient hydrogen-bond donor and acceptor cooperative heterogeneous catalyst for the synthesis of 2-substituted benzimidazoles. Chemistryselect. **8**(42), e202303092 (2023)

A Novel Analytical Modeling and Simulation of Dual-Metal Double Gate Tunnel FET Using Finite Difference Method

V. Karutharaja[1(✉)], B. S. Durkesh[1], N. B. Balamurugan[1], and S. Raj Kumar[2]

[1] Department of Electronics and Communication Engineering, Thiagarajar College of Engineering, Madurai, India
karutharaja@gmail.com, nbbalamurugan@tce.edu

[2] Department of Electronics Engineering (VLSI), Velammal College of Engineering and Technology, Madurai, India
srk@vcet.ac.in

Abstract. Highly promising devices are the Tunnel Field Effect Transistors (TFETs), which have a subthreshold swing of less than 60 mV/decade. Multi-gate devices are introduced in order to mitigate the short channel impact. This research investigates the efficiency of Double Metal Dual Gate Tunnel FET (DMDG-TFET) by altering the gate metals' work functions and analyzing the impact on the electric field, surface potential, transfer characteristics, and threshold voltage. Decomposing the 2D Poisson Equations into two 1-D equations is the aim of the analytical study. With appropriate boundary conditions, the 2D Poisson equation is solved using the finite differentiation method (FDM). The proposed analytical model was validated by a simulation study of the previously reported DMDG-TFET properties. By changing the gate bias and demonstrating conformance with the results of the TCAD simulation, the effects of the surface potential and electric field are investigated. The outcomes of the analysis and the TCAD simulation are identical. Tunneling rate and tunneling width reduction have enhanced our transfer characteristics, and we have concluded that the ON current of the DMDG-TFET is higher than that of the SMG TFET.

Keywords: Tunnel FET · Poisson Equation · FDM · Surface potential · Threshold voltage

1 Introduction

In semiconductor device technology, the necessity for high-density chips necessitates device scaling down. Scaling increases the amount of components per channel and decreases the channel length to boost processing speed. A novel design for a double-gate tunnel field-effect transistor (DG tunnel FET) that, in simulations [1]–[3], exhibits significant performance improvements over single-gate devices with a gate dielectric. It is shown that the current flow in tunnel FETs is not restricted to a channel at the gate-dielectric surface by two-dimensional analysis [4, 5]. Tunnel FETs with a high-k gate

A. Antony et al. (Eds.): icONMAT 2025, SPPHY 328, pp. 113–123, 2025.
https://doi.org/10.1007/978-981-95-2697-0_15

dielectric exhibit a reduced threshold voltage shift with temperature variation compared to conventional MOSFETs, while maintaining a nearly constant subthreshold slope (SS) for fixed values of Vg [6]. A triple gate n-channel junction less transistor (JLTFET) and a junction less tunnel field effect transistor were combined in [7]–[10]. The JLTFET is a strongly doped junction less transistor that uses the tunneling principle to narrow the barrier between the source and channel of the device to switch on and off [11].

Additionally, at ambient temperature, a simulated JLTFET with a gate length of 50 nm has a SS of 47 mV/decade, suggesting that JLTFET is a promising technology for switching efficiency. There have been suggestions regarding tunnel FETs' short-channel effects [8, 12]. This paper investigates short-channel effects (SCEs) in double-gate tunnel FETs (TFETs) using an analytical model involving depletion in the source. When the channel length is less than twice the scale length, the potential is significantly affected by the drain bias at the source [13].

The SCE mimics the Drain-Induced Barrier Lowering effect in conventional MOSFETs, showing up as finite-output conductance in the saturation field. Concepts such as electric flux and surface potential have been presented [2, 14]. Analytical modeling of the electric flow and surface potential is achieved by meeting the 2-D Laplace condition. The analytical and MATLAB data were compared with the Dual Material Gate Tunnel Field Effect Transistor (DMG-TFET) to determine the ion enhancement and lower Subthreshold Swing [15]. Using the finite difference method (FDM), the analytical model of the double metal double gate TFET's surface potential, electric field, drain current, and threshold voltage has been developed in this research. The computer simulator 2-D Sentaurus TCAD is used to simulate the analytical models.

2 Device Structure

With two gate metals, GM1 of length L1 and GM2 of length L2, the innovative proposed structure is depicted in Fig. 1. It is known as a double material dual gate tunnel-field effect transistor (DMDG-TFET) and is analytically modeled using FDM. Since electrons have a tendency to move from metals with lower work functions to those with higher work functions, we have chosen that metal 1's work function (Ø1) is lower than metal 2's work function (Ø2). Provide a way to maximize transistor scaling and lower the threshold voltage as a result. When multi-gate devices are narrowed, control over the source and drain channels becomes extremely challenging, and the creation of a junction complicates the device. By raising the surface potential at the metal interface, the usage of double material improves the device's efficiency. Table 1 compiles the physical parameters of the device.

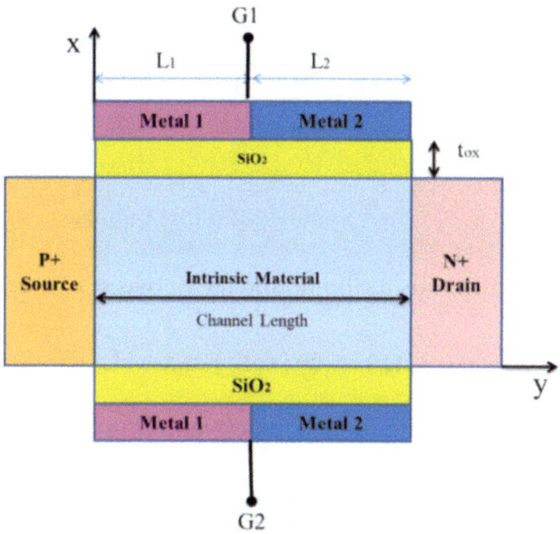

Fig. 1. The structure of proposed DMDG TFET

Table 1. The physical parameters of DMDG TFET

Quantities	Symbols	Values
Gate Work functions	$\emptyset_1$	4 eV
Gate Work functions	$\emptyset_2$	4.4 eV
Thickness of oxide	^{t}ox	2 nm
Thickness of Silicon	^{t}si	10 nm
Channel length	L_1 and L_2	10 nm
Doping level of Source	N_A	10^{20}cm-3
Doping level of Drain	N_D	5x1018cm^{-3}
Doping level Channel	NC	1017cm-^{3}

3 Analytical modeling derivation

The 2-D Poisson's equation for DMDG-TFET is given by:

$$\frac{d^2\phi(x, y)}{dx^2} + \frac{d^2\phi(x, y)}{dy^2} = \frac{qN_D}{\varepsilon_{si}}\ 0 \leq x \leq L \tag{1}$$

Where φ(x,y) - surface potential with coordinates of a rectangular, q – charge of electron, N_D - donor concentration in the channel, ε_{si} – Silicon permittivity.

The general equation for finite difference method is given by:

$$\frac{\phi(x_{i-1}, y_i) - 2\phi(x_i, y_j) + \phi(x_{i+1}, y_j)}{(h_x)^2} + \frac{\phi(x_i, y_{i-1}) - 2\phi(x_i, y_j) + \phi(x_i, y_{j+1})}{(h_y)^2} = f(x, y) \tag{2}$$

Boundary conditions are

1. Surface potential is continuous at the boundary between two different metals,

$$\phi_1(x, t_{si}) = \phi_2(x, t_{si}) \text{ at } x = L_1 \tag{3}$$

2. The constant electric field at the interface between Metal1 & Metal2:

$$\frac{d\#_1(x, y)}{dx} = \frac{d\#_2(x, y)}{dx} \text{ at } x = L_1 \tag{4}$$

3. Surface potential under Metal1:

$$\phi_{L1}(x, 0) = v_{g1}, \text{ where } v_{gi} = v_{gs} - v_{FB1} \tag{5}$$

4. Surface potential under Metal2:

$$\phi_{L2}(x, 0) = v_{g2}, \text{ where } v_{g2} = v_{gs} - v_{FB2} \tag{6}$$

$$\frac{d\#_{L1}(x, y)}{dy}|_{y=t_{si}} = \frac{\varepsilon_{ox}}{\varepsilon_{si}} \frac{\phi_{g1} - \phi_1}{t_{ox}}, \text{ where } \#_{g1} = V_{GS} - \varnothing_1 + \$ + \frac{E_g}{2} \tag{7}$$

$$\frac{d\#_{L2}(x, y)}{dy}|_{y=t_{si}} = \frac{\varepsilon_{ox}}{\varepsilon_{si}} \frac{\phi_{g2} - \phi_2}{t_{ox}}, \text{ where } \#_{g2} = V_{GS} - \varnothing_2 + \$ + \frac{E_g}{2} \tag{8}$$

5. Surface potential at the source side:

$$\phi_1(0, d) = v_{bi} = \phi_{L1}(0) \tag{9}$$

6. Surface potential at the drain side:

$$\phi(L_1 + L_2, d) = v_{bi} + v_{ds} = \phi_{L2}(L) \text{ where } L = L_1 + L_2 \tag{10}$$

Where V_{GS} - Gate to Source voltage, V_{FB} - Flat band voltage, q - Electric charge, ε_{ox} - Permittivity of SiO_2,

E_g - band gap energy, V_{ds} - Drain to Source voltage and V_{bi} - built-in potential.

3.1 Model of Surface potential

By comparing boundary conditions (1) and (2), the following three discrete points are taken along the x-axis and y-axis in order to solve 2D Poisson's equation using the finite difference method.

To calculate surface potential, the above values are substituted into Eq. (1):

$$\frac{\phi(0,y) - 2\phi(L_1,y) + \phi(L_1 + L_2, y)}{L_1^2} + \frac{\phi\left(x, \frac{d}{2}\right) - 2\phi(x,0) + \phi\left(x, -\frac{d}{2}\right)}{\frac{d^2}{4}} = \frac{qN_D}{\varepsilon_{si}} \tag{11}$$

In order to separate the DMDG TFET into two distinct single material TFETs and compute the surface potential for each one independently, let's suppose that the electric field and surface potential under M1 and M2 do not interface with one another.

Using FDM approach the Eq. (11) can be decomposed into:

$$\frac{\phi_i(0) - 2\phi_i(L_1) + \phi_i(L_1 + L_2)}{L_1^2} = \frac{qN_D}{\varepsilon_{si}} \text{ where } i = 1, 2 \tag{12}$$

$$\frac{\phi\left(x, \frac{d}{2}\right) - 2\phi(x,0) + \phi\left(x, -\frac{d}{2}\right)}{\frac{d^2}{4}} = 0 \tag{13}$$

Where the surface potentials $\varphi(0)$ and $\varphi(L_1 + L_2)$ are represented as φ_S and $\varphi(L_1)$ is represented as φ_0.

So, Eq. (12) becomes:

$$\phi_s = \phi_0 + \frac{qN_D L^2}{2\varepsilon_{si}} \tag{14}$$

With V_{th} being the DMDG TFET's threshold voltage, φ_0 can now be substituted with V_{gs}-V_{th}.

Eq. (14) becomes:

$$\phi_s = V_{gs} - V_{th} + \frac{qN_D x_d^2}{2\varepsilon_{si}} \tag{15}$$

Where x_d is the depletion width given by:

$$x_d = \sqrt{\frac{2\varepsilon_{si}\left(v_{gs} - v_{th}\right)}{qN_D}} \tag{16}$$

In the partially depleted channel, the density of charge (Qsc) is:

$$Q_{sc} = 2qN_D x_d = -2C_{ox}\left(V_{gs} - V_{fb} - \phi_{si}\right) \tag{17}$$

Substituting Eq. (16) into (17):

$$\left(v_{gs} - v_{th}\right) = \frac{C_{ox}^2\left(v_{gs} - v_{fb} - \phi_{si}\right)^2}{2\varepsilon_{si} qN_D}, \text{ where } C_{ox} - \text{gate oxide capacitance} \tag{18}$$

The surface potential for both regions 1 and 2 can be produced by replacing Eq. (15) with Eq. (18):

$$\text{For region 1}: \#_{s1} = \#_0 + \frac{C_{ox}^2}{2+_{si}qN_D}\left(v_{gs} - v_{fb} - \#_{s1}\right)^2 \tag{19}$$

$$\text{For region 2}: \#_{s2} = \#_0 + \frac{C_{ox}^2}{2+_{si}qN_D}\left(v_{gs} - v_{fb} - \#_{s2}\right)^2 \tag{20}$$

3.2 Electric Field Model

The DMDG-TFET device's performance with regard to short channel effects is determined by the electric field, which is a crucial parameter. Surface potential differentiation can be used to determine the electric field parameter for each location. As a consequence, the electric fields for each region are as follows:

$$\text{For region 1}: E_1(z) = \frac{C_{ox}^2}{+_{si}qN_D}\left(v_{gs} - v_{fb} - \#_{s1}\right) \tag{21}$$

$$\text{For region 2}: E_2(z) = \frac{C_{ox}^2}{+_{si}qN_D}\left(v_{gs} - v_{fb} - \#_{s2}\right) \tag{22}$$

3.3 Drain Current Model

Based on the band-to-band tunneling of charge carriers from the source's valance band to the intermediate region's conduction band, the DMDG-TFET drain current model was developed. Calculating the overall drain current involves integrating the Band-to-Band generation rate (Gkane) over the device's volume:

$$I_{DS} = q \iint G_{kane}\,dxdy \tag{23}$$

The definition of the band-to-band generation rate (Grate) is:

$$G_{kane} = \alpha(E)^D \exp\left(-\frac{\beta}{E}\right) \tag{25}$$

$$\alpha = \frac{q}{\pi h^2}\sqrt{\frac{m_r}{E_g}}.\beta = \frac{\pi^2\sqrt{m_r E_g^2}}{qh}, \text{ where } m_r = \frac{m_e m_h}{m_e + m_h} \tag{26}$$

3.4 Threshold Voltage Model

The threshold voltage is the lowest voltage that is put between the gate and the source in order to produce current flow. For the lowest surface potential, the analytical threshold voltage model was developed.

$$v_T = \frac{Q_{sc}}{C_{ox}} + \phi_s,\ when\ \phi_s\ is\ minimum,\ \phi_s = 2 \mid \phi_F \mid \tag{27}$$

$$v_T = (-2C_{ox}(v_{gs} - v_{fb} - 2|\varnothing_F|, \textit{where}, \phi_F - \textit{Fermi potential} \tag{28}$$

4 Result and Discussion

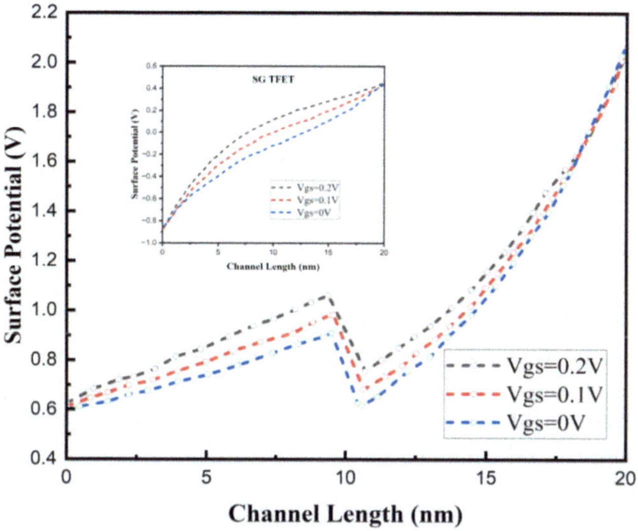

Fig. 2. Surface Potential at Vgs = 0 V, Vgs = 0.1 V and Vgs = 0.2 V.

For different gate-to-source voltages, Fig. 2 shows the surface potential curve along the channel length of the suggested TFET. The surface potential rises in tandem with an increase in gate voltage, indicating robust gate control over the channel. Band-to-band tunneling (BTBT) in TFETs depends on the channel-to-drain junction, which is indicated by a clear transition in the potential profile close to the mid-channel.

The electric field variation along the channel length for the suggested Tunnel Field Effect Transistor (TFET) structure at various gate-to-source voltages (VGS = 0 V, 0.1 V, and 0.2 V) is shown in Fig. 3. It is found that the electric field's magnitude increases as the gate voltage rises. The electric field profile of a traditional Single-Material (SM) TFET and the suggested construction are contrasted in the inset.

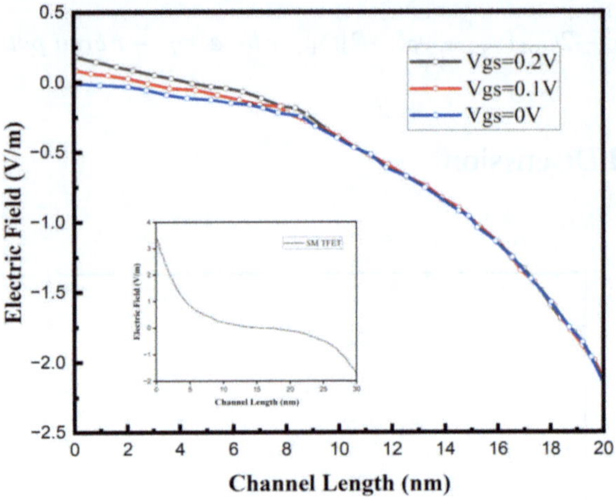

Fig. 3. Electric Field (V/m) at Vgs = 0 V, Vgs = 0.1 V and Vgs = 0.2 V

Analytical modeling and TCAD simulation of the electric field and surface potential along the channel at VGS = 0 V are contrasted in Fig. 4. The high degree of consistency confirms that the obtained models are accurate. Along the channel, the surface potential rises, and the electric field peaks close to the source-channel junction, which is essential for effective band-to-band tunneling.

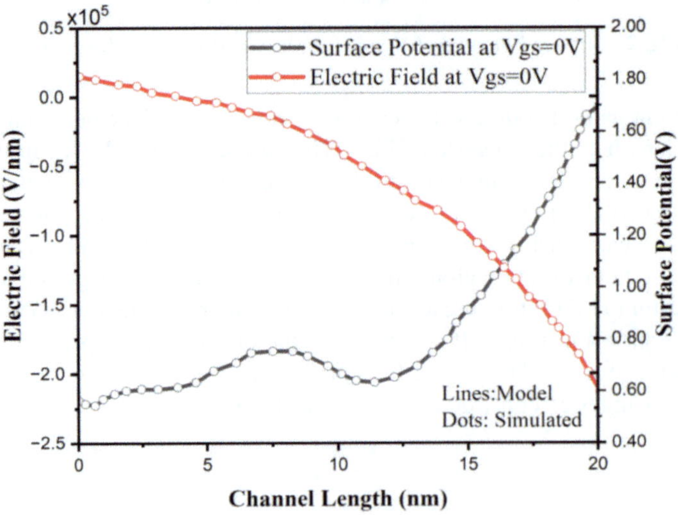

Fig. 4. Surface Potential and Electric Field (V/nm) at Vgs = 0 V.

The transfer characteristics, or the change in drain current (ID) with gate voltage (VGS), are shown in Fig. 5 for both the SMG-TFET and DMDG-TFET architectures. The accuracy and robustness of the established analytical expressions for the drain current are confirmed by the close match between the analytical model's results and the TCAD simulation data.

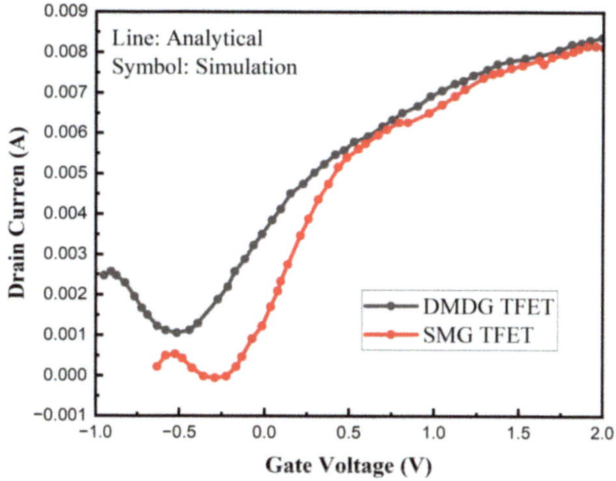

Fig. 5. Comparison of Drain Current (I_D) between DMDG TFET and SMG TFET

The surface potential fluctuation along the channel length of a Double-Material Double-Gate (DMDG) TFET for three gate-source voltages (VGS = 0 V, 0.1 V, and 0.2 V) is depicted in Fig. 6. The accuracy of the created surface potential model is validated by plotting the results of the TCAD simulation and the analytical model, which demonstrate close agreement.

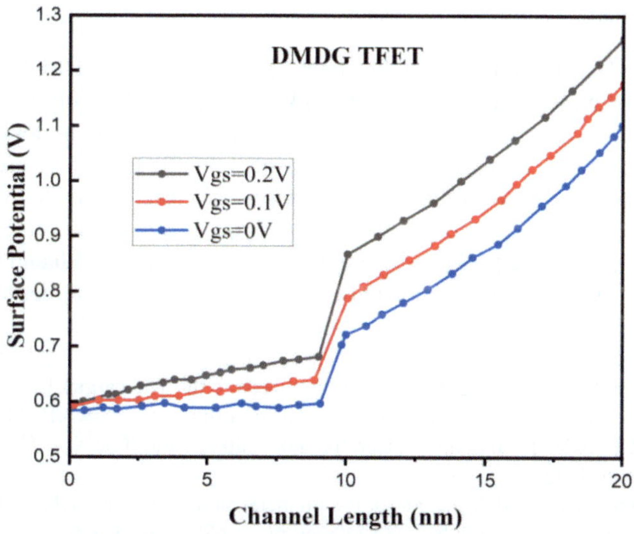

Fig. 6. Surface Potential of DMDG FET at different Vgs

The dependence of threshold voltage (VT) on gate length (L) and drain voltage (VD) for a Tunnel FET (TFET) under different top gate bias and oxide dielectric constant (εox) conditions is shown in Fig. 7. The suggested threshold voltage model is validated by plotting the results of the TCAD simulation and the predictions of the analytical model, which demonstrate a significant correlation.

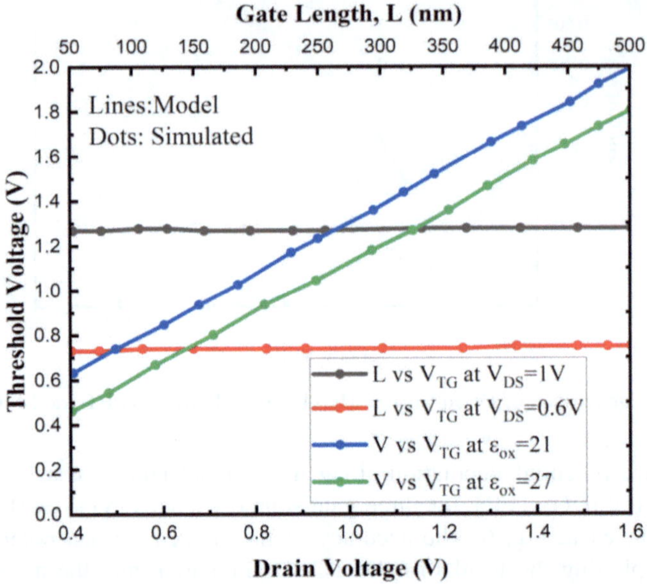

Fig. 7. Gate Length (L) Vs Threshold Voltage (V_T) for various drain voltage (V_{DS})

5 Conclusion

In this presented work, we looked at the Double Metal Dual Gate (DMDG) TFET's construction and examined important variables like the electric field, drain current, threshold voltage, and surface potential. The DMDG TFET's analytical model was created by applying the finite differentiation approach to solve the 2D Poisson expression. TCAD simulations were used to verify the validity of this structure, and the outcomes demonstrated a high degree of agreement with the analytical data. Enhancements in transfer characteristics were made possible by the decreased tunneling width and increased tunneling velocity. As such, the ON current of the DMDG TFET is higher than that of the SMG TFET.

Acknowledgement. The authors express their gratitude to the Thiagarajar College of Engineering (TCE), Madurai, India for supporting us to carry out this research work. Also, the financial support from TCE under Thiagarajar Research Fellowship scheme (File No.: TRF/July–2023/05) is gratefully acknowledged.

We also extend our heartfelt thanks for the financial support provided through the TCE Seed Money Scheme (TCE-SMS) (File No.: R&D/Seed money/2024/8), which has been instrumental in the successful execution of this study.

References

1. Saleem, S., Anirudh, M.D., Bharath, K.R., Pavan, R.R., Vimala, P.: Design and analysis of dual material double gate tunnel-field effect transistor (DMDG-TFET) with gate oxide stack. In: 3rd International Conference on Intelligent Technologies (CONIT), pp. 1–4. Institute of Electrical and Electronics Engineers, Hubli India (2023)
2. Saini, G., Miriyala, D.P.: A Systematic Investigation of Dual Material Gate TFET for Improved Performance. Indian J. Pure Appl. Physis (IJPAP). **62**, 5888 (2024)
3. Raju, V., Suresh, E., Shashikanth, B., et al.: Performance analysis of double material gate (DG) -TFET with channel doping. SILICON. **14**, 10775–10779 (2022)
4. Koppolu, K., Samuyelu, B., C B Rao K.: The effect of a dual oxide - dual gate material and a sensitivity analysis on the performance of a junctionless tunnel FET. SILICON. **16**, 3905–3915 (2024)
5. Kathy, B., Adrian, M.I.: Double-gate tunnel FET with high-K gate dielectric. IEEE Trans. Electron Devices. **54**(7), 1725–1733 (2007)
6. Woo, Y.C., Park, B.-G., Jong, D.L., Liu, T.-J.K.: Tunneling field- effect transistors (TFETs) with subthreshold swing (SS) less than 60 mV/dec. IEEE Electron Device Lett. **28**(8), 743–745 (2007)
7. Panigrahy, S., Sahu, P.K.: Performance enhancement and reduction of short channel effects of nano-MOSFET. In: 1st International Conference on Circuit, Power and Computing Technologies (ICCPCT). Institute of Electrical and Electronics Engineers, Nagercoil, India (2013)
8. Ghosh, B., Bal, P., Mondal, P.: A junction less tunnel field effect transistor with low subthreshold. J. Comput. Electron. **12**(3), 428 (2013)
9. Shara, M., Medhi, S., Tiwari, P.K.: A Performance Analysis of Hetero- Dielectric Dual-Material-Gate Silicon-on- Insulator Tunnel Field Effect Transistors. Annual IEEE India Conference (2014)
10. Lakshmi Priya, G., Balamurugan, N.B.: New dual material double gate junction less tunnel FET: Subthreshold modeling and simulation, AEU. Int. J. Electron. Commun. **99**, 130–138 (2019)
11. Singh, K.S., Kumar, S.: A review Tunnel field effect transistor for ultra-low power applications. In: 8th International Conference on Systems and Control (ICSC) (2019)
12. Safa, S., Noor, S.L., Khan, Z.R.: Physics-based generalized threshold voltage model of multiple material gate tunneling FET structure. In: 3rd International Conference on Electrical Engineering and Information Communication Technology (ICEEICT), pp. 1–4 (2017)
13. Helen, R., Bharathi, R., Karthikeyan, P.: Analytical modeling and simulation of dual material double gate all around tunnel field effect transistor using MATLAB. In: International Conference on Innovations in Information, Embedded and Communication Systems (2017)
14. Samantha, L.N., Safa, S., Khan, M.Z.R.: Dual- material double-gate tunnel FET: Gate threshold voltage modeling and extraction. J. Comput. Electron. **15**, 763 (2016)
15. Sunil, K., Sumit, K., Keshaw, K., Balwinder, R.: Analysis of Double Gate Dual Material Tunnel FET Device for Low Power SRAM Cell Design. American Scientific Publishers (2016)

Design and Development of MEMS-Based Cantilever with Integrated Micro-Tip

S. Vishal(✉), Megha Agrawal, and D. V. Priyanka

Sensor Technology Development Centre, Centre for Sensors and Vision Technology, Central Manufacturing Technology Institute, Bengaluru 560022, India
vishalayan108@gmail.com

Abstract. The development of Micro-Electro-Mechanical Systems (MEMS) has transformed sensor technology, leading to the creation of highly precise sensors at microscale dimensions for modern measurement capabilities. Among MEMS devices, micro-cantilevers have gained prominence due to their versatile functionality, compact design, and potential for integration into a variety of sensing platforms. This proposed work highlights the design and fabrication of silicon cantilevers along with Micro-tips using MEMS technology. It offers enhanced performance over conventional bonded-tip designs by eliminating alignment errors and ensuring mechanical robustness. Silicon is chosen as a structural material for both cantilevers and micro-tips due to its high Young's modulus and compatibility with MEMS fabrication techniques. These cantilevers with integrated tips have a wide range of applications, including nanometer-resolution surface profiling (e.g., Atomic force microscopy) and Nano indenter calibration, with the capability to achieve sub-nanometer displacement resolution. The piezo resistors can also be embedded on top of cantilevers to use them as a position sensor, which converts the cantilever deflection into an electrical signal. The designed sensors are simulated using the Covent or Ware tool to analyses the mechanical behavior of the silicon cantilever beam as well as to finalize the fabrication flow of the proposed device.

Keywords: MEMS · Silicon Cantilever · Micro-tips

1 Introduction

Metrology is the science of measurement, ensuring the accuracy and reliability of measurements crucial for the production of nanomaterials and devices. It plays a fundamental role in ensuring high precision in manufacturing, particularly for parameters like roughness, roundness, conicity, and edge rounding, which are critical for maintaining quality and consistency at the nanoscale. Various techniques for measuring at the nanoscale have been developed, with surface texture and roughness measurement methods typically categorized into contact stylus instruments and optical instruments like vertical scanning interferometry (VSI) [1]. Silicon cantilevers are widely used in MEMS devices, including position sensors, pressure sensors, and accelerometers. Single-crystal silicon is preferred due to its mechanical strength, stiffness, high sensitivity, and ease of fabrication with semiconductor processes. In strain-based position sensors, cantilevers serve as

A. Antony et al. (Eds.): icONMAT 2025, SPPHY 328, pp. 124–130, 2025.
https://doi.org/10.1007/978-981-95-2697-0_16

the probe element for measuring mechanical characteristics and surface topography at the micro to nanoscales. The cantilever's beam deflects when a force is applied, and this deflection is indirectly measured by the strain on the cantilever, which can be quantified using piezo-resistive elements directly fabricated on top of the cantilevers using MEMS technology [2].

Surface roughness measurement is critical in semiconductor technology, automotive manufacturing, and medical engineering. μ-CMMs and optical microscopy are widely used for dimensional quality control, while AFMs provide nanometer-scale resolution. For challenging measurements, such as vertical sidewalls or high-aspect-ratio microstructures, slender piezo-resistive silicon cantilevers with integrated silicon tips offer resolutions comparable to AFMs [3].

These microprobes enable roughness measurements at high traverse speeds, up to 15 mm/s, and their relatively low cost compared to optical instruments. The wearing effect of the tip is also a concern. Hence, diamond tips with 2 μm radius were glued to these microprobes. Although there is an advantage of reduced wear, improvements are needed in its dynamic response to support measurements at speeds exceeding 10 mm/s [4]. The integrated silicon tips still offer enhanced performance over conventional bonded-tip designs by eliminating alignment errors and ensuring mechanical robustness. Alves et al. developed a mixed etching process that combines wet etching, dry etching, and thermal oxidation to fabricate silicon micro tips [5]. This paper presents the design, fabrication, and analysis of a MEMS-based cantilever with an integrated micro-tip.

2 Design Considerations and Fabrication

A traditional cantilever-beam sensor consists of two primary components:-

1. Cantilever Beam Structure - The mechanical element that experiences an applied force, leading to deflection.
2. Micro-Tip - The silicon tip converts surface variations into mechanical forces, which are detected by strain-sensitive resistors on the cantilever.

A cantilever experiences deflection when the scanning tip encounters variations in surface topography during scanning, where the interaction forces (like van der Waals, electrostatic, or mechanical contact forces) between the tip and sample surface cause the beam to bend upward or downward.

The magnitude of deflection is proportional to the force exerted at the tip, bending the beam either upward or downward. As the cantilever moves from its neutral position, strain develops on its surface. To measure this strain, a piezo-resistive Wheatstone bridge is positioned near the cantilever beam's anchored end, where strain is maximized. The bridge becomes unbalanced in response to applied force, generating an electrical signal proportional to the strain.

The deflection of the beam (δ) is proportional to the applied force (F) and is given by:

$$\delta = \frac{4Fl^3}{Ewh^3} \tag{1}$$

The strain generated at the fixed end is given by:

$$\varepsilon = \frac{3h}{2l^2}\delta \tag{2}$$

where F is the Force applied at the free end, l is the Length of the cantilever, w is the width of the beam, h is the beam thickness, and E = 170 Gpa is Young's modulus of the material.

The schematic (Fig. 1) illustrates the complete design, including piezo resistors intended for strain sensing. However, in the present work, the piezo resistors have not yet been diffused onto the cantilever structure.

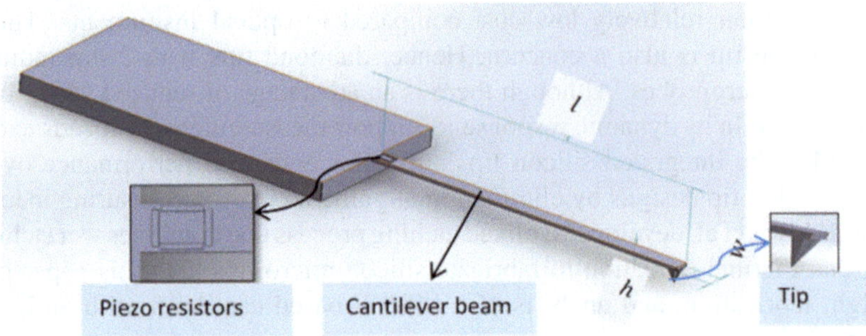

Fig. 1. Shows a schematic representation of a silicon cantilever with an integrated tip

Fabrication Flow The fabrication of the cantilever with an integrated tip involves a four-step process, as shown in Fig. 2:

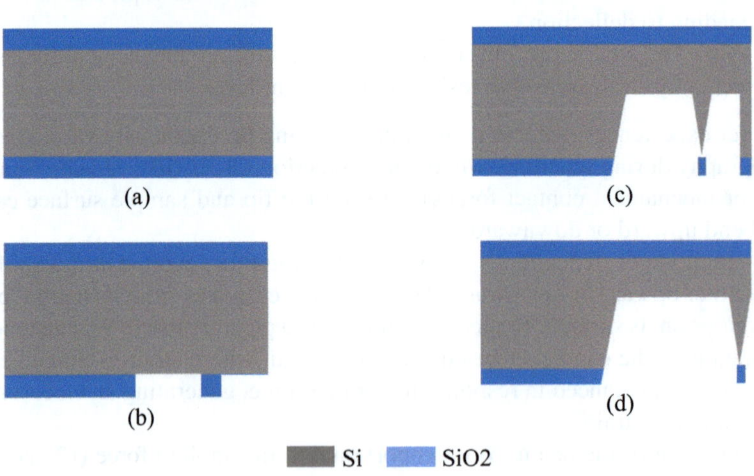

Fig. 2. Fabrication Flow for Cantilever with Integrated Micro-tip

(a) **Oxide Coating**:

A 4-inch. p-type silicon wafer with a (100) orientation is coated with a 1 µm-thick silicon dioxide (SiO_2) layer by thermal oxidation. This oxide layer acts as a hard mask, protecting selected regions of the silicon during subsequent etching steps.

(b) **Lithography and Oxide Etch:**

Photolithography is used to define the tip pattern. This involves spin-coating a photoresist, exposing it to UV light through a photomask, and developing the resist to reveal the pattern. The exposed SiO_2 areas are then etched using dry etching, creating openings that expose the underlying silicon.

(c) **Wet Etching with KOH:**

The wafer is subjected to anisotropic wet etching in potassium hydroxide (KOH) solution. KOH preferentially etches silicon along the (100) and (110) crystallographic planes while etching slowly along the (111) plane, which results in well-defined features with characteristic angled sidewalls.

(d) **Cantilever Release via ICP RIE:**

The final device singulation is carried out using Deep Reactive Ion Etching (DRIE). This through-etch process releases the cantilever, completing the fabrication of the structure.

The fabrication process, which combines thermal oxidation, photolithography, and etching, produced sharp micro-tips. As shown in Fig. 3, which includes (a) mask layout designed using Clew in software, (b) fabricated cantilevers on a silicon wafer, (c) cross section image of cantilever beam taken in optical microscope, (d) microscopic images of micro-tip, (e) prototype cantilever images and (e) SEM image of the micro-tip.

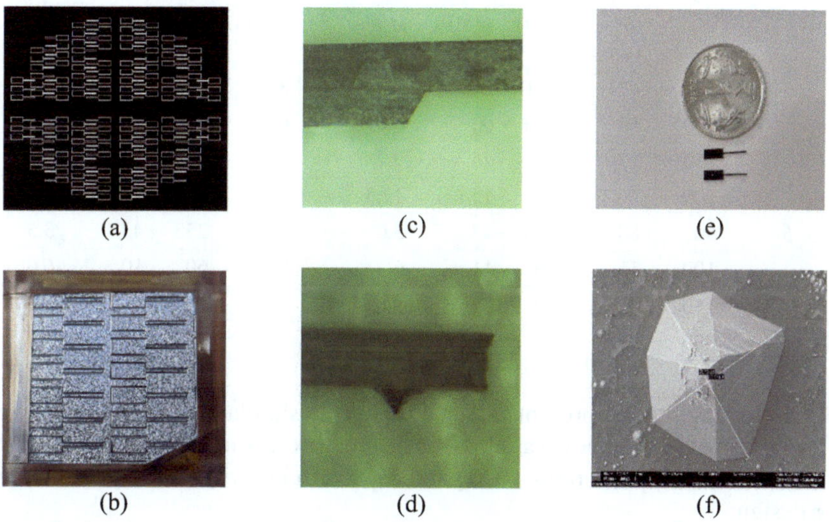

Fig. 3. Fabrication results of the MEMS cantilever with integrated micro-tip

3 Results and Discussions

3.1 Finite Element Analysis (FEA)

Covent or Ware is used for finite element modelling (FEM) to analyses the load-deflection behavior of the silicon cantilevers. Covent or Ware is a comprehensive simulation and design platform widely used in MEMS research and development. The design process in Covent or Ware generally consists of three primary stages: defining the process steps, creating the layout design, and generating the 3D model.

Table 1 presents the design parameters considered for the simulation of the beam.

Table 1. Design parameters

Sl no	Parameter	Dimensions of the Cantilever beam
1	Beam Length (l)	5 mm
2	Beam Width (w)	200 μm
3	Beam Thickness (h)	50–200 μm
4	Force (F)	750 μN–4000 μN

Table 2 compares theoretical and simulated results, highlighting the impact of cantilever thickness on deflection, strain, stress, and the strain-to-deflection ratio. For clarity, only the minimum and maximum force cases (750 μN and 4000 μN) are included.

Table 2. Calculated and Simulated deflection and strain values

F (μN)	**l**(mm)	**h**(μm)	**CALCULATED**				**SIMULATED**			
			δ (μm)	ε (μ)	σ(MPa)	$\varepsilon/\ \delta\ (m^{-1})$	δ (μm)	ε (μ)	σ(MPa)	$\varepsilon/\ \delta\ (m^{-1})$
4000	5	50	471	1410	240	2.9	390	1350	230	3.4
		100	58	353	60	6.0	48	324	55	6.7
		150	17	157	26	9.2	14	153	26	10.9
		200	7	88	15	12.5	6	94	16	15.6
750	5	50	88	265	45	3.0	72	253	43	3.5
		100	11	66	11	6	9	60	10	6.6
		150	3	29	5	9.6	3	28	5	9.3
		200	1	16	2	12	1	15	2	15

The following section presents an analysis of the simulated and theoretical results from Table 2, focusing on how variations in beam thickness influence deflection, strain, and the strain-to-deflection ratio—a key indicator of sensitivity—with implications for sensor design.

It presents a comparison of calculated and simulated values for deflection, strain, and the strain-to-deflection ratio under varying force conditions. A force range of 750 μN to 4000 μN was applied at the free end of the cantilever tip. In this study, the beam length was fixed at 5 mm and the width at 200 μm to accommodate piezo resistors for strain measurement. To analyses the effect of thickness on sensor response, the beam length and width were kept constant, while the thickness was varied from 50 μm to 200 μm in steps of 50 μm. This choice was based on the fact that thickness is the primary controllable parameter during fabrication, whereas the width was selected to accommodate the piezo-resistive Wheatstone bridge network.

3.2 Effect of Thickness on Deflection and Strain

The deflection and strain of a cantilever beam are highly dependent on its thickness. Deflection decreases as thickness increases, while strain increases with decreasing thickness. A 50 μm thick beam exhibited the highest strain, while a 200 μm thick beam showed the lowest. From the table, for an applied load of 750 μN, a 50 μm thick beam deflected by 88 μm, while a 200 μm thick beam is deflected by a micron. The strain-to-deflection ratio, a key metric for sensitivity, increases with thickness, suggesting that thicker beams are more sensitive per unit deflection.

Fig. 4(a) presents the FEA-based displacement profile of the cantilever subjected to a tip-applied load, obtained using Covent or ware. The deformation distribution along the beam indicates maximum deflection at the free end. Fig. 4(b) shows a linear relationship between cantilever thickness and the strain-to-deflection ratio (ε/δ), validated through simulated and analytical results.

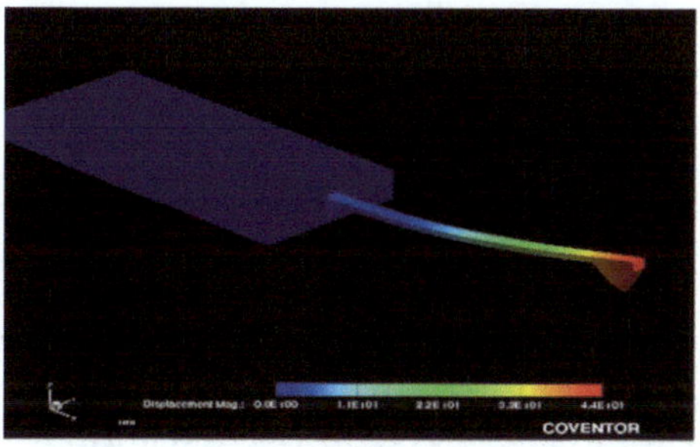

(a)

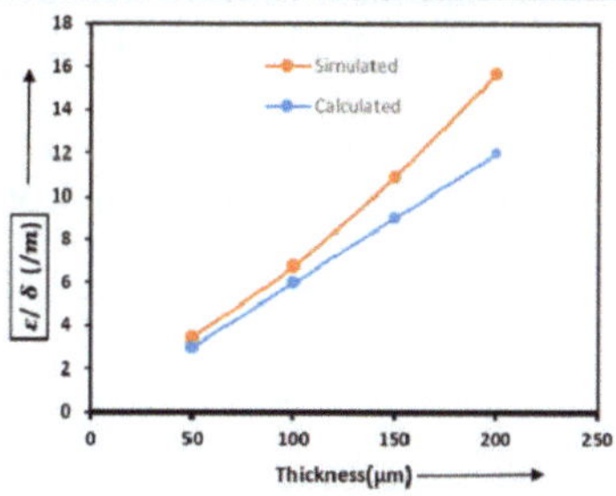

(b)

Fig. 4. **(a)** Simulated Deflection of a MEMS Cantilever under applied load in CoventorWare **(b)** Strain to deflection ratio vs. thickness of the beam for load of 750–4000 μN

3.3 Trade-offs and Design Implications

Thinner beams offer higher deflection and strain, ideal for applications like surface roughness measurement requiring a higher range. However, they are more prone to mechanical failure and handling challenges. Thicker beams offer greater robustness and stability, suitable for durability-critical applications. But they exhibit lower deflection and strain. This trade-off is critical in applications such as surface roughness measurement, where sensitivity and robustness must be balanced.

4 Conclusion

This study successfully designed and fabricated a MEMS-based silicon cantilever with an integrated micro-tip. Simulations and experiments demonstrated the impact of beam thickness on deflection and sensitivity, with thinner beams offering higher sensitivity and thicker beams providing stability. Future research will focus on integrating piezo resistive (PZR) elements to enable real-time electrical readout of cantilever deflection.

Acknowledgement. The authors gratefully acknowledge the financial support provided by the Ministry of Heavy Industries (MHI) under the Industry Accelerator Project, which enabled the successful execution of this research. We also extend our sincere thanks to the staff of the Sensor Technology Development Center (STDC) at CMTI for their valuable technical support throughout the course of this work.

References

1. Fleming, A.J.: A review of nanometer resolution position sensors: Operation and performance. Sens. Actuators Phys. **190**, 106–126 (2013). https://doi.org/10.1016/j.sna.2012.10.016
2. Teir, L., et al.: In-Line measurement of the surface texture of rolls using long slender piezoresistive microprobes. Sensors. **21**(17), 5955 (2021). https://doi.org/10.3390/s21175955
3. Brand, U., et al.: Long Slender Piezo-Resistive Silicon Microprobes for Fast Measurements of Roughness and Mechanical Properties inside Micro-Holes with Diameters below 100 μm. Sensors. **19**(6), 1410 (2019). https://doi.org/10.3390/s19061410
4. Xu, M., Li, Z., Fahrbach, M., Peiner, E., Brand, U.: Investigating the trackability of silicon microprobes in high-speed surface measurements. Sensors. **21**(5), 1557 (2021). https://doi.org/10.3390/s21051557
5. Alves, M.A.R., Takeuti, D.F., Braga, E.S.: Fabrication of sharp silicon tips employing anisotropic wet etching and reactive ion etching. Microelectron. J. **36**(1), 51–54 (2005). https://doi.org/10.1016/j.mejo.2004.10.004

Performance of ZnO – Epoxy Nanocomposite for Electrical Application

N. K. Basil Kumar and Siny Paul(✉)

Electrical and Electronics Engineering, MA College of Engineering, Kerala Kothamangalam, India
p22ee001@mace.ac.in

Abstract. The polymer Nanocomposite are advanced material made by incorporating Nano fillers in the polymer matrix, resulting in a new hybrid material with advanced unique properties. This paper emphasizes the development of ZnO (Zinc Oxide) Epoxy Nanocomposite material. The dielectric and insulating property of ZnO-Epoxy Nanocomposite, were studied. This help to contribute to the development of new material for electrical application. The simulation studies and hardware implementation of ZnO Nanoparticle with size ranges from 0-25 nm and 30-50 nm are used. Simulation studies for electric field distribution with various weight percentage of ZnO in Epoxy were analyzed in COMSOL Multiphysics software. ZnO- Epoxy Nanocomposite with various filler distribution samples are created. The high voltage AC & DC tests for samples were conducted and results were obtained.

Keywords: ZnO(Zinc Oxide) · Nanocomposites · dielectric breakdown · epoxy resins

1 Introduction

Polymer nanocomposites [1] are the polymers in which small amounts of nanometer sized fillers are homogeneously dispersed. The polymer dielectric material has very low dielectric constant but very high dielectric strength. The polymers exhibit low dielectric losses. The consolidation of materials of large permittivity with polymer of high breakdown strength may benefit the energy storage capability of new materials which can be used in electrical application. Epoxy resins [2] are thermosetting materials in high-voltage apparatus as insulation due its good mechanical and electrical properties and chemical stability. Nano epoxy resin (Nano composites) has enhanced the epoxy's properties. Due to the high electric breakdown field, low loss, light weight, and low cost, these nanocomposites are the primary choice of materials for, capacitors [3], insulating materials etc. In order to enhance the dielectric properties such as permittivity and breakdown strength of epoxy, combining epoxy with nanoparticles shows some promising result. The addition of Nano fillers should not deteriorate the electrical properties composite such as breakdown strength. It is necessary to make a new nanocomposite, having enhanced dielectric constant and high breakdown strength. In this paper an Epoxy

A. Antony et al. (Eds.): icONMAT 2025, SPPHY 328, pp. 131–143, 2025.
https://doi.org/10.1007/978-981-95-2697-0_17

nanocomposite with ZnO as Nano fillers were prepared [4, 6] and experiments were conducted to evaluate the AC and DC breakdown strength of the new composite [7]. The electrical field distribution of Nanocomposites with different weight percentage of ZnO Nano fillers were simulated in COMSOL and break down strength of new composites were conducted in High voltage lab.

2 Theoretical Model of Nanocomposite

To create and study a new dielectric material that shows advanced insulation property and check both AC and DC breakdown strength. Epoxy Nanocomposites are favored for insulation due to their extremely good dielectric overall performance, good mechanical strength, chemical stability and manufacturability. In high voltage transmission, which include GIS(Gas-insulated switchgears) traditional epoxy-based materials are inadequate. To overcome this insulation problems, epoxy resin with ZnO nanoparticles as fillers. This addition is predicted to enhance insulation property of the material and their thermal characteristics.

2.1 Simulation Studies in COMSOL

The electric field distribution of Epoxy- ZnO nanocomposite was simulated using COMSOL Multiphysics software. The simulation studies reveal electric field distribution within the nanocomposite material when subjected to high voltage. The studies carried out the use of the Finite Element Method (FEM), which intricate the behavior of the nanocomposite under high voltage conditions. In COMSOL, the geometry of the nanocomposite undergoes discretization into finite symmetrical entities, usually known as mesh factors. The problem-fixing process includes the approximation of derivatives thru finite variations. Specifically, the distribution of the electric discipline in the nanocomposite became computed using the electrostatics module embedded in COMSOL. In this simulation, each polymers and fillers have been considered isotropic substances. The polymer matrix to uniformly disperse nanoparticles, assuming a perfect connection between the particles and the epoxy. If the filler particles are round and organized in an easy cubic lattice, gap among these particles with in the epoxy nanocomposite can be obtained by using the equation [2]

$$D = \left\langle \left\{ \frac{\pi}{6} \left(\frac{\rho m}{\rho n} \right) \frac{100}{\omega t\%} \left[1 - \frac{\omega t\%}{100} \left(1 - \frac{\rho m}{\rho n} \right) \right] \right\}^{\frac{1}{3}} - 1 \right\rangle \tag{1}$$

Where, D - is the inter particle distance and d – is the diameter of the nanoparticle. The application of voltage takes place at the top face, while the bottom face is grounded within the block. The remaining side faces are left untouched. The geometry is similar to a parallel plate capacitor. Meshing is an important step in the technique that is dividing the geometry into a finite wide variety of elements. Table 1 and 2 provides the details of inter particle distances within different weight percentages of Epoxy ZnO nanocomposite. The inter particle distance serves as a metric for the spatial separation among the debris embedded within the composite. The calculated values explain the correlation between the inter particle distance and the Weight Percentage of the ZnO filler of size 0-25 nm and 30-50 nm. The particles within the composite tend to get closer to each other diminishing inter particle distance is directly linked to the increasing Weight Percentage of the ZnO filler.

Table 1. Percentage weight Vs Inter particle Distance of 0-25 nm

WEIGHT PERCENTAGE ZnO Nano Fillers (0-25 nm) (%)	INTERPARTICLE DISTANCE, D(M)
1	2.346×10^{-7}
2	1.810×10^{-7}
5	1.268×10^{-7}
10	9.547×10^{-8}
15	7.059×10^{-8}

Table 2. Percentage weight Vs Inter particle Distance of 30-50 nm

WEIGHT PERCENTAGE ZnO Nano Fillers (30-50 nm) (%)	INTERPARTICLE DISTANCE, D(M)
1	3.753×10^{-7}
2	2.896×10^{-7}
5	2.028×10^{-7}
10	1.527×10^{-7}
15	1.283×10^{-7}

2.2 Electric Field Distribution Analysis

While analyzing the electric field in pure epoxy, the field is distributed uniformly all over. After adding ZnO Nano particles, the field distribution become non uniform. The electric field distribution of ZnO Nano Composite is shown in the Fig. 1.

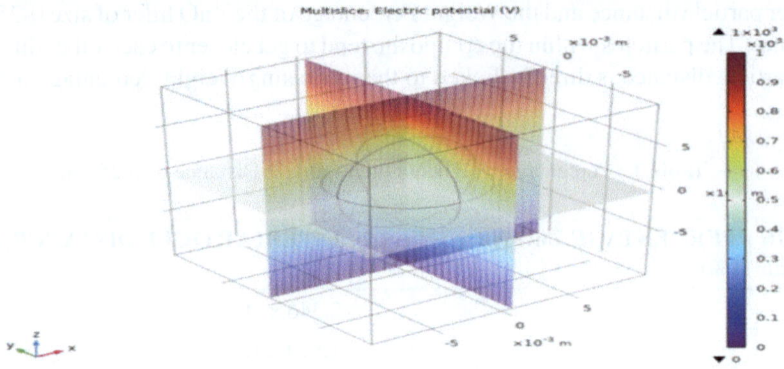

Fig. 1. Electrical Potential Distribution in Nanocomposites

2.3 Electric Field Distribution in ZnO Nano Composites with Nanoparticle Size Varies from 10-25 nm with Different Weight Percentage

1% WEIGHT RATIO.

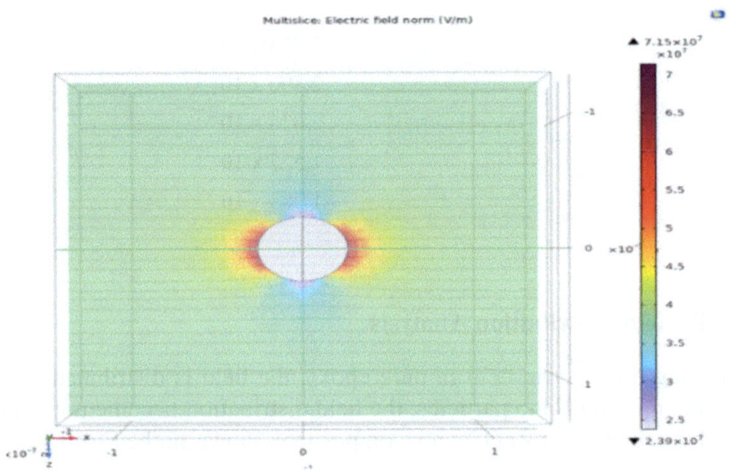

Fig. 2. Electric field of nanocomposite at 1%wt

Applying a voltage of 10 V to the top face of the unit cell, the maximum electric field intensity achievable from the applied potential is measured at 7.15 x 10^7 volts per meter Fig. 2.

2% WEIGHT RATIO.

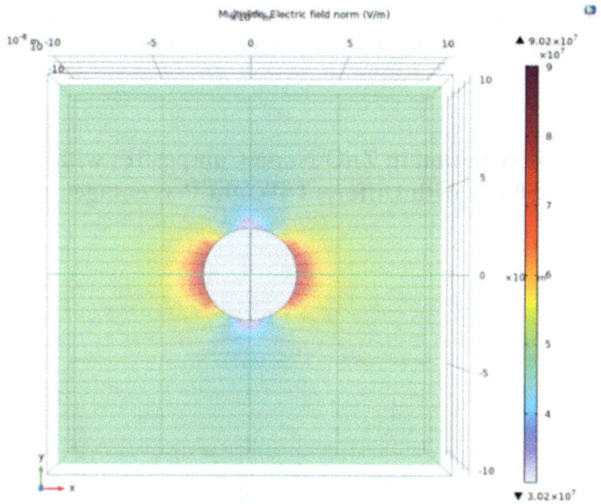

Fig. 3. Electric field of nanocomposite at 2%wt

Applying a voltage of 10 V to the top face of the unit cell, The maximum electric field intensity achievable from the applied potential is measured at 9.02 x 10^7 volts per meter as shown in Fig. 3

5% WEIGHT RATIO.

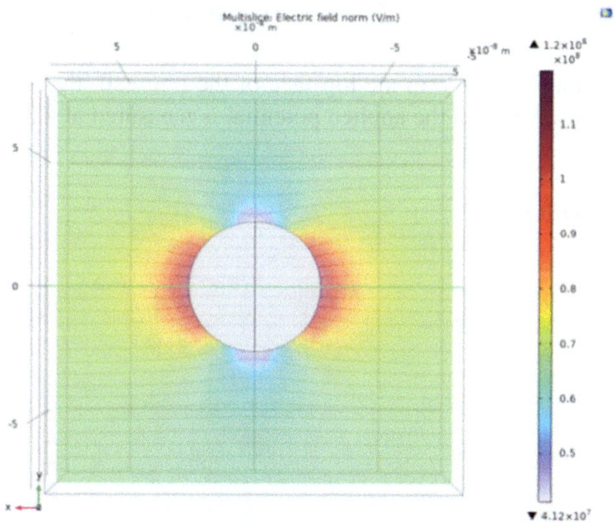

Fig. 4. Electric field of nanocomposite at 5%wt

Applying a voltage of 10 V to the top face of the unit cell, the maximum electric field intensity achievable from the applied potential is measured at 1.2 x 10^8 volts per meter Fig. 4.

2.4 Electric Field Distribution in ZnO Nano Composites with Nanoparticle Size Varies from 30-50 nm with Different Weight Percentage

1% WEIGHT RATIO.

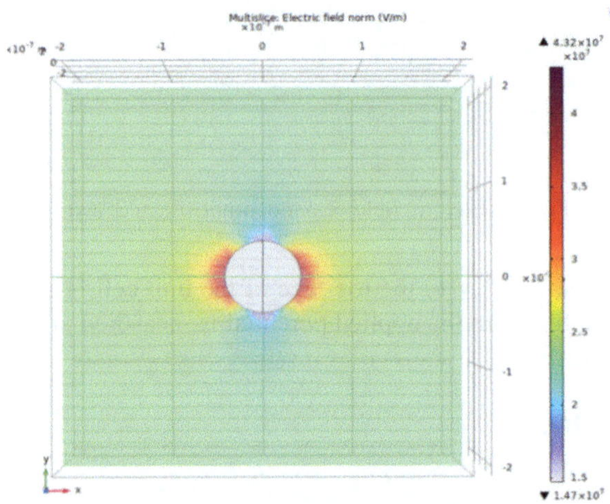

Fig. 5. Electric field of nanocomposite at 1%wt

Applying a voltage of 10 V to the top face of the unit cell, the maximum electric field intensity achievable from the applied potential is measured at 4.32 x 10^7 volts per meter Fig. 5.

2% WEIGHT RATIO.

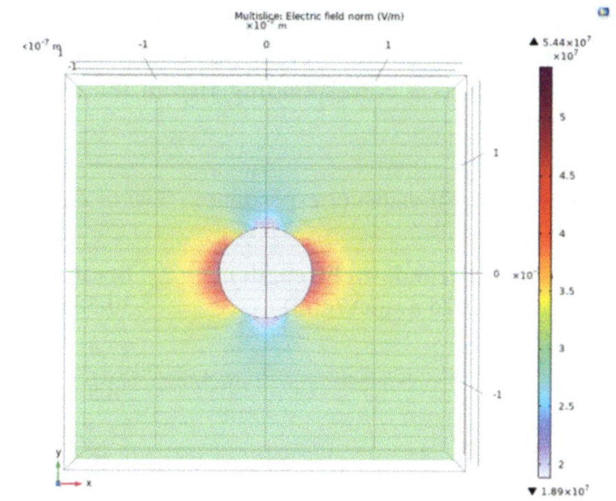

Fig. 6. Electric field of nanocomposite at 2%wt

Applying a voltage of 10 V to the top face of the unit cell, The maximum electric field intensity achievable from the applied potential is measured at 5.44 x 10^7 volts per meter Fig. 6.

5% WEIGHT RATIO.

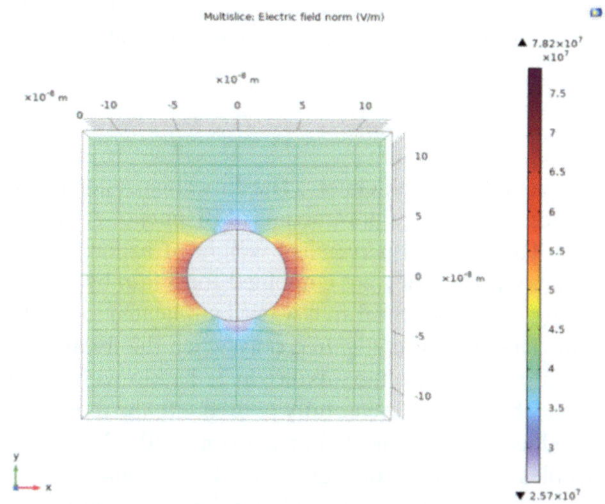

Fig. 7. Electric field of nanocomposite at 5%wt

Applying a voltage of 10 V to the top face of the unit cell, The maximum electric field intensity achievable from the applied potential is measured at 7.82 x 10^8 volts/m Fig. 7.

When examining the simulation of the nanocomposite material conducted in COMSOL software, it was noted that as the weight percentage of ZnO increases in the nanocomposite unit cell, there is a corresponding rise in the maximum electric field intensity (Table 3).

Table 3. Breakdown Strength corresponding to %wt of ZnO in Epoxy

Percentage Weight (%)	Breakdown Strength of 0-25 nm(kV/mm)	Breakdown Strength of 30-50 nm(kV/mm)
1	26.2	27.4
2	27.4	28.5
5	29.8	31.2

3 Sample Preparation

Bisphenol-A-Epoxy resin (CY1300) along with hardener (HY956) supplied by Haksons were the base polymers used for preparing the samples. The filler used for study is ZnO Nano powder provided by Nano Research Lab. The average particle size of the fillers is 10–25 & 30–50 nm. To create a sample in the shape of a round disc with the requisite dimensions, a stainless-steel mould was built. The testing device is designed to accommodate a disc with a circular form since it is both affordable and functional. This mould is made up of three sections that, when put together, make a circle with a 6 cm diameter and a 2 mm thickness. The volume of the mould is calculated using the formula,

$$Volume = \pi \times r^2 \times h \tag{2}$$

where ‘r’ is the radius and ‘h’ is the thickness. The direct dispersion method is used to create pure epoxy samples. The appropriate amounts of Bisphenol-A-epoxy resin (CY1300) is taken in a beaker. The required percentage weight of nanomaterial is weighed and added to the pure epoxy resin. They are combined for 5 min by hand stirring and then for 8–10 min by using magnetic stirrer in a water bath heated to 40 °C. Then ultrasonicated for 8–10 min. The heating is required only when the mixture is highly viscous. The Hardener (HY956) is added to the mixture in 3: 1 ratio with the mixture and hand stirrer for 2 min. Then the mixture is carefully poured into the mould without spilling or overflowing, the produced liquid is heated in an oven at 40 °C for four hours to release any trapped air bubbles. The procedure for preparation of pure epoxy samples is same as above. The difference is that we are not adding the nanomaterials in the preparation steps, that is 0%wt of nanoparticles in the sample. The weighing of Bisphenol-A-epoxy resin (CY1300) and Hardener (HY956) in 3: 1 ratio. A magnetic stirrer is a device widely used in laboratories and consists of a rotating magnet or a stationary electromagnet that creates a rotating magnetic field. This device is used to make a stir bar, immerse in a liquid, quickly spin, or stirring or mixing a solution. The moulds are assembled in a OHP sheet. Grease is applied in the mould for the separation of the moulds. Then the mixture is poured into the mould and it taken in to the preheated oven of 40 °C for four hours in an oven. After four hours the samples are taken out from the oven (Fig. 8).

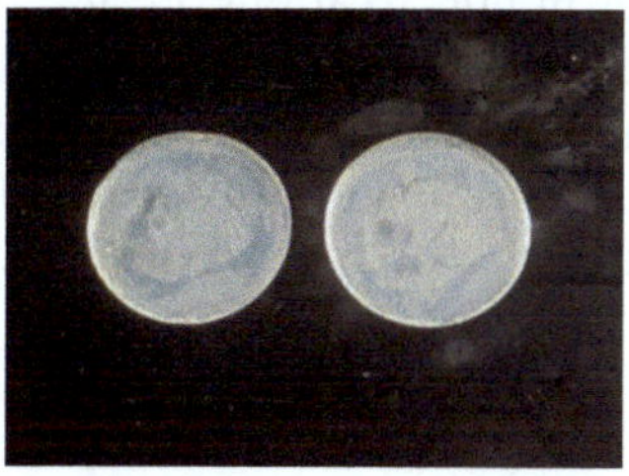

Fig. 8. Pure epoxy samples are taken out from OHP sheet

This procedure helps the mixture to harden and also removes the bubbles from the mixture. The final product that is the pure epoxy samples are obtained after the series of processes. If the procedures are done carefully, then the perfect sample will be obtained. If we do not add required amount of the hardener in the mixture, then the material would get slimy, and it can't be used for the further test procedures. The other problem may occur in the heating face in oven. If the material didn't get the required temperature, it won't get set. So, there is also a chance for the material to become slimy. All these problems took care of, then the samples can be easily taken out from the OHP sheet. For creating ZnO Nano composite with varying weight ratios such as 1% and 5%. Specifically, ZnO nanoparticles with Average Particle Sizes (APS) of 10-25 nm and 30-50 nm are used. Fig. 9. shows the Epoxy-ZnO. Nanocomposite at different weight percentages that is 1%, 5% respectively. These samples are prepared for HVDC & HVDC breakdown analysis.

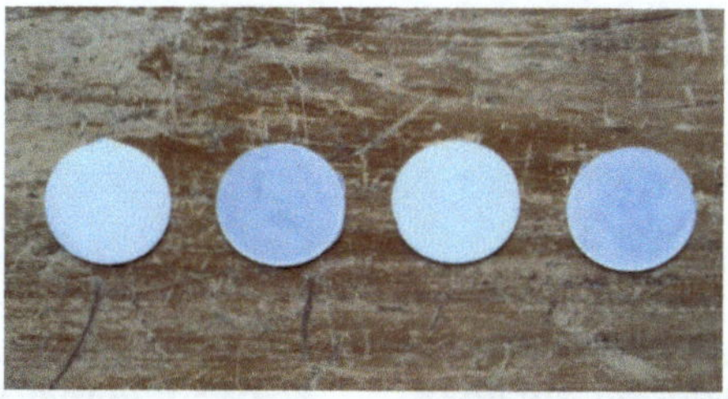

Fig. 9. Samples for HVDC & HVAC Breakdown Analysis

4 Scanning Electron Microscopy

SEM stands out as a highly versatile and extensively employed technique for analyzing material surfaces in modern science. Its capability to examine both morphology and composition makes it invaluable. SEM offers a relatively quick, cost-effective, and minimally invasive method for surface analysis.

In the SEM analysis, we examined the dispersion of ZnO particles within epoxy. This allowed us to observe the morphologies of nano-ZnO composites, providing insight into how nanoparticles are distributed within the matrix material. Initially, the samples were placed under vacuum conditions to prepare them for analysis. Subsequently, we observed the microstructures of the composites. Specifically, we tested a sample containing 1% and 5% mass fraction of ZnO particles of 0-25 nm and 30-50 nm range.

Based on the SEM images in Fig. 10 & Fig. 11, it is evident that agglomeration is occurring within the samples. This phenomenon leads to the clustering of ZnO nanoparticles, resulting in a non-uniform distribution of particles within the epoxy matrix. Consequently, this non-uniform distribution adversely affects the performance of the nanoparticles, ultimately leading to a decrease in dielectric properties.

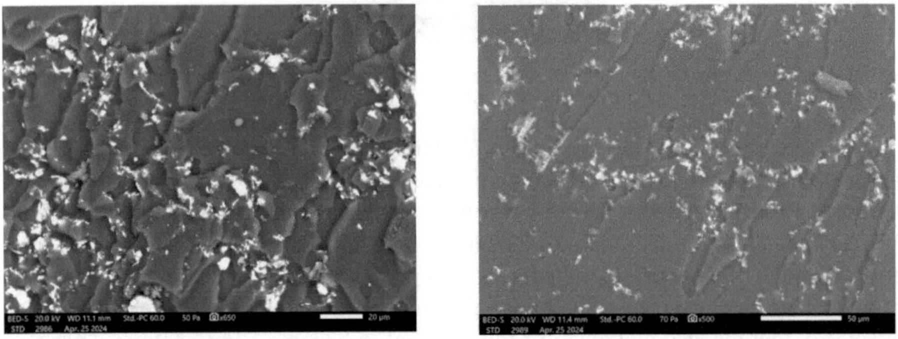

Fig. 10. SEM Images of 10-25 nm ZnO nanocomposite of 5 wt%

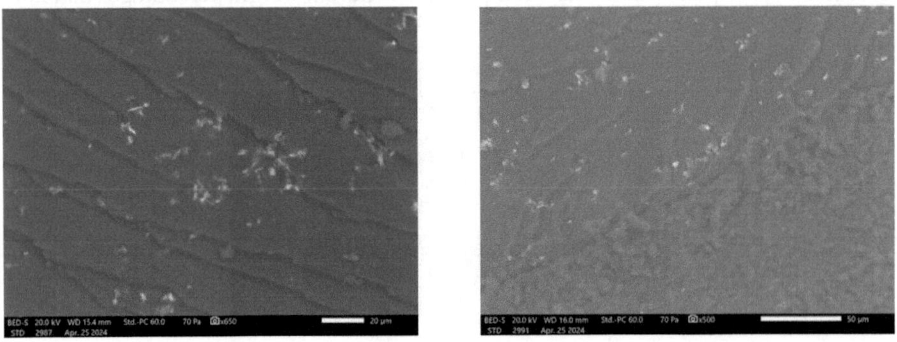

Fig. 11. SEM Images of 30-50 nm ZnO nanocomposite of 5 wt%

The observed agglomeration highlights a critical issue in the dispersion of nanoparticles within the epoxy matrix. The non-uniform distribution compromises the intended functionalities of the nanoparticles, impacting the overall performance of the composite material. To mitigate this issue and enhance the performance of the composite, further optimization of the dispersion process is recommended. Strategies such as optimizing processing parameters, incorporating dispersing agents, or employing alternative mixing techniques could potentially improve the dispersion of nanoparticles within the epoxy matrix.

5 HVAC and HVDC Analysis on Nanocomposites

Table 4 presents the results obtained from HVAC breakdown testing, demonstrating a clear trend: as the weight percentage increases in the pure epoxy, the breakdown strength also increases. Notably, the sample containing 5%wt of ZnO in epoxy exhibits the highest breakdown strength among all tested compositions. These findings lead to the conclusion that there is a direct correlation between the breakdown strength and the percentage weight of the filler material, indicating that higher weight percentages result in increased breakdown strength.

Table 4. HVAC Breakdown Analysis of Epoxy-ZnO nanocomposite

AC BREAKDOWN ANALYSIS					
Percentage weight of ZnO	Points	Thickness (mm)	Breakdown Voltage (kV)	Breakdown Strength (kV/mm)	Average Breakdown Strength (kV/mm)
Pure epoxy	1	1.98	26.1	23.8	22.95
	2	1.99	25.2	23.4	
1%(0-25 nm)	1	2.01	29.8	25.8	26.05
	2	2.00	29.3	26.3	
5%(0-25 nm)	1	1.99	39.7	29.5	30.15
	2	2.00	40.2	30.8	
1%(30-50 nm)	1	2.02	31.2	27.1	27.35
	2	2.00	32.2	27.6	
5%(30-50 nm)	1	2.03	40.1	31.6	32.15
	2	1.99	41.1	32.2	

The findings from the DC breakdown analysis of epoxy-ZnO are summarized in Table.5. Initially, the pure epoxy with 0%wt of ZnO demonstrates a breakdown tolerance of only 25.95 kV/mm. However, as the percentage weight of ZnO in the epoxy increases, there is a corresponding rise in breakdown strength. Notably, when the percentage weight of ZnO in the epoxy reaches 5%, the sample exhibits promising results in the context of the research.

Table 5. HVDC Breakdown Analysis of Epoxy-ZnO nanocomposite

DC BREAKDOWN ANALYSIS					
Percentage weight of ZnO	Points	Thickness (mm)	Breakdown Voltage (kV)	Breakdown Strength (kV/mm)	Average Breakdown Strength (kV/mm)
Pure epoxy	1	2.01	29.5	25.7	25.95
	2	1.98	30.2	26.2	
1%(0-25 nm)	1	2.03	37.8	29.8	30.05
	2	2.01	38.3	30.3	
5%(0-25 nm)	1	1.98	45.7	37.5	37.65
	2	2.02	46.2	37.8	
1%(30-50 nm)	1	2.02	38.2	30.1	30.35
	2	2.00	39.5	30.6	
5%(30-50 nm)	1	2.01	49.7	39.6	39.9
	2	1.99	50.1	40.2	

Based on our analysis of HVAC and HVDC tests conducted on various samples of nanocomposites, a significant finding emerges: the breakdown strength of nanocomposites escalates proportionally with the increase in filler loading weight percentage within the composites. For AC breakdown tests, we see an increase from 22.95 kV/mm to 32.15 kV/mm, and for DC breakdown tests, from 25.9 kV/mm to 39.9 kV/mm. These results confirm that nanocomposites offer better insulation properties compared to other materials.

6 Conclusion

The study aimed at enhancing the insulation properties of epoxy through the incorporation of ZnO nanoparticles, renowned for their high electrical insulation and thermal stability. By conducting DC and AC breakdown tests on pure epoxy and epoxy-nanocomposite samples, alongside simulation using COMSOL Multiphysics V6.0 software, we aimed to develop high-quality insulation tailored for high voltage applications.

Simulation results provided invaluable insights into the electric field distribution within the nanocomposite for varying percentages of ZnO nanoparticle filler. Encouragingly, HVDC and HVAC breakdown tests underscored the promising breakdown strength of nanocomposite materials. However, further studies are necessary to assess their reliability thoroughly.

The SEM images revealed a significant challenge: agglomeration within the samples, leading to the clustering of ZnO nanoparticles and a non-uniform distribution within the epoxy matrix. This adversely affects the nanoparticles' performance, ultimately decreasing dielectric properties.

Addressing this agglomeration issue is crucial for optimizing epoxy-based nanocomposites. Further optimization of the dispersion process through strategies such as adjusting processing parameters, employing dispersing agents, or alternative mixing techniques is recommended. ZnO is a conducting nano material, so that the breakdown occurs at slightly higher voltages for nanocomposite than the pure epoxy. The agglomeration also creates a low breakdown strength for the samples as compared to theoretical samples in COMSOL.

References

1. Kadim, E.J., Noorden, Z.A., Adzis, Z., Azis, N.: Nanoparticles application in high voltage insulation systems. IEEE Trans. Dielectr. Electr. Insul. **28**(4), 1380–1399 (2021)
2. Paul, S., Sindhu, T.K.: A Neural network model for predicting the dielectric permittivity of epoxy aluminum nanocomposite and its experimental validation. IEEE Trans, oncomponents packag. manuf. technol. **5**(8), 1122–1128 (2015)
3. Paul, S., Sindhu, T.K.: Effect of filler particle size on energy density of epoxy- aluminum Nanocomposites. IEEE Trans, Dielectr. Electr. Insul. **23**(5), 2786–2794 (2016)
4. Kaur, D., Bharti, A., Sharma, T., Madhu, C.: Dielectric properties of ZnO-based nanocomposites and their potential applications. Hindawi Int. J. Opt. **20**, 9950202 (2021)
5. Hardoň, Š., Kúdelčík, J., Hornak, J., Michal, O., Totzauer, P.: The Influence of ZnO nanoparticles in the epoxy resin on the complex permittivity and dissipation factor. Transp. Res. Procedia. **40**, 30–33 (2019)
6. Anandraj, J., Joshi, G.M.: Fabrication, performance and applications of integrated nanodielectric properties of materials - a review. Compos. Interfaces. **25**(5–7), 455–489 (2018)
7. Liu, M., Weilin, S., Qin, X., Cheng, K.: Mechanical/electrical characterization of ZnO nanomaterial based on AFM/nanomanipulator embedded in SEM. Micromachines. **12**, 12030248 (2021)

Recent Developments in Luminescent Materials for Anti-counterfeiting Technologies

P. D. Belsare[1(✉)], H. S. Bhoyar[1], B. T. Kumbhare[2], and S. V. Moharil[3]

[1] Ramdeobaba University, Katol Road, Nagpur 440013, India
{belsarepd,bhoyarhs}@rknec.edu
[2] SSESA'S Science College, Nagpur 440012, India
[3] R.T.M. Nagpur University, Nagpur 440033, India

Abstract. The act of creating, distributing, or selling fake or copied products, documents, currency, or medicines to mislead others into believing they are genuine is known as counterfeiting. Advances in technology, such as high-resolution equipment, advanced 3D printers, and high-tech software, are easily available, influencing people to engage in counterfeiting activities. The growth of counterfeiting has now increasing day by day so there is much need to develop a solution to counter the rapidly growing counterfeiting activities. Counterfeiting activities are a major concern for governments, companies, and individuals worldwide, with significant consequences for financial security, safety, and health. Luminescent materials offer effective techniques to detect the counterfeited product. Luminescent paints and inks, made from phosphors, are highly adaptable, easy to apply, and convenient for reading across different environments. Visual inspection remains a simple and cost-efficient method for identification while combining both visible and invisible inks provides an additional layer of security against counterfeiting. Different types of ink technologies used for anti-counterfeiting such as UV-sensitive, IR-sensitive, magnetic, optically variable and thermally sensitive inks are used for anti-counterfeiting. This paper discusses advanced dual mode and multimode luminescence techniques employed in anticounterfeiting technologies.

Keywords: Luminescence · Phosphor · Smartphone based anticounterfeiting

1 Introduction

Counterfeiting involves creating or selling an illegal copy of a legitimate product, designed with the intention to mislead people into thinking it is real. The widespread circulation of counterfeit materials adversely impacts individuals, governments, and society. It affects a wide range of domains, including secure documents, currency, genuine branded company products, medicines, etc. [1]. Advanced developments in technology such as advanced 3D printers, high-resolution devices, high-tech software, etc., are easily available that can be used to perform counterfeiting activities. In recent years some modern techniques were developed to counter counterfeiting activities. Luminescent materials played a vital role in anti-counterfeiting and attracted researchers because of their unique characteristics and broad range of applications such as low cost, high accuracy, and easy synthesis techniques etc.

A. Antony et al. (Eds.): icONMAT 2025, SPPHY 328, pp. 144–152, 2025.
https://doi.org/10.1007/978-981-95-2697-0_18

1.1 Anti-counterfeiting Methods

The EUIPO, in collaboration with the European Observatory on Infringements of Intellectual Property Rights, has recently introduced a technological guide to counter the counterfeiting of genuine products. This guide promotes the use of electronic, chemical, mechanical, and digital methods as shown in Fig. 1. Anticounterfeiting (ANTC) technologies can be deployed through human inspection or using automated systems, but all rely on unique, secure markers often embedded with secret codes—affixed to genuine products. This enables accurate identification and enhanced protection against forgery.

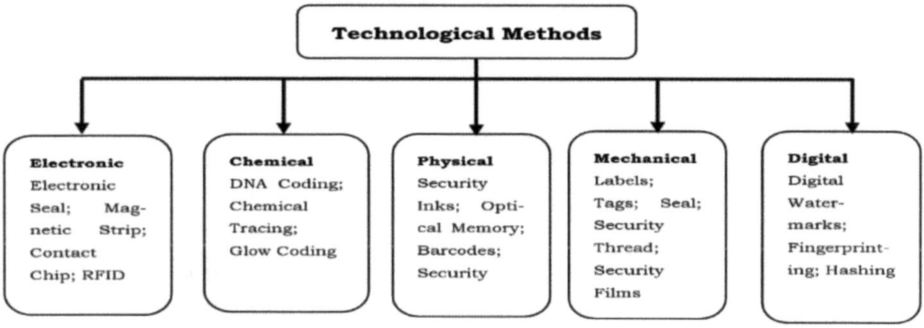

Fig. 1. Schematic diagram showing different methods of anti-counterfeiting.

1.2 Luminescent Materials for Anticounterfeiting

For anticounterfeiting (ANTC) there are a wide range of technologies developed but especially the ink-based ANTC technology that relies on the use of phosphor provides a high-level security against counterfeiting. Phosphors mostly used for the synthesis of luminescent paints are highly in demand because of their ease of use and can be read in any environment. Visible or invisible anticounterfeiting inks, play a very crucial role in identifying counterfeited goods [2]. Different types of ink-technologies UV-sensitive inks, IR-sensitive inks, magnetic inks, optical variable inks, and thermally-sensitive inks are now available for anticounterfeiting [3].

2 Printing Techniques for Security Inks

The three methods that are used in preventing counterfeiting of products are

1. Screen printing method
2. Inkjet printing method
3. Aerosol printing method

2.1 Screen Printing Method

Screen printing is one of the most commonly used methods, allowing thick deposition of luminescent inks through a fine mesh stencil onto the substrate. In screen printing, a stencil or mesh screen with permeable and impermeable areas defines the pattern to be printed. The stepwise process of screen printing is shown in Fig. 2. Luminescent ink is applied over the screen, and a squeegee is forced to press the ink through the open mesh areas, transferring it onto the underlying substrate. A reverse stroke is then applied, causing the screen to touch the substrate momentarily along a line of contact. The ink wets the substrate and needs to be pulled out of the mesh screen as the screen springs back after the blade has passed. The screen printing method is suitable for high-durability security elements, such as seals, logos, certificates, and official documents, especially when strong luminescent brightness is required. Screen printing is a slower process and may not be suitable for metal surfaces or electronic items. The luminescent ink used in screen printing must be optimized to prevent clogging.

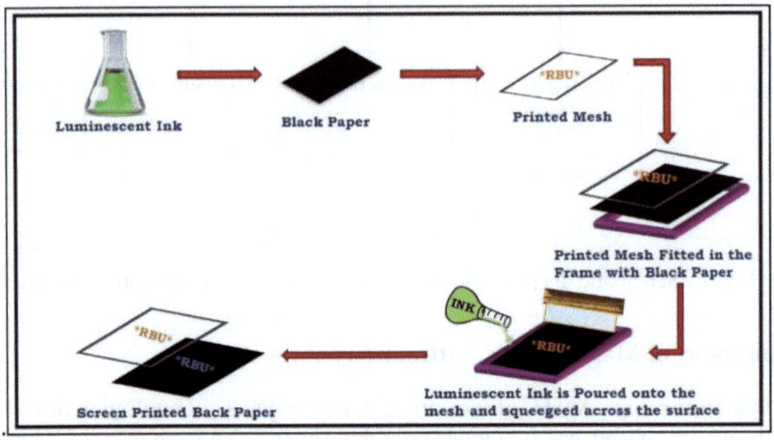

Fig. 2. Schematic diagram for screen printing process

2.2 Inkjet Printing Method

Inkjet printing is a digital technique that deposits small ink droplets of functional ink onto a substrate without direct contact. When used with luminescent materials such as phosphors, quantum dots, or fluorescent dyes it enables the creation of security features with high resolution and customizable designs. In inkjet printing, a computer-controlled printhead ejects tiny ink droplets through nozzles onto a surface. In the configuration of a piezoelectric printhead, the actuator is integrated around or behind a small ink chamber. When a voltage pulse is applied, the piezoelectric material contracts or expands, generating a pressure wave that displaces the ink within the chamber and forces a droplet through the nozzle as shown in Fig. 3. As the voltage is removed, the actuator returns to its original shape, and the ink refills the chamber via capillary action or through controlled ink channels. This entire process occurs on the order of microseconds, enabling printing frequencies exceeding 20 kHz, and droplet sizes ranging from 1 to 100 pL depending on the nozzle diameter, waveform, and fluid properties. Piezoelectric inkjet (PIJ) printing enables printing of fine patterns, microtexts, QR codes, or invisible barcodes that are hard to replicate. There are certain limitations of PIJ method such as it requires inks with specific viscosity and surface tension.

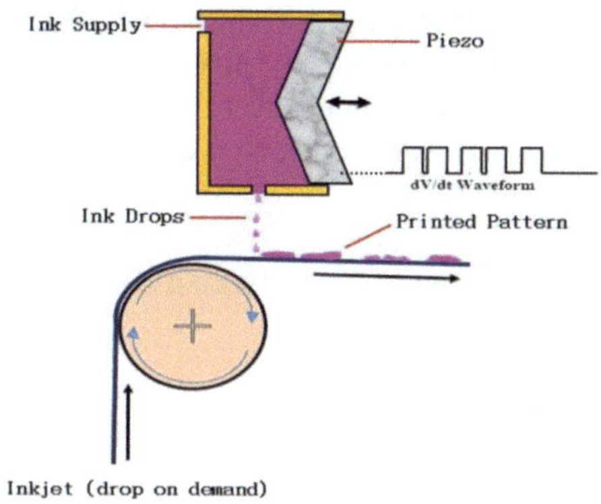

Fig. 3. Schematic diagram for screen printing process

2.3 Aerosol Jet Printing Method

The aerosol jet printing method employs an ultrasonic or pneumatic atomizer to generate a fine aerosol mist of ink droplets (1–5 μm), which are then focused via a sheath gas into a tightly collimated stream and deposited onto a substrate with high precision. The focused aerosol can be deposited on flat or 3D surfaces in programmable patterns, with control over line width, thickness, and feature geometry. This method accommodates a wide range of ink viscosities and particle sizes, making it compatible with complex formulations such as metal–organic frameworks (MOFs) doped with luminescent centers, quantum dots, rare-earth complexes, and other functional materials.

3 Multimode Anticounterfeiting Techniques

Single-mode phosphors can be recognized by counterfeiters and easily duplicated. The blending of two or more phosphors results in inconsistent luminescence and uneven dispersion of materials. It is therefore necessary to develop and authenticate phosphors with additional security features. One technique could involve using phosphors that emit multiple wavelengths under varying excitations. Other luminescence characteristics of phosphors, such as long-lasting emission even after the excitation is stopped, emission lifetime, and afterglow colors, can be used to provide more reliable authentication. Additionally, luminescence properties such as mechanoluminescence, thermoluminescence, and photostimulated luminescence of the material can be utilized to authenticate the phosphors.

3.1 $Na_2CaGe_2O_6$: Ln^{3+}, Yb^{3+}, Tb^{3+} Phosphor

$Na_2CaGe_2O_6$ (NCG): Ln^{3+}, Yb^{3+}, Tb^{3+} ($Ln^{3+} = Ho^{3+}$, Er^{3+}, Tm^{3+}) is a promising phosphor for multilevel anti-counterfeiting applications [4]. Upconversion, downconversion, and persistent luminescence in $Na_2CaGe_2O_6$ doped with lanthanide ions have been reported. Under 980 nm near-infrared excitation, upconversion luminescence results in green emission for NCG:Er^{3+}, Yb^{3+}; yellow emission for NCG:Yb^{3+}; and blue emission for NCG:Tm^{3+}, Yb^{3+} as shown in Fig. 4. The upconversion emission is found to be power-dependent, increasing with the intensity of the excitation source. Additionally, interesting luminescence properties have been observed in $Na_2CaGe_2O_6$ doped with Tb^{3+} [5].

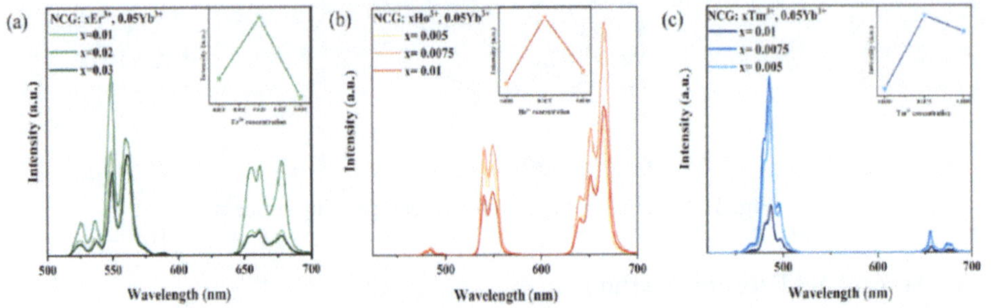

Fig. 4. Upconversion emission spectra of (a) $Na_2CaGe_2O_6$: Er^{3+}, Yb^{3+} (b) $Na_2CaGe_2O_6$: Ho^{3+}, Yb^{3+} and (c) $Na_2CaGe_2O_6$: Tm^{3+}, Yb^{3+} under 980 nm excitation. (Reprinted from Journal of Luminescence,249, X. Jin et al., 118937–118945, Copyright (2022), with permission from Elsevier).

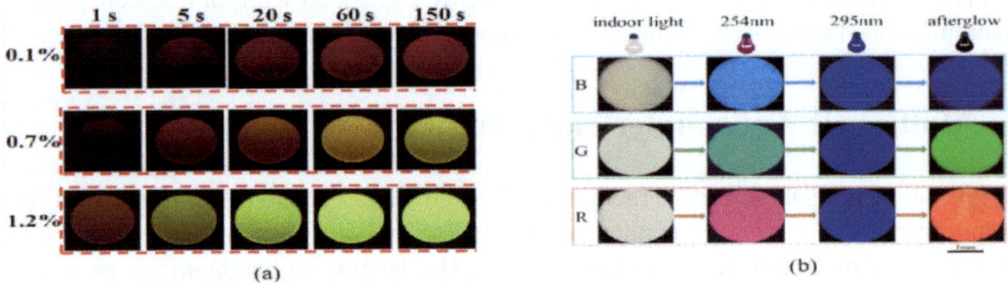

Fig. 5. (a) PL photos of the NCG:Tb^{3+} materials at different irradiation times for $\lambda_{ex} = 254$ nm. (Reprinted with permission from ACS Applied Materials & Interfaces **2019** 11 (39), 35871–35878 Copyright (2019), American Chemical Society) (b) Dependence of the PL spectra on the excitation wavelengths $\lambda_{ex} = 254$ nm and $\lambda_{ex} = 295$ nm in $Na_2CaGe_2O_6$:Pb^{2+},Y^{3+} (blue), $Na_2CaGe_2O_6$:Pb^{2+},Tb^{3+} (green), $Na_2CaGe_2O_6$:Pb^{2+},Mn^{2+},Yb^{3+} (red) phosphors. (Reprinted with permission from ACS Appl. Mater. Interfaces 2019, 11, 18548–18554 Copyright (2019), American Chemical Society)

Dynamic photoluminescence (PL) as demonstrated in Fig. 5 was observed when the material was excited with 254 nm UV light for varying irradiation times. At a Tb^{3+} concentration of 1.2%, the material emitted red after 1 s of irradiation, yellow after 5 s, and green after 150 s of exposure to 254 nm UV light. The observed dynamic photoluminescence (PL) behavior is primarily attributed to the presence of electronic trap states and the cross-relaxation mechanisms involving Tb^{3+} ions. $Na_2CaGe_2O_6$ (NCG) has also been investigated with Pb^{2+} and Er^{3+} doping [6]. Excitation-dependent dual-mode luminescence has been reported in NCG:Pb^{2+},Y^{3+}; $Na_2CaGe_2O_6$:Pb^{2+},Tb^{3+}; and $Na_2CaGe_2O_6$:Pb^{2+},Mn^{2+},Yb^{3+} phosphors [7].

3.2 Ca_2MgWO_6: Er^{3+},Yb^{3+}, Eu^{3+} Phosphor

Multimode luminescence has also been reported in triply doped Ca_2MgWO_6 CMW):Er^{3+}/Yb^{3+}/Eu^{3+} phosphors [8]. Doping with Er^{3+} in Ca_2MgWO_6 results in weak upconversion luminescence at 533 nm, 544 nm, and 663 nm under 980 nm excitation. The addition of Yb^{3+} to this host improves the luminescence performance, as Yb^{3+} absorbs more NIR light and efficiently transfers its energy to the Er^{3+} luminescent center, Eu^{3+} ions, acting as activators, were then doped into the CMW:Er^{3+} (2%)/(5%) phosphor to obtain upconversion and downconversion emission. The CMW:Er^{3+} (2%)/Yb^{3+} (5%)/Eu^{3+} (3%) sample exhibits multiple downconversion emission bands at 533 nm, 544 nm, 592 nm, and 617 nm under 365 nm excitation. Compared to the sample without Eu^{3+} doping, this sample shows significant downconversion luminescence from Eu^{3+} ions and a sharp decrease in the emission intensity associated with Er^{3+}.This shift in emission intensity suggests that energy transfer occurs from Er^{3+} to Eu^{3+}. The overall energy transfer process $Yb^{3+} \rightarrow Er^{3+} \rightarrow Eu^{3+}$ is illustrated in Fig. 6.

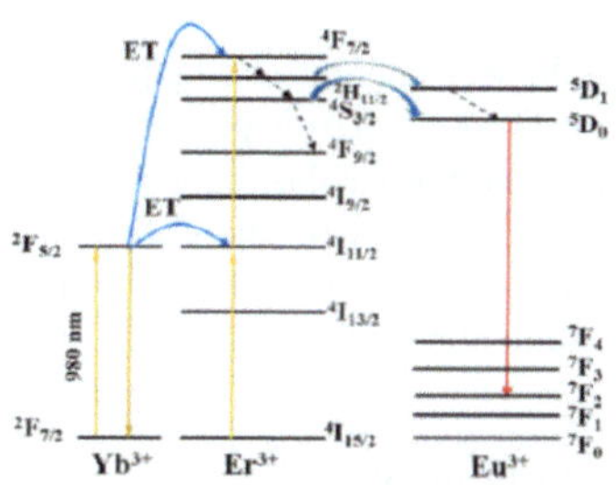

Fig. 6. Energy transfer process from Yb^{3+}➔ Er^{3+}➔ Eu^{3+} in Ca_2MgWO_6 (Reprinted from Journal of Luminescence, Volume 246, H. Cheng et al., 118804, Copyright (2022), with permission from Elsevier).

3.3 Multilevel Anticounterfeiting Using Smartphone

Due to the requirement of specialized hardwares such as use of different excitation sources, lifetime measuring instruments etc. multimode anticounterfeiting technologies cannot be used everywhere. Recently ANTC labels were developed that can be excited and detected using only a smartphone. In smartphone flashlight a broadband emitting phosphor usually YAG:Ce^{3+} is coated on blue (450 nm) emitting LED chip. Phosphors based ANTC labels that can be detected using only a smartphone shall have sufficient absorption as well as emission in the visible region. Efficiency of emission should be high enough to be detected by the smartphone. Also luminescence of such phosphor must last longer and shift in the emission color with time is required. $SrAl_2O_4$:Eu^{2+}, Dy^{3+} is a known long lasting phosphor. Green emitting $SrAl_2O_4$:Eu^{2+}, Dy^{3+} (SAED), and red-emitting CaS:Eu^{2+} phosphors with varying lifetimes (milliseconds to seconds) for developing ANTC labels are reported recently [9].

Fig. 7 illustrates a significant overlap between the excitation spectrum of the SAED phosphor and the 450 nm emission wavelength of smartphone LED flashlights. Emission spectra of SAED with emission peak at 514 nm detectable by smartphone cameras without external filters. By annealing SAED at varying temperatures (700–860 °C), its persistent emission lifetime was finely tuned between 0.5 and 11.7 s. Similarly, the Eu^{2+} doping level in CaS and SrS phosphors was modulated to achieve red-emitting lifetimes in the 0.1–0.6 s range. This tuning is critical to establishing dynamic color-shifting behavior upon excitation.

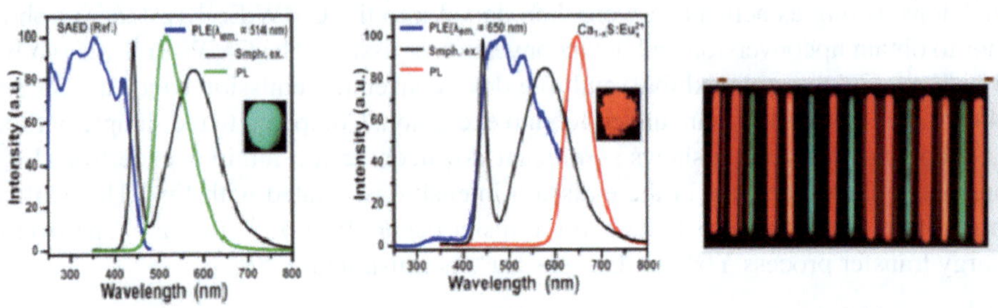

(a) (b) (c)

Fig. 7. (a) PL spectra of smartphone flashlight overlapping with excitation spectra of SAED alongside emission spectra of SAED with emission peak at 514 nm (b) Good overlap between the emission spectra of CaS:Eu^{2+} and smartphone flashlight. Emission spectra with emission peak at 640 nm. (c) raw barcode lines made from SAED and CaS:Eu^{2+}. (Katumo, N., Li, K., Richards, B.S. et al. Dual-color dynamic anti-counterfeiting labels with persistent emission after visible excitation allowing smartphone authentication. Sci Rep 12, 2100 (2022)).

The PL mechanism in SAED involves Eu^{2+} as the luminescent center and Dy^{3+} ions and oxygen vacancies as electron traps. After visible light excitation, electrons are trapped and then thermally released, producing persistent afterglow. In CaS/SrS phosphors, the red emission arises from Eu^{2+} $4f^6 5d^1 \rightarrow 4f^7$ transitions, and the lifetime is strongly dependent on non-radiative deactivation influenced by Eu^{2+} concentration. Phosphors were blended in precise ratios (e.g., 95:5 wt% SAED:red phosphor) to achieve dynamic dual-color transitions, with red emission fading rapidly and green emission persisting. This blending leads to visually observable transitions from red to green post-excitation, a feature essential for optical coding and time-resolved security applications.

4 Challenges in Using Anti-counterfeiting Methods

Advanced ANTC techniques based on optical materials and smartphone-assisted authentication present significant advantages in terms of security and functionality compared to conventional methods, but they also introduce several scientific and practical challenges. One of the foremost issues is the complexity involved in the synthesis of the functional materials themselves. Many modern systems employ rare-earth-doped phosphors, quantum dots, or persistent luminescent materials that require highly controlled synthesis environments, precise stoichiometry, and thermal treatment processes. In contrast, conventional methods such as watermarks, holograms, and barcodes rely on mature manufacturing techniques that are easily scalable and economically feasible. Another critical concern is the long-term stability and environmental durability of luminescent anti-counterfeiting materials. These materials may undergo photobleaching, oxidation, or hydrolysis under ambient conditions, especially in the presence of UV light, moisture, or elevated temperatures, thereby degrading their optical performance over time. Ensuring that the security features remain functional throughout the product's lifecycle necessitates additional encapsulation strategies or protective coatings. Furthermore, while these materials offer multi-level security through features such as excitation-dependent emission, time-gated luminescence, and multichannel detection (e.g., RGB separation), the complexity of these systems introduces potential barriers for end-users, particularly when relying on consumer devices like smartphones for verification. The reliance on smartphones as both excitation and detection tools, while advantageous, also imposes technical constraints. The flashlight LEDs in smartphones typically emit around 450 nm, which limits excitation compatibility to materials with absorption bands in the visible region. Consequently, designing phosphors that can be efficiently excited at these wavelengths while maintaining high quantum efficiency and persistent emission remains an ongoing challenge in materials science. Moreover, the cameras in smartphones have varying sensitivities and exposure settings, which may affect the consistency of luminescence detection across different devices. The absence of standardized imaging protocols can compromise the reliability of verification.

5 Conclusions

This paper highlights the role of luminescent materials in anti-counterfeiting techniques. The rare earth-doped $Na_2CaGe_2O_6$ phosphor offers several advantages, exhibiting up-conversion (UC), down-conversion (DC), persistent luminescence, and photostimulated luminescence (PSL), making it suitable for multilayered anti-counterfeiting applications. This multimodal luminescence behavior provides enhanced security compared to conventional single-mode luminescent materials. In particular, the $Na_2CaGe_2O_6$:Tb^{3+} system displays a novel dynamic photoluminescence (PL) response, wherein the emission color transitions from red to green under continuous UV irradiation. This temporal evolution in PL color significantly increases the difficulty of replication using static luminescent inks. Similarly, Ca_2MgWO_6 doped with Er^{3+}, Yb^{3+}, and Eu^{3+} exhibits efficient UC and DC emissions. Energy transfer between these dopant ions enhances emission intensity and enables dual-mode excitation using both 980 nm (NIR) and 365 nm (UV) sources, which is highly advantageous for anti-counterfeiting applications.

However, several challenges remain in employing these materials for advanced non-traditional counterfeiting (ANTC) purposes. The dynamic luminescence behavior of $Na_2CaGe_2O_6$:Tb^{3+} is highly dependent on trap states within the host lattice. Variability in trap depth and distribution can lead to inconsistent optical responses, particularly under uncontrolled environmental conditions. Furthermore, multicolor emissions—especially in mixed-doped systems such as $Tb^{3+}/Er^{3+}/Yb^{3+}$—may suffer from spectral overlap, reducing color purity and complicating the accurate detection or decoding of encoded security patterns.

References

1. Linsner, M., Bristows, L.L.P., EUIPO Observatory publishes Anti-counterfeiting Technology Guide. Lexology. https://www.lexology.com 2021. Accessed 18 Mar 2021
2. Ren, W., et al.: Optical nanomaterials and enabling technologies for high-security-level anticounterfeiting. Adv. Mater. **32**(18), 1901430 (2019)
3. Abdollahi, A., Roghani-Mamaqni, H., Razavi, B., Salami-Kalajahi, M.: Photoluminescent and chromic nanomaterials for anticounterfeiting technologies: Recent advances and future challenges. ACS Nano. **14**(11), 14417–14492 (2020)
4. Jin, X., et al.: Dual-mode multicolor luminescence based on lanthanide-doped $Na_2CaGe_2O_6$ phosphor for anticounterfeiting application. J. Lumin. **249**, 118937 (2022)
5. Wang, J., et al.: Advanced Dynamic photoluminescent material for dynamic anticounterfeiting and encryption. ACS Appl. Mater. Interfaces. **11**(39), 35871–35878 (2019)
6. Sun, Z., et al.: Spy Must Be Spotted: A Multistimuli-Responsive Luminescent Material for Dynamic Multimodal Anticounterfeiting and Encryption. ACS Appl. Mater. Interfaces. **10**(25), 21451–21457 (2018)
7. Shi, C., et al.: Excitation wavelength-dependent dual-mode luminescence emission for dynamic multicolor anticounterfeiting. ACS Appl. Mater. Interfaces. **11**(20), 18548–18554 (2019)
8. Cheng, H., et al.: Simultaneous achievement of sensitivity enhancement and dual-mode luminescence through co-doping Eu^{3+} in $Ca_2MgWO_6:Er^{3+}/Yb^{3+}$ phosphor. J. Lumin. **246**, 118804 (2022)
9. Katumo, N., et al.: Dual color dynamic anticounterfeiting labels with persistent emission after visible excitation allowing smartphone authentication. Sci. Rep. **12**, 1–14 (2022)

Role of H_3BO_3 as Flux on Enhancing the Luminescence of CMS: Eu^{3+} Phosphor

Navya Sara Kuriyan(✉), R. S. Nikhil Raj, and M. Sabeena

Department of Physics, Cochin University of Science and Technology, Kochi 682022, Kerala, India

navyasarakuriyan@cusat.ac.in

Abstract. In this study, Eu^{3+} doped calcium magnesium silicate (CMS: Eu^{3+}) phosphors were synthesized with and without H_3BO_3 to investigate the role of the flux in altering their structural, morphological and optical properties. The incorporation of H_3BO_3 as a flux facilitated a phase transformation from a combination of phases to a single phase merwinite ($Ca_3MgSi_2O_8$) with monoclinic crystal structure and improved particle formation. Enhanced excitation and emission intensities were observed with flux addition, along with an increase in color purity in the red spectral region, from 85.53% to 92.46%. These modifications reflect improved luminescent behavior and enable fine color tunability. Such tunability in photoluminescence (PL) properties underscores the potential of flux assisted CMS: Eu^{3+} phosphors for biomedical applications.

Keywords: Eu^{3+} doped calcium magnesium silicate · Flux addition · Fine color tunability

1 Introduction

Recent advancements in luminescent materials have significantly broadened the application of phosphors, making them indispensable in diverse areas such as solid state lighting like Light Emitting Diodes (LEDs) and biomedical technologies, including bioimaging and biosensing [1, 2]. Among various extrinsic phosphors, Eu^{3+} doped calcium magnesium silicate (CMS: Eu^{3+}) is recognized as a highly efficient red emitting phosphor, due to its strong crystal field effects, sharp and narrow emission lines due to $f-f$ transitions, wide bandgap and low phonon energy [1]. CMS: Eu^{3+} can crystallize in structural phases such as diopside ($CaMgSi_2O_6$) and merwinite ($Ca_3MgSi_2O_8$) of monoclinic structure, akermanite ($Ca_2MgSi_2O_7$) of tetragonal structure and monticellite ($CaMgSiO_4$) of orthorhombic structure [3, 4]. These structural variations directly influence the luminescent behavior, including parameters like emission intensity, wavelength, lifetime, quantum yield (QY), color purity (CP), and chromaticity [3]. Apart from structural phase tuning, the synthesis methodology plays a crucial role in tailoring the photoluminescence (PL) properties [1]. One effective approach is the use of flux assisted synthesis. Flux helps to modify the crystal structure and promotes the reduction in melting temperature of CMS: Eu^{3+} phosphor [3].

A. Antony et al. (Eds.): icONMAT 2025, SPPHY 328, pp. 153–157, 2025.
https://doi.org/10.1007/978-981-95-2697-0_19

In the present study, CMS: Eu^{3+} phosphors were synthesized via a solid state reaction method with and without the addition of H_3BO_3 as a flux. The influence of flux on the phase optimization and particle morphology was systematically examined. Notably, the use of H_3BO_3 led to improved crystallinity and particle formation, as well as enhanced PL performance. This work provides a detailed experimental investigation into the role of H_3BO_3 in modulating the structural and luminescent characteristics of CMS: Eu^{3+} phosphor, highlighting its potential for biomedical and photonic applications.

2 Experimental

Solid state reaction method was employed for the preparation of CMS: Eu^{3+} phosphor. For that, CaO (90% pure, Loba Chemi), MgO (95% pure, Emplura), SiO_2 (99.5% pure, Spectrochem) were taken in the ratio 3:1:2. Along with these, 1 mol% of Eu_2O_3 (99.99% pure, Spectrochem) was added. CMS: Eu^{3+} was synthesized without adding H_3BO_3 (flux) and with adding 0.1 g of H_3BO_3 as flux. The precursors were stoichiometrically weighed and ground for 12 h separately (with and without flux) using a mortar and pestle to produce homogeneous powders. The prepared powders were annealed at 1100 °C for 4 h in air for the preparation of phase pure samples of CMS: Eu^{3+}.

The phase identification was done by Rigaku Miniflex Benchtop powder X-ray diffractometer of 40 kV and 15 mA Cu-Kα source with 2°/minute speed. The morphology studies were carried out using Field Emission Scanning Electron Microscopy (FESEM) of the make Carl Zeiss Sigma Supra 40VP. PL excitation and emission were obtained using HORIBA CANADA QM-8075-11-C spectrofluorometer with a mercury xenon lamp source. Using Commission Internationale d'Eclairage (CIE) coordinates diagram 1931, color chromaticity and CP values were calculated.

3 Results and Discussion

3.1 Phase Identification

Figure 1(a) and (b) represent the XRD pattern of CMS: Eu^{3+} without and with the addition of H_3BO_3. XRD analysis reveals that CMS: Eu^{3+} synthesized without flux contains a mixture of phases such as diopside (ICDD 00–017-0318), merwinite (ICDD: 00–035-0591) of monoclinic crystal structure, akermanite (ICDD: 00–035-0592) of tetragonal crystal structure and monticellite (ICDD: 00–035-0590) of orthorhombic crystal structure [3, 4]. This multiphase formation results from uneven precursor reactions. However, incorporating H_3BO_3 as a flux leads to a single-phase merwinite structure with improved homogeneity. The flux acts as a medium to enhance precursor interaction and ion diffusion, promoting phase purity. As reported earlier, phase pure CMS: Eu^{3+} was also obtained using H_3BO_3 with $CaCO_3$, $MgCO_3$ and SiO_2 precursors [3].

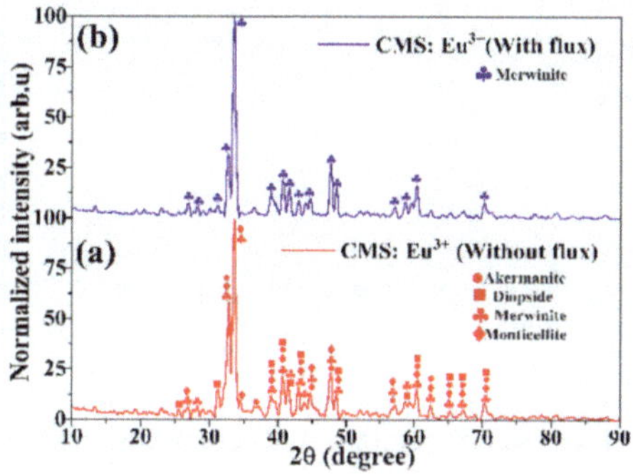

Fig. 1. XRD pattern of CMS: Eu^{3+} (a) without H_3BO_3 and (b) with H_3BO_3.

3.2 Morphology Studies

The morphology varies from the porous structure with small agglomerations to clear particles with no pores and agglomerations with flux addition and given as Fig. 2(a) and (b), respectively. All CMS: Eu^{3+} phosphors show clear particle formation with distinct boundaries. Flux reduced the presence of small agglomerations coexisting with the particles irrespective of phases. The agglomerations could be intermediate phases. However, confirmation necessitates an Energy Dispersive X-ray (EDX) analysis, which was not conducted in this study. The reduction of minute traces of intermediate phases likely contributes to the absence of agglomerations with flux.

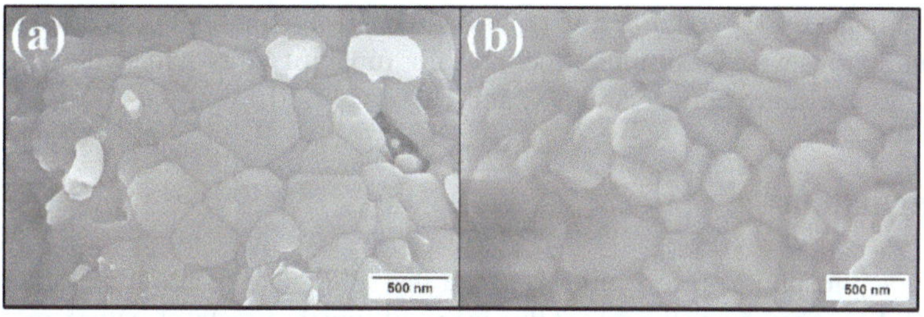

Fig. 2. FESEM images of CMS: Eu^{3+} (a) without H_3BO_3 and (b) with H_3BO_3.

3.3 Photoluminescence Studies

To analyze the effect of H_3BO_3 as a flux on the photoluminescence behavior of CMS: Eu^{3+} phosphors, excitation and emission spectra were recorded in the 350–550 nm and 550–750 nm ranges, respectively, as shown in Fig. 3(a) and (b). The excitation transitions include ^7F_J (J = 0, 1) $\leftarrow$ ^5D_J (J = 1, 2, 4) and ^7F_J (J = 0, 1) $\leftarrow$ ^5L_J (J = 6, 7) with maximum excitation at 394 nm in the UV region. The emission transitions are ^5D_J (J = 0, 1) $\rightarrow$ ^7F_J (J = 0–5) with maximum emission in the pure red region of the spectra at 614 nm with an induced electric dipole (ED) transition [4, 5].

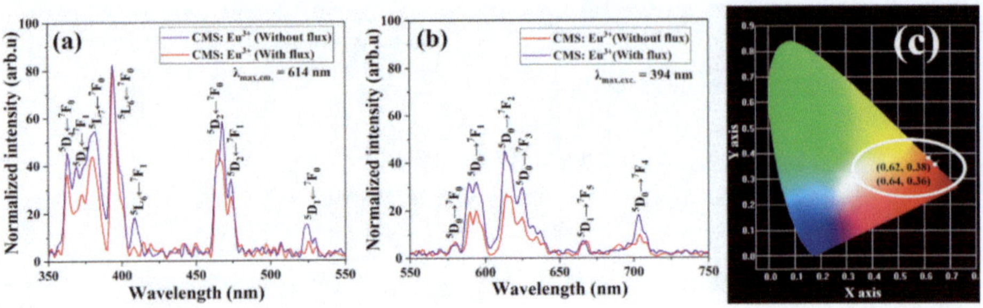

Fig. 3. PL (a) excitation, (b) emission spectra and (c) color chromaticity diagram of CMS: Eu^{3+} without and with H_3BO_3.

A noticeable enhancement in PL intensity is observed upon flux addition, independent of excitation or emission spectra. This improvement is attributed to the elimination of multiple phases, resulting in better luminescent performance due to flux addition. The incorporation of flux helps for the formation of phase pure CMS: Eu^{3+}, leading to a more favorable coordination environment for Eu^{3+} ions and improved site occupancy in the CMS host lattice factors in turn promotes the peak intensity and also helps to reduce the melting temperature of phases of CMS: Eu^{3+}. These findings are consistent with observations reported in our earlier study [4]. Notably, no significant peak splitting is detected in either the excitation or emission spectra with or without flux.

Additionally, the flux induces a fine tuning in color chromaticity coordinates from (0.62, 0.38) to (0.64, 0.36), corresponding to an increase in color purity from 85.53% to 92.46% in the red region, as illustrated in Fig. 3(c). This color tuning aligns well with the enhanced PL behavior. The equation for CP calculation is provided elsewhere [3–5]. In previous work, a similar effect on color tunability was achieved by varying the annealing temperature between 800 °C and 1200 °C [1, 5].

4 Conclusions

CMS: Eu^{3+} phosphors were synthesized via the solid state reaction method with and without H_3BO_3 to evaluate the role of flux in modifying the crystal structure, morphology and PL properties, including excitation, emission, chromaticity and CP. The addition of H_3BO_3 effectively suppresses multiphase formation, resulting in phase pure CMS: Eu^{3+}. Flux assisted synthesis also enhances particle growth and reduces the presence of smaller agglomerates and pores. PL analysis shows enhanced emission intensity across excitation and emission spectra, attributed to modification in the coordination geometry and improved site occupancy with flux addition. The CIE chromaticity diagram supports this enhancement, showing a red color fine tuning from (0.62, 0.38) to (0.64, 0.36), with CP increasing from 85.53% to 92.46%, aligning well with the PL findings.

Acknowledgements. The authors gratefully acknowledge the Department of Science & Technology, India, for providing the Innovation of Science Pursuit for Inspire Research (DST-INSPIRE) Fellowship and University Grant Commission, India for the UGC-BSR startup grant.

References

1. Kuriyan, N.S.: Luminescence studies of Eu^{3+} doped calcium magnesium silicate phosphor prepared at different annealing temperatures for fine color tunability. J. Lumin. **249**, 119038–119050 (2022)
2. Kuriyan, N.S.: Investigation to understand the role of phase variation in red emitting Eu^{3+}-doped calcium magnesium silicate phosphor for in vitro bioimaging. ACS Appl. Bio Mater. **8**, 1646–1664 (2025)
3. Kuriyan, N.S.: Site preference-based luminescence studies in Eu doped calcium magnesium silicate phosphor: A combined experimental and DFT approach. J. Lumin. **260**, 119901–119912 (2023)
4. Kuriyan, N.S.: Unraveling the role of crystal structural variations in modifying the luminescence properties of CMS: Eu^{3+} phosphor. J. Lumin. **277**, 120894–120908 (2025)
5. Navya, N.S., Sabeena, M.: Emission tuning in Eu^{3+} doped calcium magnesium silicate phosphor via annealing temperature. In: Tyagi, M., Bahadur, J., Pant, L.M., Kumar, G.R. (eds.) 67th DAE Solid State Physics Symposium 2023, AIP Conference Proceedings, vol. 3198, pp. 020028–020033 (2025)

Structural and Luminescent Properties of Cool White Light Emitting Dy^{3+} Doped $BaSr_2TeO_6$ Phosphors for Outdoor Lighting Applications

Arya Gopinath, Sincy Anna Oommen, P. B. Gayathri, and P. R. Biju(✉)

School of Pure and Applied Physics, Mahatma Gandhi University, Kottayam 686560, Kerala, India

prb.mgu@gmail.com

Abstract. A series of $BaSr_2TeO_6$:xDy^{3+} (x = 0, 0.02, 0.04, 0.06, 0.08, and 0.10) phosphors were successfully prepared via conventional solid-state reaction method. XRD analysis confirmed the formation of a single-phase structure and FTIR spectroscopy revealed the characteristic Te–O vibrational bands. SEM images show that the optimized $BaSr_2TeO_6$:0.04Dy^{3+} phosphor exhibited agglomerated morphology with an average particle size of 1.51 μm. EDS validated the elemental composition of the samples and confirmed the successful incorporation of Dy^{3+} in the host. Diffuse reflectance spectroscopy (DRS) analysis showed that the materials possess wide band gaps in the range of 5.11–5.17 eV, suggesting their suitability for optoelectronic and photonic applications. Photoluminescence emission under 351 nm excitation exhibited prominent blue (482 nm), yellow (575 nm), and far-red (675 nm) emissions attributed to Dy^{3+} transitions, resulting in near-white light emission. The phosphor with x = 0.04 displayed the highest luminescence efficiency. The observed concentration quenching beyond this concentration is attributed to dipole–dipole interactions between adjacent Dy^{3+} ions. The calculated CIE chromaticity coordinates were found to be close to those of commercial white LEDs, while the correlated color temperature values (ranging from 4100 to 4700 K) categorized the emission as cool white light. These results demonstrate that $BaSr_2TeO_6$:xDy^{3+} phosphors are promising candidates for single-phase white-light emitters, particularly for outdoor lighting applications.

Keywords: Cool white emission · $BaSr_2TeO_6$ · Dy^{3+}doping

1 Introduction

Research on rare earth (RE) doped phosphors is increasing nowadays owing to their exceptional optical properties and broad applicability across a range of emerging technologies. One such technology is in solid state lighting applications. The solid-state lighting utilizes phosphor-converted white light-emitting diodes (pc-WLEDs) stands out as a transformative lighting technology, poised to replace conventional lighting systems due to its numerous advantages such as low operating costs, extended lifespan, compact form factor, environmental friendly due to the absence of mercury vapor, low

A. Antony et al. (Eds.): icONMAT 2025, SPPHY 328, pp. 158–171, 2025.
https://doi.org/10.1007/978-981-95-2697-0_20

power consumption, and high luminous efficacy [1]. Since their inception, the commercial standard for pc-WLEDs has been based on blue-emitting InGaN chips coated with yellow-emitting $Y_3Al_5O_{12}:Ce^{3+}$ (YAG:Ce^{3+}) phosphors [2, 3]. These systems offer high luminous efficacy (up to ~150 lm·W^{-1}) and are widely used in cool white light applications. However, this blue-chip + yellow phosphor combination exhibits certain limitations. It often suffers from a low Color Rendering Index (CRI < 90) due to poor spectral coverage in the red region, and it produces high Correlated Color Temperature (CCT > 6000 K) values, leading to a cold bluish-white appearance [4]. Moreover, recent studies have raised health concerns regarding excessive exposure to blue light, particularly its potential to disrupt the circadian rhythm and adversely affect sleep patterns when used at night. To mitigate these drawbacks, researchers have turned toward an alternative strategy that employs near-UV (n-UV) LED chips in conjunction with red, green, and blue (RGB) tri-color phosphors. This approach offers improved color quality, more balanced white light emission, and greater spectral control. However, it introduces challenges such as the complexity of blending multiple phosphors, differential degradation rates among the emitters (especially red), and high manufacturing costs. In response to these challenges, current research efforts are increasingly focused on the development of single-phase white-light-emitting phosphors. When excited by a single UV or n-UV LED chip, these phosphors can generate high-quality white light with better CRI, lower CCT, and enhanced spectral stability [5, 6]. Therefore, in the current study, a single-phase $BaSr_2TeO_6:Dy^{3+}$ phosphor was successfully synthesized, which emits cool white light when excited by near-ultraviolet (n-UV) radiation. This desirable luminescent behavior highlights its potential as a promising candidate for use in white light-emitting diode (WLED) applications, particularly in energy-efficient solid-state lighting systems.

The host matrix $BaSr_2TeO_6$ is a tellurate host. Tellurate-based host materials are known for their intrinsically low phonon energy, typically ranging from 800 to 900 cm^{-1}. This characteristic is highly advantageous for RE doped phosphors, as it effectively suppresses non-radiative relaxation processes, thereby enhancing the radiative transitions that are crucial for strong luminescence [7]. Dysprosium (Dy^{3+}) ions are widely recognized as effective activators in the development of white-light-emitting phosphors due to their rich and distinct photoluminescent behavior. Upon excitation in UV, n-UV spectral regions, Dy^{3+} ions exhibit characteristic emissions in blue light from the $^4F_{9/2} \rightarrow {}^6H_{15/2}$ transition, yellow light from the $^4F_{9/2} \rightarrow {}^6H_{13/2}$ transition, and a weaker red emission from the $^4F_{9/2} \rightarrow {}^6H_{11/2}$ transition. The simultaneous presence of these three emission bands results in the generation of white light, making Dy^{3+}-doped materials ideal for solid-state lighting [8]. In the present study, a Dy^{3+} activated $BaSr_2TeO_6$ single-phase phosphor was synthesized via the conventional solid-state reaction method. A comprehensive investigation of its structural integrity, optical and photoluminescent properties, was conducted. The findings revealed that the synthesized $BaSr_2TeO_6:Dy^{3+}$ phosphor exhibits strong emission in the visible region under n-UV excitation, making it a highly promising material for use in white light-emitting diodes (WLEDs).

2 Experimental

The $BaSr_2(1\text{–}x)TeO_6{:}xDy^{3+}$ ($BST{:}xDy^{3+}$) phosphors, with x = 0, 0.02, 0.04, 0.06, 0.08, and 0.10, were synthesized via the solid-state reaction method. High-purity precursors $BaCO_3$ (99.99%), $SrCO_3$ (99.99%), TeO_2 (99.99%), and Dy_2O_3 (99.99%) were weighed in stoichiometric ratios and thoroughly ground in an agate mortar for 15 min. Then the homogeneous compounds were transferred to alumina crucibles and sintered at 750 °C for 5 h in a programmable muffle furnace. The resultant compounds were allowed to cool down to room temperature. The obtained compounds were again ground thoroughly for 15 min and heated at 950 °C for 8 h in alumina crucibles using a muffle furnace. After heating, the products were cooled down to room temperature. The obtained products are then ground into fine powder for further analysis.

The synthesized samples are subjected to different structural, morphological and optical studies. The phase purity and crystal structure of the as prepared samples were characterized by a powder X-ray diffraction (XRD) analysis using Rigaku Miniflex 600 employing Cu-Kα radiation ($\lambda = 1.5406$ Å). High-resolution surface morphology and elemental composition of the samples were examined using Scanning Electron Microscopy (SEM) and Energy Dispersive Spectroscopy (EDS). These analyses were performed with the help of a TESCAN BRONO field emission scanning electron microscope. The Diffuse Reflectance Spectra (DRS) of the phosphor samples were recorded using UV-Vis-NIR Spectrophotometer. Photoluminescence (PL) excitation and emission spectra were made with the help of spectrofluorophotometer (Horiba Fluorolog-3). The chromaticity co-ordinates were evaluated from the luminescence spectra obtained by converting the spectra into CIE-1931 chromaticity diagram using ColorCalculator7.49 software (OSRAM).

3 Results and Discussions

3.1 XRD Analysis

The phase purity, crystal structure, and crystallite size of the synthesized phosphor samples were thoroughly investigated using X-ray diffraction (XRD) analysis. The XRD patterns of $BaSr_2TeO_6{:}xDy^{3+}$ phosphors ($BST{:}xDy^{3+}$, where x = 0, 0.02, 0.04, 0.06, 0.08 and 0.10), are presented in Fig. 1, along with the standard reference pattern of $BaSr_2TeO_6$ obtained from the JCPDS number 35–0992. It is clearly observed that all the diffraction peaks corresponding to the $BST{:}xDy^{3+}$ samples are well indexed to the standard $BaSr_2TeO_6$ crystal structure without the appearance of any secondary or impurity phases. This is further corroborated by the comparable ionic radii of the dopant Dy^{3+} ions (0.912 Å) and the host matrix ions Sr^{2+} (1.18 Å), both having a coordination number of 6, which facilitates the successful substitution of Dy^{3+} into the Sr^{2+} sites within the lattice. The compound $BaSr_2TeO_6$ crystallizes in a cubic crystal system with space group F and a crystallographic multiplicity (Z) of 4. Furthermore, the lattice parameters $BST{:}xDy^{3+}$ phosphors calculated from the XRD data are summarized in Table 1. The crystallite size (d) of the $BST{:}xDy^{3+}$ phosphors were obtained from Debye Scherrer equation [9],

$$d = \frac{k\lambda}{\beta cos\theta} \tag{1}$$

where k is a constant called shape factor (k = 0.9), is the wavelength of the X-ray beam used (0.154 nm), β is the FWHM of the strongest peak in radian and θ is the diffraction angle of the diffracted peak. The calculated average crystallite size of phosphors is in the range 19 to 34 nm. The crystallite sizes are given in the Table 1 along with the lattice parameters.

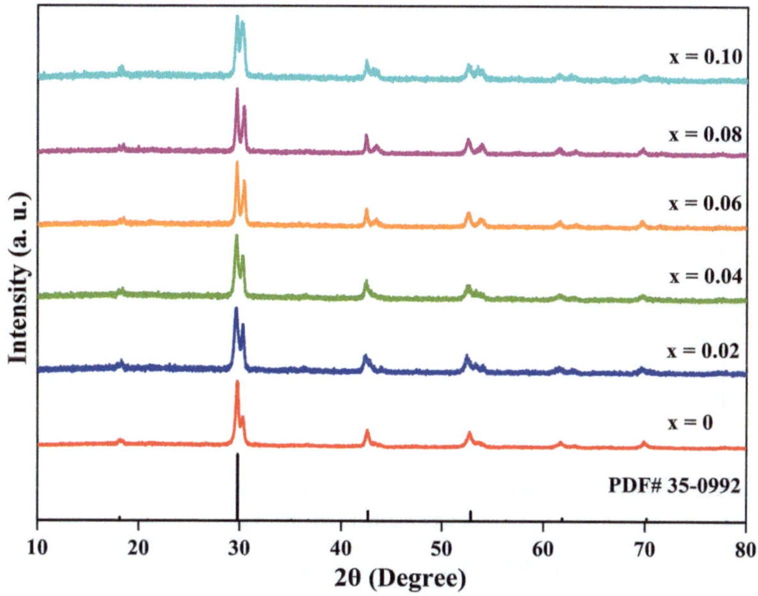

Fig. 1. XRD of BST:xDy^{3+} phosphors

Table 1. Unit cell parameters and crystallite size of BST:xDy^{3+} phosphors

Dy^{3+}Concentration	a (Å)	V (Å^3)	d (nm)
0	8.502	614.554	21.015
0.02	8.529	620.626	21.375
0.04	8.513	616.967	19.526
0.06	8.508	615.965	26.707
0.08	8.519	618.374	33.886
0.10	8.510	616.397	29.982

3.2 FTIR Analysis

The presence of Te–O bonds in the TeO_6 octahedral units is confirmed through Fourier Transform Infrared (FTIR) spectroscopy, as illustrated in Fig. 2. The FTIR spectra of both undoped and Dy^{3+}-doped samples exhibit highly similar absorption patterns, indicating that the incorporation of Dy^{3+} ions does not significantly disturb the fundamental lattice structure of the host matrix. Specifically, the prominent absorption bands observed at approximately 855 cm^{-1} and 659 cm^{-1} are attributed to the stretching vibrations of Te–O bonds within the TeO_6 groups. Additionally, the peak observed around 1469 cm^{-1} is associated with the vibrational mode of the O–H bond, which is likely due to the presence of adsorbed moisture from the atmosphere on the surface of the phosphor particles [10].

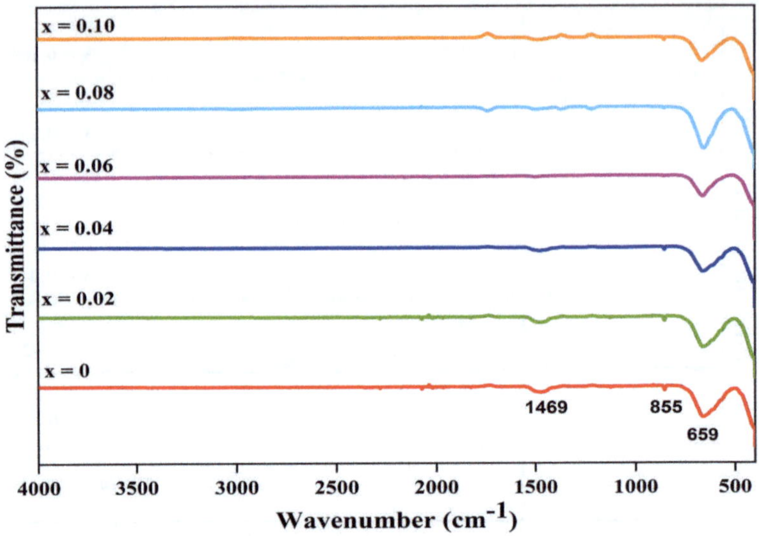

Fig. 2. FTIR spectra of BST:xDy^{3+} phosphors

3.3 SEM and EDS Analysis

The particle size and morphology of the optimized BST:0.04Dy^{3+} phosphor were comprehensively examined using Scanning Electron Microscopy (SEM) to gain insights into the surface features and microstructural characteristics of the synthesized material. Fig. 3(a–c) shows the SEM images of different magnification of BST:0.04Dy^{3+} phosphor. The particles exhibit irregular morphology, with a tendency to form clusters and agglomerates. This agglomeration is likely a consequence of the high-temperature solid-state reaction method employed during synthesis, where elevated sintering temperatures promote particle coalescence [11]. To further quantify the particle dimensions, a histogram illustrating the particle size distribution is provided in Fig. 3(d). The average particle size was estimated to be approximately 1.51 μm. This result confirms the successful formation of microcrystalline phosphor powders. The presence of micro-sized crystalline particles offers several distinct advantages for photoluminescent applications, particularly in solid-state lighting and display technologies. The micrometer-sized particles inherently possess lower surface area, which minimizes the formation of undesirable surface states. This results in enhanced radiative recombination, leading to higher luminescent output [12, 13].

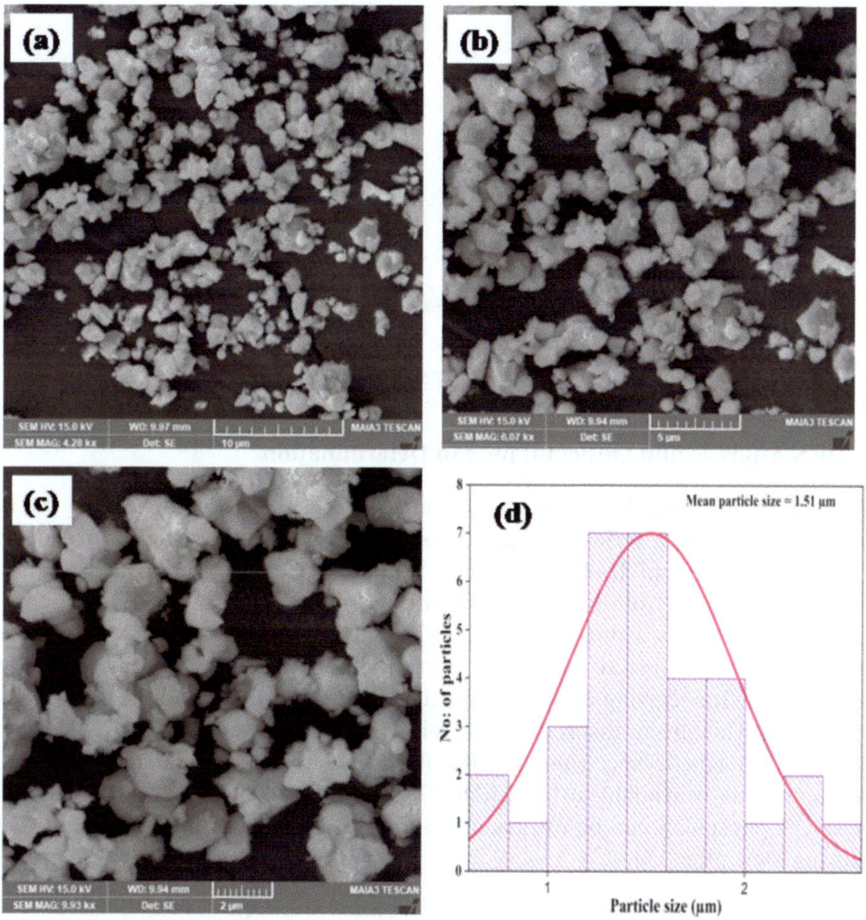

Fig. 3. (a), (b), and (c) FESEM images of BST:0.04Dy^{3+} phosphor at different magnification (d) Histogram showing average particle size

Energy Dispersive Spectroscopy (EDS) analysis was performed to examine the elemental composition of the optimized BST:0.04Dy^{3+} phosphor sample and to confirm the successful incorporation of dopant ions into the host matrix. The resulting EDS spectrum is presented in Fig. 4, which clearly shows the characteristic energy peaks corresponding to the constituent elements. The spectral data conclusively confirm the presence of strontium (Sr), bismuth (Bi), tellurium (Te), oxygen (O), and the dopant dsprosium (Dy) in the synthesized phosphor. Additionally, a quantitative analysis table, provided as an inset in the figure, lists the relative atomic percentages of the detected elements. Thus, the EDS analysis provides direct evidence supporting the successful doping of Dy^{3+} ions into the $BaSr_2TeO_6$ matrix.

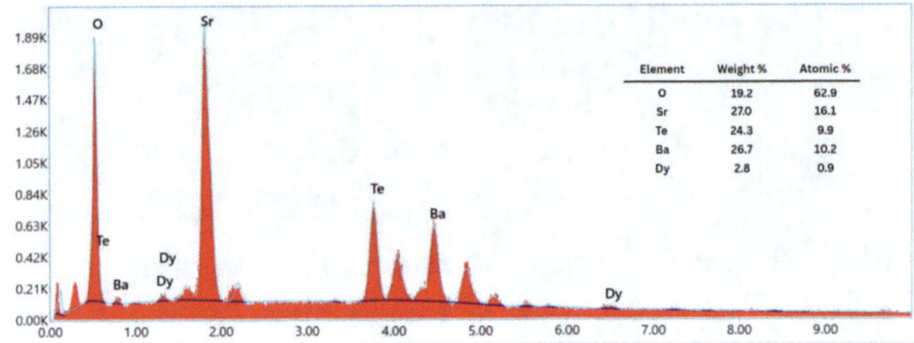

Element	Weight %	Atomic %
O	19.2	62.9
Sr	27.0	16.1
Te	24.3	9.9
Ba	26.7	10.2
Dy	2.8	0.9

Fig. 4. EDS of BST:0.04Dy^{3+} phosphor

3.4 DRS Analysis and Optical Bandgap Determination

The Fig. 5(a) shows the DRS spectra of synthesized phosphor. Upon doping the BST host with Dy^{3+} ions, a series of sharp absorption peaks appear, which correspond to the intra-configurational 4f–4f electronic transitions of Dy^{3+} [14]. The appearance of broad absorption peak from 300 nm to 380 nm confirms the substitution of Dy^{3+} ions into appropriate lattice sites within the host material, without disturbing the overall structural integrity. Among the broad absorption transitions of Dy^{3+} ions, the transition from 6H15/2 to 6P7/2, occurring at approximately 350 nm, exhibits the highest absorption intensity [14]. This strong absorption band falls within the near-ultraviolet (n-UV) region of the electromagnetic spectrum, making it particularly advantageous for excitation using commercially available n-UV LEDs. The optical band gap (Eg) of the synthesized BST:xDy^{3+} phosphors can be estimated from diffuse reflectance spectroscopy (DRS) data using Tauc plot [15].

$$\alpha = \frac{K}{S} = \frac{1 - \mathrm{R}^2}{2\mathrm{R}} \tag{2}$$

$$[\alpha h\vartheta]^{1/\gamma} = A\left(h\vartheta - E_g\right) \tag{3}$$

where R is the measured reflectance, K and S are the absorption and scattering coefficients, respectively, h is Planck's constant, ϑ is the photon frequency, and A is a proportionality constant. The exponent γ denotes the nature of the electronic transition and takes the value 1/2 for direct allowed transitions and 2 for indirect allowed transitions. To determine the band gap energy (E_g), the function $[\alpha h\vartheta]^{1/\gamma}$ is plotted as a function of photon energy and the linear portion of the curve is extrapolated to intersect the energy axis at the point where $[\alpha h\vartheta]^{1/\gamma} = 0$. This intercept gives the value of the optical band gap. Using this method, the estimated optical band gap values for the BST:xDy^{3+} phosphors were found to be 5.16, 5.11, 5.17, 5.17 and 5.11 eV respectively for x = 0.02, 0.04, 0.06, 0.08, and 0.10. These values fall within the typical range for wide-band-gap materials, indicating that the synthesized phosphors are capable of efficiently absorbing n-UV radiation.

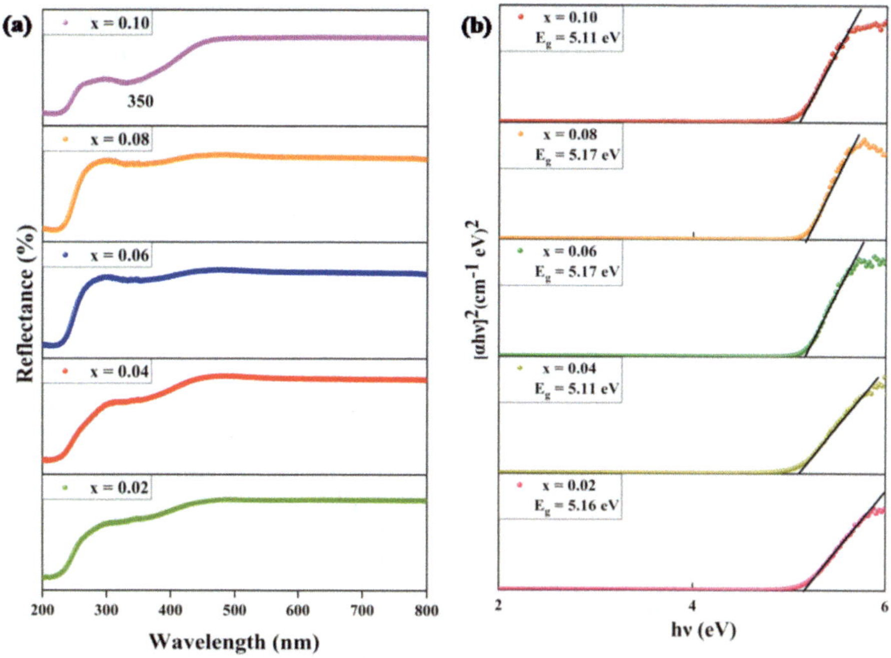

Fig. 5. (a) DRS of BST:xDy^{3+} phosphors (b) Tauc plot showing optical bandgap energy

3.5 Photoluminescence Excitation Spectra Analysis

The photoluminescence excitation (PLE) spectra of the synthesized BST:xDy^{3+} phosphors were recorded by monitoring the emission at 575 nm is shown in Fig. 6. The obtained excitation spectra reveal several well-defined peaks at wavelengths of 328, 351, 365, 390, 425, 454, and 472 nm. These peaks are associated with transitions originating from the ground state $^6H_{15/2}$ to various excited states $^6P_{3/2}$, $^6P_{7/2}$, $^6P_{5/2}$, $^4I_{13/2}$, $^4G_{11/2}$, $^4I_{15/2}$ and $^4F_{9/2}$ respectively of the Dy^{3+} ion [16]. Among these transitions, the excitation band at 351 nm was found to be the most intense. Thus, BST:xDy^{3+} phosphors can be efficiently excited by commercially available n-UV LED chips, making them highly promising candidates for application in pc-WLEDs. The result that obtained from PLE is in agreement with the results that obtained from DRS analysis.

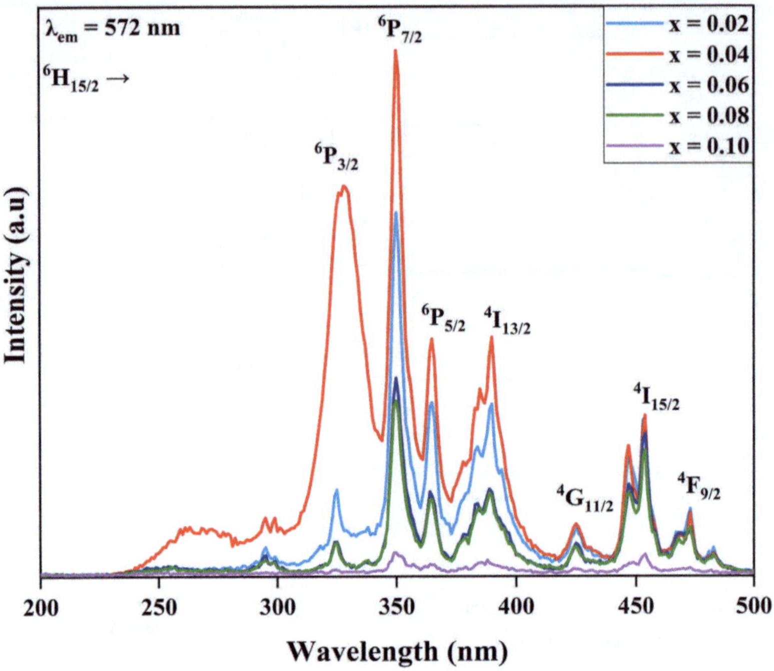

Fig. 6. PL excitation spectra of BST:xDy^{3+} phosphors

It is important to emphasize that rare-earth (RE) ion excitation processes inherently involve intra-configurational 4f–4f transitions. In RE ions, the 4f orbitals are shielded by outer filled 5 s and 5p orbitals, which protect them from the influence of the surrounding ligand field. Despite this shielding, interactions with ligands in the host lattice can still slightly affect the 4f energy levels through a phenomenon known as the Nephelauxetic effect**.** This effect refers to the expansion of the 4f electron cloud as a result of reduced electron–electron repulsion in the presence of ligands, effectively indicating increased covalent character in the metal–ligand bond. The extent of the Nephelauxetic effect is influenced by the nature and electronegativity of the ligands and can be quantitatively described using the Nephelauxetic ratio (β) and the bonding parameter (δ) [17, 18].

$$\beta = \frac{1}{n}\sum \frac{\vartheta_{complex}}{\vartheta_{free\ ion}} \tag{4}$$

$$\delta = \frac{1-\overline{\beta}}{\overline{\beta}} \tag{5}$$

Where $\vartheta_{complex}$ the energy (wavenumber) of a particular transition is in the Dy^{3+}-doped complex, $\vartheta_{free\ ion}$ is the energy of the same transition in the free Dy^{3+} ion, and n is the number of transitions considered. In the present study, the calculated values of the Nephelauxetic ratio and bonding parameter were found to be $\beta = 0.998$ and $\delta = 0.002$, respectively. The positive value of β and δ shows the covalent character in the $Dy^{3+} - O^{2-}$ bond.

3.6 Photoluminescence Emission Spectra Analysis

The photoluminescence (PL) emission spectra were recorded by monitoring at excitation 351 nm is shown in Fig. 7. Upon excitation at these wavelengths, the Dy^{3+}-doped phosphors exhibit characteristic emission bands in the visible region at 482 nm (blue), 575 nm (yellow) and 675 nm (far-red). These peaks correspond to the transition of $^4F_{9/2} \rightarrow {}^6H_{15/2}$, $^6H_{13/2}$ and $^6H_{11/2}$, respectively, of Dy^{3+} ions [19]. The PL emission intensity of BST:xDy^{3+} phosphors exhibits a strong dependence on the doping concentration of Dy^{3+} ions. As shown in Fig. 7, the emission intensity initially increases with increasing Dy^{3+} concentration, reaching a maximum at x = 0.04. However, upon further increasing the Dy^{3+} concentration beyond this optimal value, a gradual decrease in emission intensity is observed. This is due to concentration quenching effect, which is commonly encountered in RE ion-doped phosphor materials. The concentration quenching phenomenon arises primarily due to non-radiative energy transfer processes between neighboring Dy^{3+} ions [20]. At higher doping concentrations, the average interionic distance between Dy^{3+} ions decrease, facilitating increased interactions. These interactions lead to energy migration among excited ions, eventually resulting in non-radiative relaxation or energy loss through defects or quenching centers. The dominant mechanism responsible for this energy loss depends on the critical distance (R_c) between activator ions, which can be estimated using Blasse's equation [21],

$$R_c \approx 2\left(\frac{3V}{4\pi X_c N}\right)^{\frac{1}{3}} \tag{6}$$

For the synthesized $BaSr2TeO_6$:0.04Dy^{3+} phosphor, the unit cell volume is V= 616.967 Å3, the optimum doping concentration is Xc = 0.04, and N = 4. Substituting these values into Blasse's equation yields a critical distance Rc = 19.45 Å. This value is significantly greater than 5 Å, which indicates that exchange interactions are not dominant; instead, multipolar interactions are responsible for the observed concentration quenching behavior. To further investigate the nature of the multipolar interaction, the relationship between PL intensity and Dy^{3+} concentration was analyzed using Van Uitert's model, which describes how the emission intensity varies as a function of dopant concentration [22],

$$\frac{I}{x} = k\left[1 + \beta(x)^{\frac{\theta}{3}}\right]^{-1} \tag{7}$$

A plot of log (1/x) versus log (x), shown in Fig. 8. From the linear fit, the slope was determined to be −1.9, which results in a calculated θ ≈ 5.7. Since θ values of 6, 8, and 10 correspond to dipole–dipole, dipole–quadrupole, and quadrupole–quadrupole interactions, respectively. The obtained value of θ is closest to 6, indicating that the dipole–dipole interaction is the dominant mechanism responsible for concentration quenching in BST:xDy^{3+} phosphors.

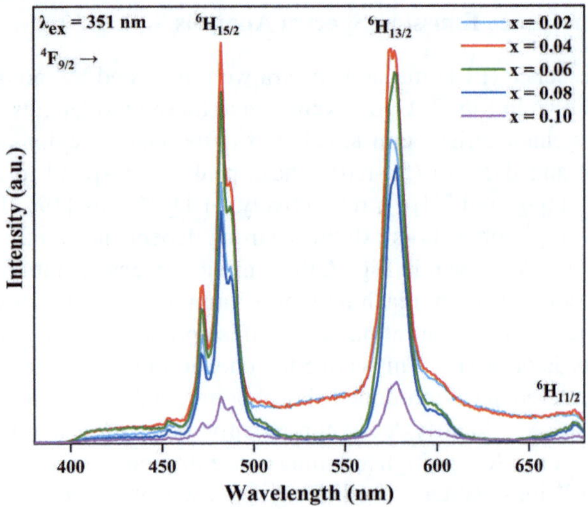

Fig. 7. PL emission spectra of BST:xDy^{3+} phosphors

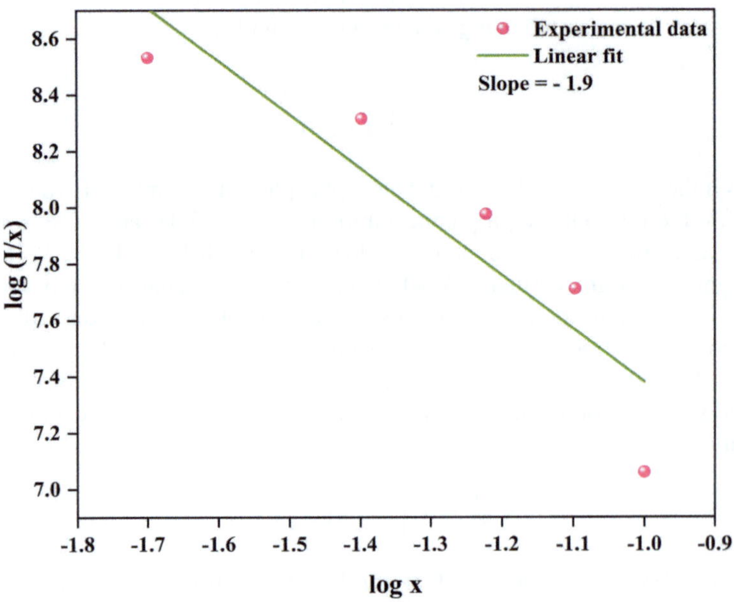

Fig. 8. Plot of log (I/x) versus log x graph

3.7 CIE Coordinates and CCT

Fig. 9 illustrates the Commission Internationale de l'Éclairage (CIE) chromaticity coordinates of the BST:xDy^{3+}phosphors, derived from the photoluminescence emission spectra under 351 nm excitation. These coordinates represent the perceived color of the emitted light and are plotted on the CIE 1931 chromaticity diagram to assess the color quality of the luminescence. The calculated CIE coordinates of optimized BST:xDy^{3+}phosphors are shown in Table 2. Notably, the obtained coordinates are very close to those of commercially available white-light LEDs, particularly the InGaN-based LED combined with YAG:Ce^{3+} phosphor, which typically exhibits CIE coordinates around (0.32, 0.32). Furthermore, they are also comparable to the National Television System Committee (NTSC) standard for white light emission, designated at (0.310, 0.316) [23]. This proximity indicates that the color output of the synthesized phosphor lies within the desirable white-light region, supporting its potential utility in lighting technologies.

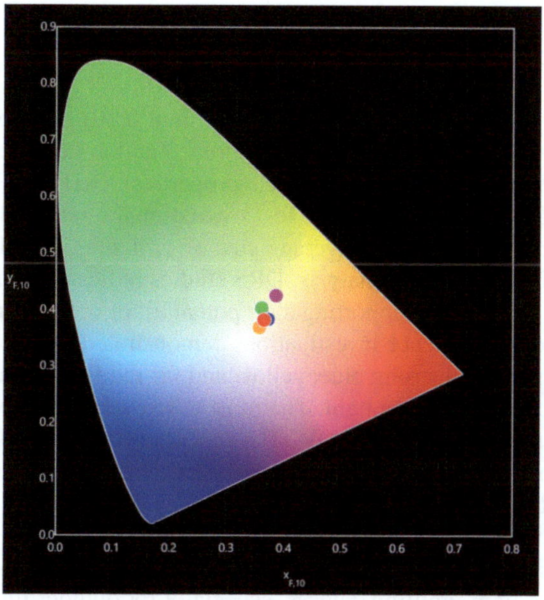

Fig. 9. CIE coordinates of BST:xDy^{3+} phosphors

Table 2. CIE coordinates and CCT values of BST:xDy^{3+} phosphors

Concentration of Dy^{3+} (x)	CIE coordinates	CCT (K)
0.02	(0.385, 0.425)	4179
0.04	(0.360, 0.403)	4461
0.06	(0.356, 0.369)	4676
0.08	(0.371, 0.383)	4317
0.10	(0.364, 0.384)	4494

Furthermore, the Correlated Color Temperature (CCT) is a critical photometric parameter that describes the hue of a light source by comparing its color appearance to that of an ideal blackbody radiator (also known as a Planckian radiator) heated to a specific temperature. It provides an estimate of whether the light emitted appears warm (yellow-red hue) or cool (blue-white hue). The CCT is especially relevant in lighting applications, as it directly influences the visual comfort, mood, and functional suitability of illumination. The CCT values of the synthesized BST:xDy^{3+} phosphors were calculated using the McCamy empirical relation [24]. The computed CCT values for the optimized Dy^{3+}-doped BST phosphors are presented in Table 2, and are found to be in the range of 4100–4700 K. These values indicate that the emitted light falls in the cool white region, which is desirable for applications requiring high visual acuity and brightness. Cool white light, typically characterized by higher CCT (> 4000 K) of BST:xDy^{3+} phosphors, make it suitable for outdoor illumination, street lighting etc.

4 Conclusions

$BaSr_2TeO_6$:xDy^{3+} (x = 0, 0.02, 0.04, 0.06, 0.08, and 0.10) phosphors were successfully synthesized using solid-state reaction approach and systematically characterized for their structural, morphological, and optical properties. XRD and FTIR analyses confirmed phase purity and the presence of Te–O vibrational modes, while SEM revealed agglomerated morphology with an average size of 1.51 μm. The elemental composition was validated through EDS analysis. DRS studies indicated wide band gap values ranging from 5.11 to 5.17 eV, supporting their potential for optoelectronic and photonic applications. Photoluminescence investigations demonstrated that under 351 nm excitation, the phosphors emitted strong blue, yellow and far-red light corresponding to Dy^{3+} transitions, producing near-white light. The most efficient emission was achieved at an optimal dopant concentration of x = 0.04. The concentration quenching phenomenon is found to be due to dipole-dipole interactions of neighboring Dy^{3+} ions. The CIE chromaticity coordinates of synthesized phosphors are notably close to that of commercial white LEDs, and the CCT values (4100–4700 K) classify the emission as cool white light. These findings affirm that $BaSr_2TeO_6$:xDy^{3+}phosphors are suitable for use in single-phase, white-light-emitting devices, especially for outdoor lighting applications.

References

1. Senden, T., Van Dijk-Moes, R.J.A., Meijerink, A.: Quenching of the red Mn4+ luminescence in Mn^{4+}-doped fluoride LED phosphors. Light Sci. Appl. **7**, 1–8 (2018)
2. Nakamura, S.: InGaN-based blue light-emitting diodes and laser diodes. J. Cryst. Growth. **201–202**, 290–295 (1999)
3. Duan, Y., et al.: Photoluminescence properties of $Tb_3Al_5O_{12}$:Ce^{3+}, Mn^{2+} phosphor ceramics for high color rendering index warm white LEDs. Opt. Mater. **111**, 110670 (2021)
4. Wu, H., et al.: Synthesis, structure and optical properties of novel thermally robust Dy^{3+}-doped $Ca_9Sc(PO_4)_7$ phosphors for NUV-excited white LEDs. J. Rare Earths. **39**, 277–283 (2021)
5. Bispo-Jr, A.G., Saraiva, L.F., Lima, S.A.M., Pires, A.M., Davolos, M.R.: Recent prospects on phosphor-converted LEDs for lighting, displays, phototherapy, and indoor farming. J. Lumin. **237**, 118167 (2021)
6. Ye, S., Xiao, F., Pan, Y.X., Ma, Y.Y., Zhang, Q.Y.: Phosphors in phosphor-converted white light-emitting diodes: Recent advances in materials, techniques and properties. Mater. Sci. Eng. R Rep. **71**, 1–34 (2010)

7. Zhao, L., et al.: Synthesis and photoluminescence properties of Sm^{3+} and Dy^{3+} ions activated double perovskite Sr_2MgTeO_6 phosphors. J. Lumin. **207**, 520–525 (2019)
8. Gupta, I., Singh, S., Bhagwan, S., Singh, D.: Rare earth (RE) doped phosphors and their emerging applications: A review. Ceram. Int. **47**, 19282–19303 (2021)
9. Manohara, B.M., et al.: Synthesis and luminescent properties of Tb^{3+} activated cadmium silicate nanophosphor. J. Alloys Compd. **592**, 319–327 (2014)
10. Li, K., Lian, H., Deun, R.V.: Site occupancy and photoluminescence properties of a novel deep-red-emitting phosphor $NaMgGdTeO_6$:Mn^{4+} with perovskite structure for w-LEDs. J. Lumin. **198**, 155–162 (2018)
11. Wang, Z., Liang, H., Zhou, L., Wang, J., Gong, M., Su, Q.: $NaEu_{0.96}Sm_{0.04}(MoO_4)_2$ as a promising red-emitting phosphor for LED solid-state lighting prepared by the Pechini process. J. Lumin. **128**, 147–154 (2008)
12. Lei, F., Yan, B., Chen, H.-H.: Solid-state synthesis, characterization and luminescent properties of Eu^{3+}-doped gadolinium tungstate and molybdate phosphors: $Gd_{(2-x)}MO_6$:Eux^{3+} (M=W, Mo). J. Solid State Chem. **181**, 2845–2851 (2008)
13. Vishwakarma, A.K., et al.: Red light emitting $BaNb_2O_6$:Eu^{3+} phosphor for solid state lighting applications. J. Alloys Compd. **622**, 97–101 (2015)
14. Behera, M., Panda, R., Arun Kumar, R., Mishra, N.K., Kumar, K., Monte, T.D.: Microwave-assisted combustion synthesis and characterization studies of novel dysprosium doped yttrium calcium borate (Dy^{3+}: $Y_2CaB_{10}O_{19}$) phosphor materials for efficient white light applications. Ceram. Int. **50**, 18146–18156 (2024)
15. Singh, J., Sharma, S., Sharma, S., Singh, R.C.: Effect of tungsten doping on structural and optical properties of rutile TiO2 and band gap narrowing. Optik. **182**, 538–547 (2019)
16. Ponkumar, S., Prakashbabu, D., Parasuraman, K., Uthrakumar, R., Kaviyarasu, K.: Synthesis, characterization, and enhanced photoluminescence of ZrO2:Dy3+ phosphors by incorporating Li^+, Na^+ and K^+ ions for LED applications. Ceram. Int. **50**, 13219–13228 (2024)
17. Sekaran, S.A.R., Padhi, R.K., Yousef, E.S., Marimuthu, K.: Spectroscopic studies on Sm^{3+} ions doped modifiers incited calcium phospho-silicate glasses for photonic and optoelectronic applications. J. Non-Cryst. Solids. **623**, 122692 (2024)
18. Upasani, M.: Synthesis of Y3Al5O12:Eu and $Y_3Al_5O_{12}$:Eu,Si phosphors by combustion method: Comparative investigations on the structural and spectral properties. J. Adv. Ceram. **5**, 344–355 (2016)
19. Zhang, W., Deng, C., Zhang, M.: Synthesis, structural and optical properties of a novel double perovskite Gd2MgTiO6: Dy^{3+} phosphors for LED applications. Mater. Res. Bull. **179**, 112982 (2024)
20. Parvathy, M., Ghosh, P.S., Arya, A., Sabeena, M.: Modification of local site preference for achieving tunable color emission from blue to near-white in $Sr_3(PO_4)_2$:Dy^{3+} phosphor. Mater. Res. Bull. **178**, 112885 (2024)
21. Panda, R., et al.: Development of thermally stable, single-phased $CaAl_4O_7$:Dy^{3+} phosphor and a study on their down-conversion features for lighting applications. Adv. Powder Technol. **35**, 104672 (2024)
22. Han, C., et al.: Photoluminescence, optical transition properties and temperature-induced shift of charge transfer band and temperature sensing property of $GdNbTiO_6$: Sm^{3+} phosphors. Spectrochim. Acta A Mol. Biomol. Spectrosc. **260**, 119951 (2021)
23. Sehrawat, P., et al.: Emanating cool white light emission from novel down-converted $SrLaAlO_4$:Dy^{3+} nanophosphors for advanced optoelectronic applications. Ceram. Int. **46**, 16274–16284 (2020)
24. McCamy, C.S.: Correlated color temperature as an explicit function of chromaticity coordinates. Color Res. Appl. **17**, 142–144 (1992)

Stable Mixed Organic Tin Bromide Low Dimensional Perovskites by Solution Processing

C. V. Mary Vijila[1(✉)], Antony Sharon[2], Meenu K. Pillai[1], Aldrin Antony[2,3,4], and M. K. Jayaraj[4]

[1] Department of Physics, St. Paul's College, Kalamassery, Kochi 683503, Kerala, India
maryvijila@stpauls.ac.in

[2] Centre of Excellence in Advanced Materials, Cochin University of Science and Technology, Kochi 682022, Kerala, India

[3] Inter University Centre for Nanomaterials and Devices, Cochin University of Science and Technology, Kochi 682022, Kerala, India

[4] Department of Physics, Cochin University of Science and Technology, Kochi 682022, Kerala, India

Abstract. We report solution-processed lead-free organic-inorganic tin halide perovskite thin films for optoelectronic applications. Here, we synthesized and characterized bulk methylammonium tin bromides and mixed methylammonium-octylammonium tin bromide low dimensional perovskites. Stable tin-based organic-inorganic hybrid perovskites can be a replacement for toxic-lead-based perovskites. The hydrophobic long-chain organic cation-octylammonium presents stability to the material from atmospheric moisture. The method introduced is fundamental and simple. Solution-synthesized perovskites were suspended in dimethylformamide. The suspension is spin-coated and drop-cast to glass slides for further characterization. Perovskite films exhibit a crystalline nature and a band gap around 2.4 eV.

Keywords: Low dimensional perovskites · bulk perovskite · tin-based perovskites

1 Introduction

Rapid progress is occurring in the research field of lead-based hybrid perovskites like methylammonium lead halides and their optoelectronic applications [1]. However, a few works have been reported on alternatives to toxic lead in hybrid perovskites due to stability issues [2, 3]. Hybrid means it contains organic and inorganic cations and possesses the advantages of organic and inorganic materials. In this work, we replaced lead with a tin alternative and enhanced the stability by reducing the amount of organic cation methylammonium with long-chain organic cation such as octylammonium. Synthesis of low dimensional stable tin-based hybrid perovskite is achieved using simple solution processing. We report the synthesis of bulk (3D) methylammonium tin bromides ($MASnBr_3$), quasi two dimensional (quasi-2D) methylammonium-octylammonium tin

A. Antony et al. (Eds.): icONMAT 2025, SPPHY 328, pp. 172–176, 2025.
https://doi.org/10.1007/978-981-95-2697-0_21

bromide perovskites ($OA_2MA_{n-1}Sn_nBr_{3n+1}$), and two dimensional (2D) octylammonium tin bromides (OA_2SnBr_4) in ambient atmospheric conditions. Hydrophobic long-chain octylammonium cations stabilize the system in the presence of moisture and high temperatures.

2 Experimental Methods

2.1 Materials

Tin bromide, n-octylamine and hydrobromic acid from Sigma Aldrich, methylamine (40% in methanol) from TCI chemicals, N, N dimethylformamide (DMF) of analytical grade from Merck were used in the synthesis of methylammonium tin bromide, mixed organic methylammonium-octylammonium tin bromide and octylammonium tin bromide films.

2.2 Fabrication of Methylammonium tin Bromide, Methylammonium-Octylammonium tin Bromide and Octylammonium tin Bromide Perovskite Films

Precursors, methylammonium bromide (MABr) and octylammonium bromide (OABr) were prepared by chemically reacting methylamine and octylamine with bromic acid, respectively [4]. To synthesize $MASnBr_3$ films, 0.112 g MABr, and 0.278 g tin bromide solution are prepared in 1 mL of DMF by magnetically stirring for twelve hours at 60 °C.

Mixed MA-OA-based quasi-2D perovskite precursor solution is prepared by dissolving 0.278 g $SnBr_2$, 0.1050 g OABr, and 0.056 g MABr in 1 mL DMF. For OA_2SnBr_4, 0.21 g OABr and 0.278 g tin bromide are dissolved in 1 mL DMF. The solutions are spin-coated on cleaned glass substrates at 1000 rpm and dried at 100 °C for 30 min.

2.3 Characterization

XRD patterns of drop-cast and annealed films were measured using a PANalytical X'Pert PRO high-resolution X-ray diffractometer (Cu Kα with $\lambda = 1.5418$ Å). Field emission scanning electron microscopy (FESEM) images provide morphological information. JASCO V-570 UV-VIS-NIR spectrometer provides absorption spectra of samples. The Fluoromax-3 spectrophotometer was used for Photoluminescence (PL) measurements.

3 Results and Discussions

Perovskite thin films were fabricated by solution processing and spin coating. Fig. 1 shows images of (a) the precursor solution used to fabricate thin films and (b) a spin-coated thin film methylammonium tin bromide ($MASnBr_3$). The films were smooth and pinhole-free. However, octylammonium-based quasi-2D and 2D perovskites exhibit less adhesion towards glass substrate, and the films were made by drop casting the solution on glass substrates for characterizations.

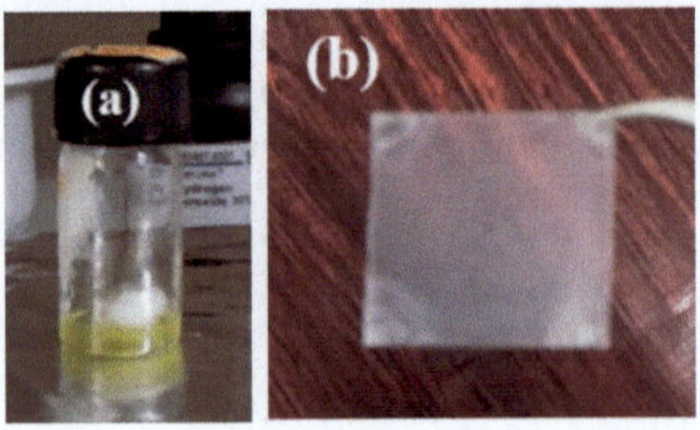

Fig. 1. Images of $MASnBr_3$ solution used for spin coating

Fig. 2 shows XRD patterns of $MASnBr_3$, mixed MA-OA-based perovskite, and OA_2SnBr_4 drop-cast films. From the X-ray diffraction patterns, it is clear that some lower-angle peaks arise when octylammonium long-chain cations replace smaller methylammonium cations due to enhanced d-spacing. An increase in d-spacing indicates intercalation of long chain cation into 3D perovskite producing quasi-2D and 2D perovskites. In addition, the quasi-2D mixed organic perovskite and 2D perovskites show improved crystallinity with sharp XRD peaks exhibiting the stability of the structure.

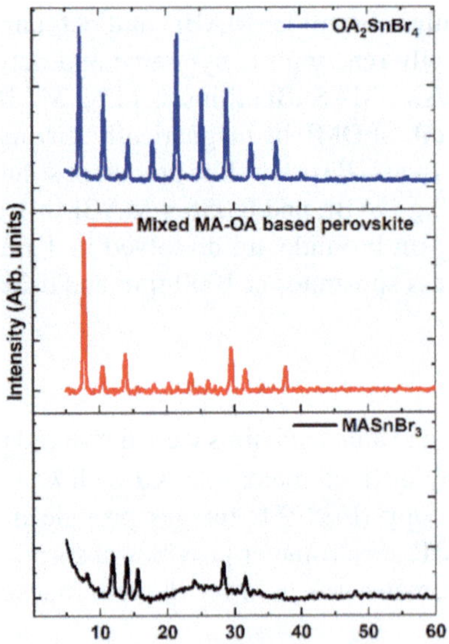

Fig. 2. XRD spectra of $MASnBr_3$, mixed MA-OA based perovskite and OA_2SnBr_4 drop cast films

Fig. 3 shows FESEM images revealing morphological evolution from **(a)** drop cast Mehtylammonium tin bromide $MASnBr_3$ to (b) MA-OA-based perovskite, then to (c) OA_2SnBr_4. The morphological evolution also indicates the formation of quasi-2D and 2D perovskites.

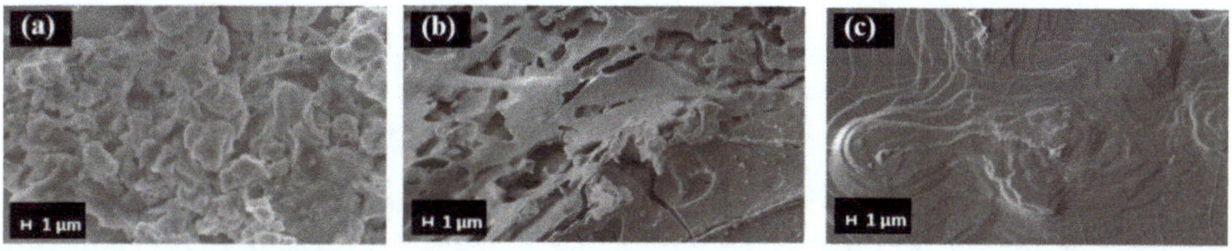

Fig. 3. FESEM Images of (a) drop cast Mehtylammonium tin bromide $MASnBr_3$ (b) MA-OA based perovskite (c) OA_2SnBr_4

Fig. 4 shows the absorption spectra of $MASnBr_3$, mixed MA-OA-based perovskite, and OA_2SnBr_4. It exhibits a blue shift in absorption peak from 326 nm for 3D $MASnBr_3$ to 294 nm for MA-OA mixed quasi-2D perovskite and 2D OA_2SnBr_4. Hence, the band gap of quasi-2D and 2D perovskites is larger. An increase in band gap or blue shift in absorption peak indicates a lowering of dimension from 3D to quasi-2D and 2D structures. Quasi-2D and 2D perovskites have almost the same peaks.

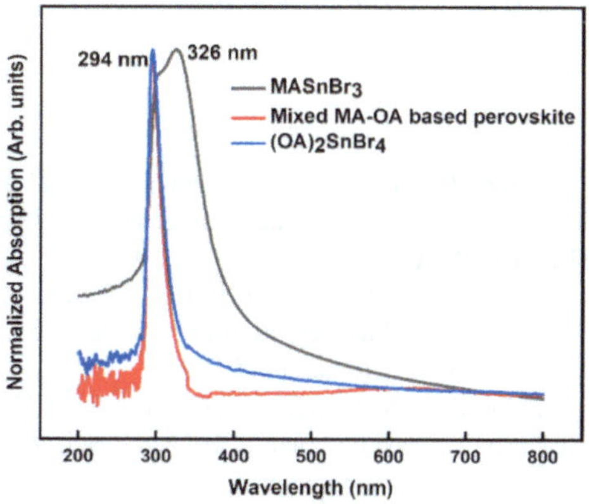

Fig. 4. Absorption spectra of $MASnBr_3$, mixed MA-OA based perovskite and OA_2SnBr_4

Fig. 5 shows PL emission spectra (a) $MASnBr_3$, (b) Mixed MA-OA-based perovskite, and (c) OA_2SnBr_4. A weak emission peak at 507 nm is present in all the samples. Also, the films are not distinctly luminescent as lead-based perovskites and are unsuitable for light-emitting device applications.

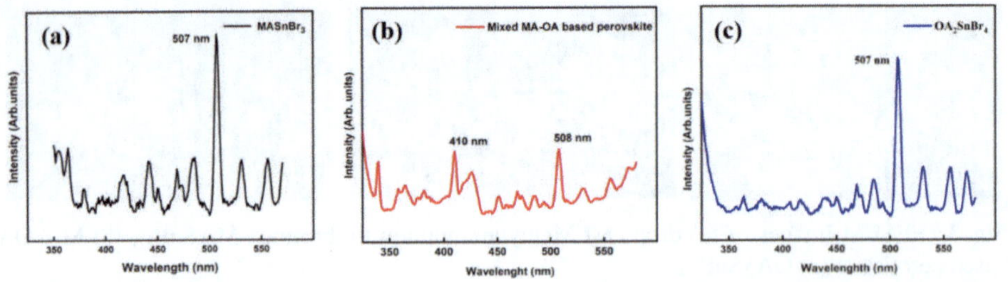

Fig. 5. Photoluminescence emission spectra of (a) $MASnBr_3$, (b) Mixed MA-OA based perovskite and (c) OA_2SnBr_4

4 Conclusions

Simple solution processing methods at ambient conditions can synthesize stable quasi-2D and 2D perovskites. The long-chain organic cation octylammonium easily replaces methylammonium cations. The hydrophobic nature of long-chain organic cations presents stability to the perovskites from moisture.

References

1. Adinolfi, V., Peng, W., Walters, G., et al.: The electrical and optical properties of organometal halide perovskites relevant to optoelectronic performance. Adv. Mater. **30**(1), 1–13 (2017). https://doi.org/10.1002/adma.201700764
2. Mosconi, E., Azpiroz, J.M., De Angelis, F.: Ab initio molecular dynamics simulations of methylammonium lead iodide perovskite degradation by water. Chem. Mater. **27**, 4885–4892 (2015). https://doi.org/10.1021/acs.chemmater.5b01991
3. Ava, T.T., Al Mamun, A., Marsillac, S., Namkoong, G.: A review: Thermal stability of methylammonium lead halide based perovskite solar cells. Appl. Sci. **9** (2019). https://doi.org/10.3390/app9010188
4. Bhaumik, S., Veldhuis, S.A., Ng, Y.F., et al.: Highly stable, luminescent core-shell type methylammonium-octylammonium lead bromide layered perovskite nanoparticles. Chem. Commun. **52**, 7118–7121 (2016). https://doi.org/10.1039/c6cc01056c

Bandgap Determination in PZT: Limitations of the Tauc Plot Method

Karthika P. Moni[1], Shreevarada Shreyas[2], Ankitha Baindla[1], P. N. Vinod[3], and Maneesh Chandran[1(✉)]

[1] Department of Physics, National Institute of Technology Calicut, Kozhikode 673601, Kerala, India
maneesh@nitc.ac.in
[2] School of Pure and Applied Physics, Mahatma Gandhi University, Priyadarsini Hills P.O., Kottayam 686 560, Kerala, India
[3] Naval Physical and Oceanographic Laboratory, Kochi 682021, Kerala, India

Abstract. Lead zirconate titanate (PZT) is a widely studied ferroelectric material known for its excellent piezoelectric properties. Despite extensive exploration of its electrical characteristics, the optical behavior, specifically bandgap estimation, has received relatively less attention. In this work, PZT powders were synthesized using the sol–gel method and subjected to varying annealing conditions to study their optical properties. The optical bandgap was determined using UV–visible diffuse reflectance spectroscopy, and analyzed through the Tauc plot method. However, the conventional approach of linear extrapolation in the Tauc plot poses limitations due to uncertainties in the nature of electronic transitions, structural disorder, compositional variations, and the presence of Urbach tails. This study emphasizes the necessity for more rigorous and objective analytical methods to accurately determine the optical bandgap in complex oxide systems like PZT.

Keywords: PZT · Tauc plot · Kubelka-Munk function

1 Introduction

Lead zirconate titanate [$Pb(Zr_xTi_{1-x})O_3$] (PZT) is a well-known ferroelectric material with excellent piezoelectric properties. PZT crystallizes in a perovskite structure, typically described by a cubic unit cell consisting of a central cation surrounded by an anion framework [1]. Depending on temperature and composition, PZT can undergo phase transitions and adopt different crystal symmetries. A key feature of PZT is the presence of the morphotropic phase boundary (MPB), which refers to a narrow composition range where two different crystal phases, rhombohedral and tetragonal, coexist. This boundary typically occurs when the Zr/Ti ratio is close to 52/48 [2]. PZT finds widespread use in various applications, including non-volatile memories, ultrasonic transducers, sensors, actuators, sonar devices, energy harvesting systems, wearable devices, acoustic and temperature sensors, biomedical devices, MEMS technologies, etc. [3–7].

This study shifts focus from the extensively studied ferroelectric properties and applications of PZT to a relatively underexplored area, primarily its optical properties,

A. Antony et al. (Eds.): icONMAT 2025, SPPHY 328, pp. 177–184, 2025.
https://doi.org/10.1007/978-981-95-2697-0_22

with a specific emphasis on bandgap determination. The bandgap of PZT was found to be in the range of 3.0–3.4 eV. This value can vary depending on the Zr/Ti ratio and can also be significantly influenced by the presence of dopants [8–10]. The optical bandgap of a material is typically determined using UV–visible diffuse reflectance spectroscopy (UV–DRS). The reflectance data are converted to absorbance using the Kubelka–Munk function, and the bandgap energy is determined by extrapolating the linear region of the Tauc plot, depending on whether the electronic transition is direct or indirect.

The Kubelka–Munk function relates the absorption coefficient (), the scattering coefficient (), and the reflectance ($_{\infty}$) of an optically thick (infinitely thick) sample and is expressed as:

$$F(R_{\infty}) = \frac{K}{S} = \frac{(1 - R_{\infty})^2}{2R_{\infty}} \tag{1}$$

The Tauc equation for the determination of bandgap of a material is

$$(\alpha h\upsilon)^{\frac{1}{\gamma}} = B\big(h\upsilon - E_g\big) \tag{2}$$

where α, h, υ and E_g are the absorption coefficient, Planck's constant, photon frequency, and bandgap, respectively; B is the proportionality constant. The value of γ depends on the nature of the electronic transition, taking a value of ½ for direct bandgap and 2 for indirect bandgap [11]. By substituting the Kubelka-Munk function from Eq. (1) into Eq. (2) for the absorption coefficient, we obtain:

$$(F(R_{\infty})h\upsilon)^{\frac{1}{\gamma}} = B\big(h\upsilon - E_g\big) \tag{3}$$

The resulting curve derived from these equations may exhibit both linear and nonlinear regions, which is indicative of allowed electronic transitions. The estimation of the optical bandgap using the Tauc plot method can be unreliable, particularly for complex metal oxide materials like PZT. This is due to several factors, including uncertainty in the nature of the electronic transitions, the presence of structural and compositional disorder, metal oxygen bond lengths, or the existence of Urbach tails. In real materials, especially those with structural or electronic disorder, the absorption edge is not sharp but instead shows a gradual increase in absorption below the actual bandgap. This phenomenon occurs due to localized states near the band edges, introduced by defects or thermal vibrations, and results in an exponential tail in the absorption spectrum. The degree of tailing can be quantitatively described by the Urbach energy, which serves as an indicator of disorder within the material [11, 12]. The Tauc method assumes a clear onset of absorption at the band edge, and the presence of Urbach tails causes a deviation from this ideal behaviour, often leading to an underestimation of the bandgap. As a result, accurately determining the bandgap energy becomes particularly challenging [13, 14].

In this study, PZT powders were synthesized using sol-gel technique under varying annealing conditions, and their optical properties were systematically investigated. The primary focus was on addressing the challenges associated with determining the bandgap energy using the Tauc plot method, specifically the extrapolation of its linear region. A detailed analysis was carried out by varying the selected energy range used for slope estimation in the Tauc plot.

2 Materials and Methods

The precursor solution was prepared using lead acetate trihydrate ($(CH_3COO)_2Pb{\cdot}3H_2O$, Sigma Aldrich), zirconium oxynitrate hydrate ($ZrO(NO_3)_2{\cdot}xH_2O$, Sigma Aldrich), and titanium n-butoxide ($Ti(OC_4H_9)_4$, Alfa Aesar), with ethylene glycol serving as the solvent. Lead acetate trihydrate and zirconium oxynitrate hydrate were dissolved in ethylene glycol and refluxed at 125 °C for 3 h. To compensate for potential lead loss during processing, a 10% excess lead precursor was added. Titanium n-butoxide was mixed with ethylene glycol and stirred for 2 h, separately. The two solutions were then combined and stirred for 3 h to ensure homogeneity. The precursors were used in stoichiometric ratios corresponding to the target composition $PbZr_{0.52}Ti_{0.48}O_3$. The solution is then dried at 90 °C to obtain the samples in powder form. The powders were annealed (500, 550, 600 °C) with different heating rates (2, 10 °C/min) to obtain phase pure sample. The structural properties were analyzed using X-ray diffraction (XRD, SmartLab SE Rigaku) and optical properties were studied using UV-vis DRS (Shimadzu UV-2600) using $BaSO_4$ as the reference.

3 Results and Discussion

The XRD patterns of PZT powder samples annealed at different annealing temperatures and annealing rates are shown in Fig. 1. From the XRD patterns, it is understood that all the experimental conditions resulted in the perovskite phase formation of PZT. But the presence of impurity peaks corresponding to ZrO_2 was observed, except for the sample annealed at 600 °C with a heating rate of 10 °C/min. The diffraction peaks of this sample refer to ICDD 00–033-0784, indicating the formation of $Pb(Zr_{0.52}Ti_{0.48})O_3$ with preferred orientation along (110) plane.

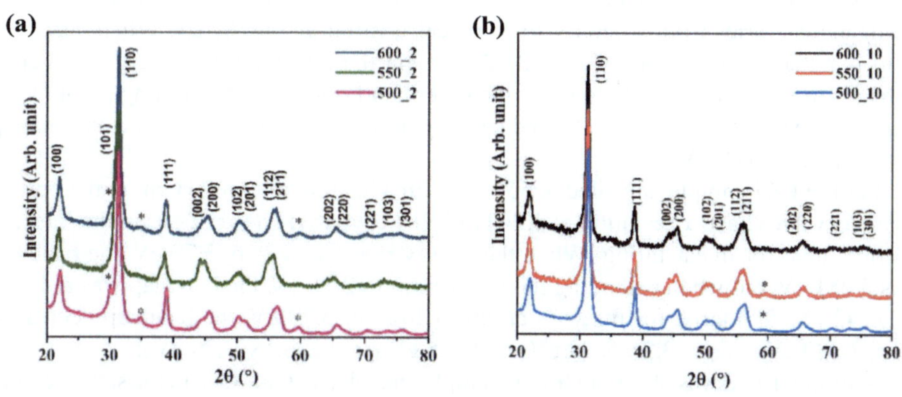

Fig. 1. XRD patterns of PZT powder annealed at 500, 550, 600 °C with different heating rates of (a) 2 and (b) 10 °C/min.

The UVDR spectra of the PZT powder in the range 200–800 nm is shown in Fig. 2. It is observed that the reflectance for all the samples starts from ~400 nm and reaches the maximum in the range of 600–800 nm. The optical bandgap was estimated by constructing a Tauc plot based on the Kubelka–Munk function derived from the diffuse reflectance data, as shown in Fig. 3. To evaluate the sensitivity and accuracy of the optical bandgap determination, a detailed analysis was conducted by varying the selected energy range within the plot. The linear region was selected by fixing one endpoint of the energy range and systematically varying the other. The specific energy range selected for the linear fitting in the Tauc plots, along with the calculated bandgap values for each sample, are summarized in Table 1.

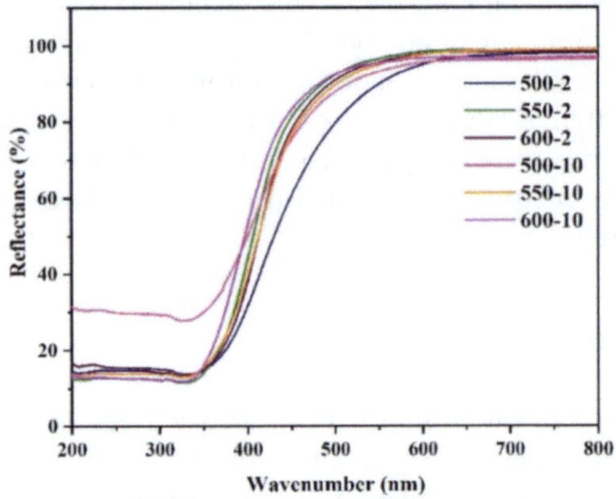

Fig. 2. UV-vis optical reflectance spectra of PZT powder annealed at 500, 550, 600 °C with a heating rate of 2 and 10 °C/min, which is labelled as 500–2, 550–2, 600–2, 500–10, 550–10 and 600–10 respectively.

The PZT powder annealed at 500 °C with a heating rate of 2 °C/min showed a bandgap that varies from 3.046 eV to 3.064 eV, a variation of 0.02 eV. When the upper limit is reduced (e.g., from 3.73 to 3.62 eV) and the lower limit is fixed, the change in bandgap was considerably low. On the other hand, lowering the starting point of the fitting range (e.g., 3.06 to 3.02 eV) and keeping the upper limit fixed lowers the bandgap in the order of 0.02 eV.

For the PZT sample annealed at 550 °C with a heating rate of 2 °C/min, the optical bandgap values have a deviation of 3.242 to 3.281 eV based on variations in the selected energy range for linear fitting. When the energy range was 3.26–3.70 eV, the bandgap was 3.281 eV, but when extending the lower limit to 3.18 eV, the bandgap was reduced to 3.242 eV. Similarly, for fits with an upper limit at 3.79 eV, the bandgap decreased from 3.280 eV (range: 3.26–3.79 eV) to 3.245 eV (range: 3.18–3.79 eV).

A similar trend was observed for the sample annealed at 600 °C with the same heating rate, where the optical bandgap values ranged from 3.105 eV to 3.166 eV, based on the chosen fitting intervals. For an energy range of 3.12–3.73 eV, the bandgap was 3.166 eV, while extending the range down to 3.00–3.73 eV resulted in lowering the bandgap to 3.113 eV. Likewise, when the upper limit was fixed at 3.62 eV, a gradual decrease in bandgap was observed from 3.162 eV (for 3.12–3.62 eV) to 3.105 eV (for 3.00–3.62 eV).

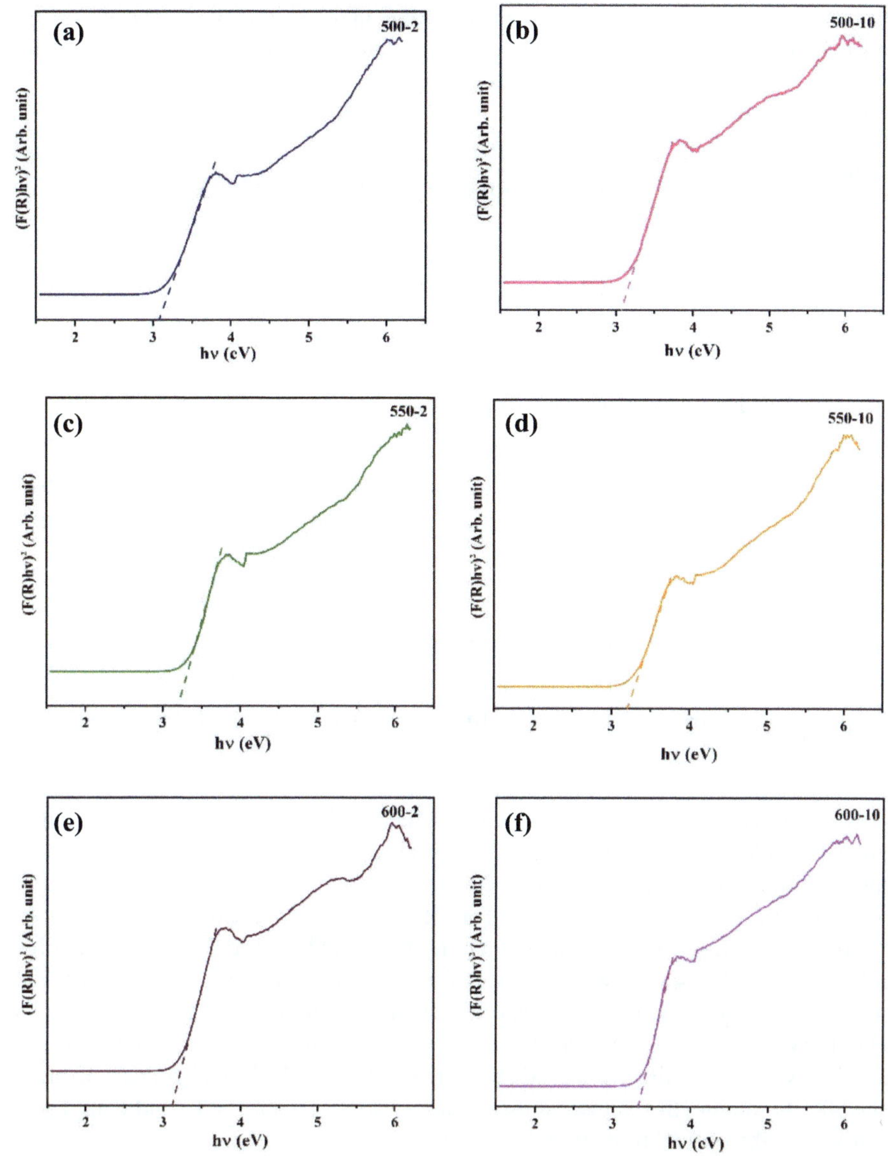

Fig. 3. Bandgap evolution of PZT powder annealed at 500, 550, 600 °C with a heating rate of 2 and 10 °C/min, shown in (a)-(f), respectively.

For the sample annealed at 500 °C with a heating rate of 10 °C/min, the calculated bandgap varied from 3.113 eV to 3.133 eV, depending on the selected energy range for Tauc plot fitting. When the fitting window was 3.10–3.76 eV, the bandgap was 3.133 eV, while narrowing the range to 3.06–3.65 eV resulted in a slightly lower value of 3.113 eV. A gradual decrease in bandgap was observed as both the lower and upper limits of the

Table 1. Selected energy ranges used for linear fitting in the Tauc plots and the corresponding estimated bandgap values of PZT samples annealed at different temperatures and heating rates.

Heating rate (°C/min)	Annealing temperature (°C)					
	500		550		600	
	Energy range (eV)	Optical bandgap (eV)	Energy range (eV)	Optical bandgap (eV)	Energy range (eV)	Optical bandgap (eV)
2	3.06–3.73	3.064	3.26–3.79	3.280	3.12–3.73	3.166
	3.06–3.65	3.061	3.26–3.70	3.281	3.12–3.62	3.162
	3.04–3.73	3.057	3.22–3.79	3.263	3.08–3.73	3.150
	3.04–3.65	3.053	3.22–3.70	3.261	3.08–3.62	3.144
	3.02–3.73	3.051	3.18–3.79	3.245	3.00–3.73	3.113
	3.02–3.65	3.046	3.18–3.70	3.242	3.00–3.62	3.105
10	3.10–3.76	3.133	3.18–3.76	3.224	3.34–3.78	3.359
	3.10–3.65	3.130	3.18–3.65	3.215	3.34–3.67	3.356
	3.08–3.76	3.125	3.14–3.76	3.206	3.32–3.78	3.321
	3.08–3.65	3.121	3.14–3.65	3.195	3.32–3.67	3.345
	3.06–3.76	3.117	3.10–3.76	3.186	3.30–3.78	3.341
	3.06–3.65	3.113	3.10–3.65	3.086	3.30–3.67	3.338

energy range were reduced to a lower energy value. Although the variation is relatively small (~0.02 eV), the trend again reflects the sensitivity of the optical bandgap to small changes in the selected linear region.

In the case of the sample annealed at 550 °C with a heating rate of 10 °C/min, the estimated optical bandgap varied, ranging from 3.224 eV to 3.086 eV. The highest bandgap value (3.224 eV) was obtained for the fitting interval 3.18–3.76 eV, while extending the range toward lower energies, such as 3.10–3.65 eV, led to a lower value of 3.086 eV. As observed in earlier cases, lowering both the lower and upper limits of the fitting window resulted in a gradual decrease in the calculated bandgap. Compared to the 500 °C sample, this dataset shows a larger variation (~0.14 eV), indicating greater sensitivity of the optical transition.

When the sample was annealed at 600 °C with a heating rate of 10 °C/min, the calculated optical bandgap values showed relatively minor variation, ranging from 3.359 eV to 3.338 eV across different fitting ranges. When the range 3.34–3.78 eV was selected, a higher bandgap value of 3.359 eV was obtained, while extending the fitting range as 3.30–3.67 eV resulted in a reduced bandgap of 3.338 eV.

By constructing the Tauc plot and varying the selected energy range used for linear fitting, noticeable inaccuracies were observed in the calculated bandgap of the sample. These discrepancies arise from the sensitivity of the intercept method to the choice of the energy range.

4 Summary

In summary, PZT powder samples were synthesized at different annealing conditions, and their optical bandgap was estimated in detail. A thorough analysis was carried out by systematically varying the energy range selected for slope fitting in the Tauc plot. Small shifts in the selected energy range, lead to noticeable variations in the calculated bandgap, ranging from ~0.02 eV to ~0.14 eV. This variability introduces a level of uncertainty in the accuracy and reproducibility of bandgap values. The visual identification of the linear region in the Tauc plot, combined with assumptions regarding the type of electronic transitions (direct or indirect), further highlights the limitations of this method for accurately evaluating the bandgap in complex materials. While Tauc analysis remains a widely used and convenient approach, these findings emphasize the importance of maintaining a consistent and well-justified fitting strategy. The careful selection and clear reporting of the fitting interval are essential to improve the reliability of bandgap estimations, and complementary techniques may be required to validate the results obtained. Therefore, more rigorous, objective, and reproducible approaches, such as derivative analysis or cumulative distribution function-based methods, are essential for accurate bandgap determination in complex oxides.

Acknowledgements. The authors gratefully acknowledge the Department of Physics, National Institute of Technology Calicut, for providing access to the XRD facility supported under FIST (SR/FST/PSI-113/2019). The authors acknowledge Centre for Innovation, Entrepreneurship & Incubation (CIEI), National Institute of Technology Calicut for funding this work through Student Innovative Projects (SIP) scheme.

References

1. Eledath, M., Chandran, M.: Status review of the science and technology of PZT/diamond heterostructures and their applications. J. Mater. Res. **36**, 4725–4745 (2021). https://doi.org/10.1557/s43578-021-00329-9
2. Heywang, W., Lubitz, K., Wersing, W.: Piezoelectricity: Evolution and Future of a Technology. Springer, Berlin, Heidelberg (2008). https://doi.org/10.1007/978-3-540-68683-5
3. Wang, Y., Cheng, H., Yan, J., Chen, N., Yan, P., Yang, F., Ouyang, J.: Large piezoelectricity on Si from highly (001)-oriented PZT thick films via a CMOS-compatible sputtering/RTP process. Materialia. **5**, 100228 (2019). https://doi.org/10.1016/j.mtla.2019.100228
4. Muralt, P.: Piezoelectric thin films for mems. Integr. Ferroelectr. **17**, 297–307 (1997). https://doi.org/10.1080/10584589708013004
5. Lee, C., Itoh, T., Suga, T.: Micromachined piezoelectric force sensors based on PZT thin films. IEEE Trans. Ultrason. Ferroelectr. Freq. Control. **43**(4), 553–559 (1996). https://doi.org/10.1109/58.503715
6. Izyumskaya, N., Alivov, Y.-I., Cho, S.-J., Morkoç, H., Lee, H., Kang, Y.-S.: Processing, structure, properties, and applications of PZT thin films. Crit. Rev. Solid State Mater. Sci. **32**, 111–202 (2007). https://doi.org/10.1080/10408430701707347
7. Xiang, W., Yin, Z., Yang, X.: Study and application of PZT in the Field of Wearable Devices. Adv. Mater. Technol. **10**(14) (2025). https://doi.org/10.1002/admt.202401692

8. Pandey, S.K., James, A.R., Raman, R., Chatterjee, S.N., Goyal, A., Prakash, C., Goel, T.C.: Structural, ferroelectric and optical properties of PZT thin films. Physica B: Condens. Matter. **369**, 135–142 (2005). https://doi.org/10.1016/j.physb.2005.08.024
9. Samanta, S., Sankaranarayanan, V., Sethupathi, K.: Band gap, piezoelectricity and temperature dependence of differential permittivity and energy storage density of PZT with different Zr/Ti ratios. Vacuum. **156**, 456–462 (2018). https://doi.org/10.1016/j.vacuum.2018.08.015
10. Khorrami, Gh.H., Khorsand Zak, A., Kompany, A., Yousefi, R.: Optical and structural properties of X-doped (X=Mn, Mg, and Zn) PZT nanoparticles by Kramers–Kronig and size strain plot methods. Ceram. Int. 38, Issue 7 5683–5690 (2012). doi:https://doi.org/10.1016/j.ceramint.2012.04.012.
11. Jubu, P.R., et al.: Comment about the use of unconventional Tauc plots for bandgap energy determination of semiconductors using UV–Vis spectroscopy. Results Opt. **14**, 100606 (2024). https://doi.org/10.1016/j.rio.2024.100606
12. Balakrishnan, R., Sahoo, P., Karuppannan, B., Dixit, A.: Probing electronic transitions and defect-induced Urbach tail bands in functional perovskite oxides using diffuse reflectance spectroscopy and the Kubelka-Munk function. arXiv preprint arXiv:2504.00033
13. Klein, J., Kampermann, L., Mockenhaupt, B., Behrens, M., Strunk, J., Bacher, G.: Limitations of the Tauc Plot Method. Adv. Funct. Mater. **33** (2023). https://doi.org/10.1002/adfm.202304523
14. Makuła, P., Pacia, M., Macyk, W.: How To Correctly Determine the Band Gap Energy of Modified Semiconductor Photocatalysts Based on UV–Vis Spectra. J. Phys. Chem. Lett. **9**, 6814–6817 (2018). https://doi.org/10.1021/acs.jpclett.8b02892

Tuning of Band Gap Through Transition Metal Doping in ZnO Nanostructures

A. Abinshan, Shibi Krishnan, S.Suchithra, and P. M. Aswathy(✉)

St. Stephen's College, Maloor College P.O., Pathanapuram, Kollam 689695, India
pmaswathy@ststephenspathanapuram.ac.in

Abstract. ZnO nanostructures are wide band gap semiconductors well established especially for optoelectronic applications. The addition of transition metals to the ZnO crystal lattice is known to create new energy levels below the conduction band, which changes the electronic structure of ZnO. The band gap tuning of ZnO can be achieved through dopant concentration variation. Apart from the notable changes in optical properties, the amount of dopant is found to affect the structural, and photocatalytic properties of the doped oxide. Doping ZnO with transition metals can also improve its photocatalytic activity. Some transition metals that can be used to dope ZnO include Mn, Fe, Co, Ni, and Cu. The present study shows the synthesis of nanostructured ZnO doped with transition metals such as Mn, Co and Cu. The structural, morphological and optical studies are discussed. The transition metal doping was found to tune the bandgap of ZnO towards a red shift. The band gap narrowing and observed red shift is due to the exchange interactions, confirming the substitution of dopants into the ZnO lattice.

Keywords: Nanostructures · Band Gap · Metal Oxide

1 Introduction

Metal oxides are a diverse class of materials, especially in their nano forms, that exhibit a wide range of electrical, optical, and catalytic properties, making them useful in applications such as sensors, catalysis, energy storage, and optoelectronics. Among the known nanostructured metal oxide semiconductors, zinc oxide (ZnO) is widely researched and highly compatible oxide due to its unique properties. The wide direct band gap of ~ 3.37 eV, exciton binding energy of ~ 60 meV and assymetric wurtzite crystalline structure are plausible reasons for piezoelectric and pyroelectric properties [1]. ZnO is non-toxic, abundant, and chemically stable, making it a promising material for a wide range of applications, including UV photodetectors, light-emitting diodes (LEDs), transparent conductive oxides, gas sensors and biomedical devices [2, 3].

The nanostructured ZnO exhibits size and shape dependent properties due to quantum confinement, high surface-to-volume ratio, and surface effects. Band gap tuning is a key strategy to optimize ZnO for specific applications, especially in optoelectronics and photocatalysis. The band gap can be tuned by doping i.e. introducing foreign atoms (e.g., Al, Ga, In, N, or transition metals) into the ZnO lattice that can create new energy levels or

A. Antony et al. (Eds.): icONMAT 2025, SPPHY 328, pp. 185–189, 2025.
https://doi.org/10.1007/978-981-95-2697-0_23

modify the conduction/valence bands, leading to changes in optical and electronic properties [4]. The localized states introduced within the band gap using transition metal or rare earth dopants can cause red or blue shifts depending on the dopant type and concentration. In nanostructures, reducing the particle size to the nanoscale (<10 nm) can cause the band gap to widen due to quantum confinement, shifting absorption and emission to shorter wavelengths [5]. Lattice strain and intrinsic defects such as oxygen vacancies or zinc interstitials can alter the local electronic structure, effectively influencing the band gap [6]. Forming heterostructures or decorating ZnO with organic or inorganic materials can also modify the effective band structure at interfaces, enabling band alignment for specific applications [7]. Understanding and controlling these mechanisms is vital for tailoring ZnO nanostructures for specific applications. Ongoing research continues to explore novel synthesis techniques and theoretical models to optimize band gap tuning strategies and improve device performance.

In the present work, we have tried to tune the band gap of ZnO nanostructures prepared via co-precipitation technique using a few transition metal dopants such as Manganese (Mn), Cobalt (Co) and Copper (Cu). We have done structural characterizations and absorption studies to confirm the phase formation and to understand the band gap dependence on transition metal doping.

2 Experimental Section

Zinc acetate dihydrate $Zn(CH_3COO)_2.2H_2O$, 99.99%), sodium hydroxide (NaOH), cobalt nitrate ($Co(NO_3)_2.6H_2O$, 99.99%), cupric nitrate ($Cu(NO_3)_2{\cdot}2H_2O$, 99.99%) and manganese nitrate ($Mn(NO_3)_2{\cdot}4H_2O$, 99.99%), were used as received for the preparation of the products. 0.5 M zinc acetate solution is prepared in 50 ml of deionized water, taken in a beaker and stirred using a magnetic stirrer. At the same time, 1 g of NaOH pellets is added to 50 ml of deionized water to prepare 50 ml of NaOH solution. While stirring zinc acetate solution, NaOH solution is added drop by drop and continuously stirred for 2 h. Using centrifuge, the white precipitate obtained is filtered and washed several times with distilled water and ethanol to remove unwanted impurities. The product obtained after washing is kept in the oven at 100 °C. The final product was dried in a muffle furnace at 300 °C for 3 h and crushed into powder. In a similar manner, copper, manganese and cobalt doped (1% and 5%) zinc oxide nanostructures were synthesized by adding adequate weight of copper, manganese and cobalt nitrates after dilution of zinc acetate.

3 Results and Discussion

The XRD spectra of pure, copper, manganese and cobalt doped (1% and 5%) zinc oxide nanostructures are shown in Fig. 1. The diffraction patterns for pure and doped ZnO exhibit good crystalline nature. All the peaks are indexed in agreement with that of the standard datasheet (JCPDS-036–1451) corresponding to the hexagonal wurtzite [8, 9]. The ionic radii of zinc ($Zn^{2+} = 0.74$ Å), copper ($Cu^{2+} = 0.69$ Å), cobalt ($Co^{2+} = 0.79$ Å) and manganese ($Mn^{2+} = 0.80$ Å) in the host matrix are comparable. No traces of any impure phase of metallic copper/manganese could be seen in the Cu/Mn doped ZnO samples. Thus, substitution of Cu and Mn ions at the Zn ion lattice site without any noticeable crystal structure can be confirmed [10–12]. The average crystalline size (D) was calculated using the Debye–Scherrer's formula [12, 20]. The average crystalline size of the samples was 20.52 nm, 21.2 nm, 22.3 nm, 24.2 nm and 25.1 nm, respectively. The larger crystallite size observed after copper, cobalt and manganese doping in the samples indicates better crystallization [12–14].

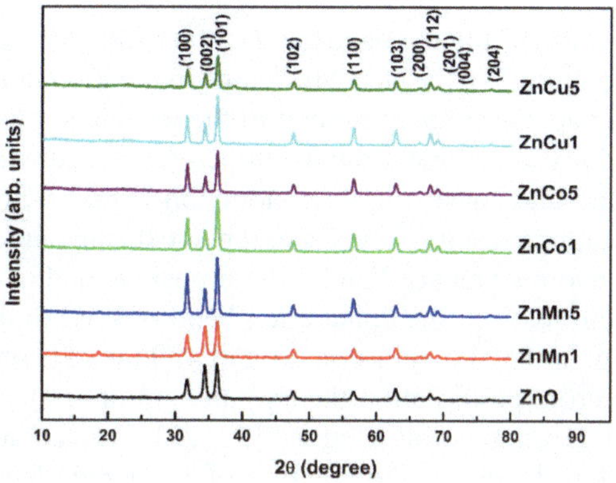

Fig. 1. XRD spectra of pure, copper, manganese and cobalt doped (1% and 5%) zinc oxide nanostructures.

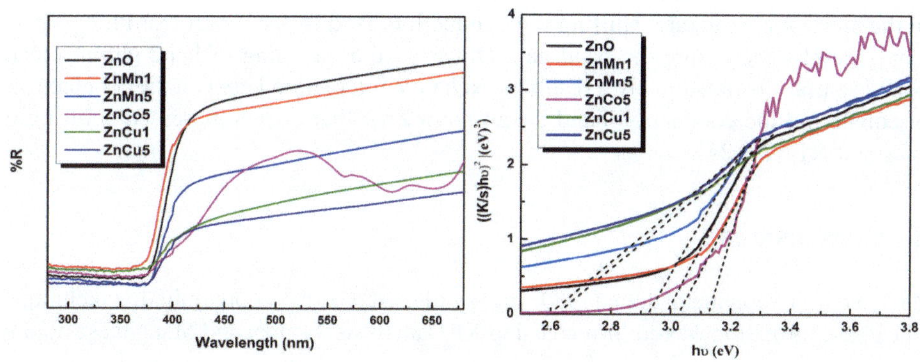

Fig. 2. A) UV-Vis spectra and b) Tauc plots of pure, copper, manganese (1% and 5%) and cobalt doped (5%) zinc oxide nanostructures.

The percentage of reflectance (% R) obtained from DRS (Fig. 2a) was used to estimate k/s (equivalent to the absorption coefficient, α) following the Kubelka - Munk expression. A plot between $((k/s)h\nu)^2$ and hν (photon energy), known as Tauc plot was drawn as given in Fig. 2b. From Tauc plot, the band gap for the pure ZnO sample was estimated to be about 2.96 eV. Then the bandgap slightly increases for 1% Mn doping and decreases steadily with doping to 2.85 eV for the 5% Mn doped ZnO and 2.65 eV and 2.62 eV for 1% and 5% Cu doped samples respectively. The highest band gap in this set of samples was observed for the 5% cobalt doped ZnO which is around 3.15 eV. The variation of diffuse reflectance near the band edge for the doped samples as compared to the pure sample is an indication of incorporation of the dopant ions in the ZnO lattice. The change in band gap is a signature for confirming whether Zn2 + ions are replaced by Cu^{2+} ions/Mn^{2+} ions in ZnO lattice. Nevertheless, regarding the cobalt doped samples, we have presented only the 5% doped ZnO, and it shows an increased band gap which is quite interesting.

Figure 2b shows a red shift in the band gap i.e. the band gap energy values decrease. Affirming the incorporation of Cu^{2+} and Mn^{2+} ions in the ZnO lattice [9]. The red-shift observed in the band gap energy is attributed to the exchange interaction of sp-d ZnO band electrons with localized d-electrons of doped Cu^{2+} and Mn^{2+} ions which raise the O 2p valence band and lower the 4s conduction band of Zn [15–17]. Band gap narrowing is the cumulative result of the s–d and p–d exchange interactions that modify the conduction band and the valence band in both positive and negative ways [18–20]. The transition of electrons from the conduction band of ZnO to the Zn vacant sites of the ZnO lattice results in the UV emission around 420 nm. Theoretically explanations based on the second-order perturbation theory can help understanding the mechanism behind the s-d and p-d exchange interactions [21, 22]. We can attribute the increased band gap in cobalt doped ZnO to the creation of localized electronic states through cobalt doping within the ZnO band structure, which can shift the Fermi level and thus alter the electronic properties. Cobalt ions ($Co2^{+}$) may introduce new impurity states above the valence band, causing an increase in the band gap due to the strong interactions between the dopants and the ZnO matrix. Cobalt doping can distort the crystal symmetry of ZnO, which can lead to an increase in strain. This strain may lead to modifications in the electronic structure, shifting the conduction band further apart from the valence band, thereby increasing the band gap. However, the widening of band gap in cobalt doped samples is presumably a result of exchange interaction between the 3d-electrons of cobalt and the conduction band electrons in ZnO that causes increased splitting of energy levels [23, 24].

4 Conclusions

ZnO and transition metal doped ZnO were synthesized using co-precipitation technique. The phase formation was confirmed using XRD analysis. Copper and Manganese doping was found to tune the bandgap of ZnO towards a red shift. Bandgap was found to decrease up to 2.62 eV. In contrast to these metal doped samples, cobalt doping showed an opposite trend showing a blue shift in band gap which needs to be studied further for promising applications.

References

1. Özgür, Ü., Alivov, Y.I., Liu, C., et al.: A comprehensive review of ZnO materials and devices. J. Appl. Phys. **98**(4), 041301 (2005)

2. Wang, Z.L.: Zinc oxide nanostructures: growth, properties and applications. J. Phys.: Condens. Matter **16**(25), R829 (2004)
3. Baruah, S., Dutta, J.: Hydrothermal growth of ZnO nanostructures. Sci. Technol. Adv. Mater. **10**(1), 013001 (2009)
4. Look, D.C.: Recent advances in ZnO materials and devices. Mater. Sci. Eng., B **80**(1–3), 383–387 (2001)
5. Pearton, S.J., et al.: Recent progress in processing and properties of ZnO. Prog. Mater Sci. **50**(3), 293–340 (2005)
6. Janotti, A., Van de Walle, C.G.: Fundamentals of zinc oxide as a semiconductor. Rep. Prog. Phys. **72**(12), 126501 (2009)
7. Bera, A., Basak, D., Lahiri, D.: Photoluminescence and band gap tuning of ZnO nanorods. J. Appl. Phys. **100**(3), 034309 (2006)
8. Chanda, A., Gupta, S., Vasundhara, M., Joshi, S. R., Mutta, G.R., Singh, J.: RSC Adv. **7**, 50527 (2017)
9. Al-Ghamdi, M.A., et al.: Computational and experimental study of wurtzite phase ZnO nanoparticles. Mater. Today: Proc. **84**, 1–8 (2023)
10. Gandhi, V., Ganesan, R., Abdulrahman Syedahamed, H.H., Thaiyan, M.: J. Phys. Chem. C **118**, 9717 (2014)
11. Bylsma, R.B., Becker, W.M., Kossut, J., Debska, U.: Phys. Rev. B **33**, 8207 (1986)
12. Djaja, N.F., Montja D.A., Saleh, R.: The effect of Co incorporation into ZnO nanoparticles, Adv. Mater. Phys. Chem. **03**, 33 (2013)
13. Basith, N.M., Vijaya, J.J., Kennedy, L.J., Bououdina, M., Jenefar, S., Kaviyarasan, V.: J. Mater. Sci. Technol. **30**, 1108 (2014)
14. Sharma, D., Jha, R.: Transition metal (Co, Mn) co-doped ZnO nanoparticles: effect on structural and optical properties. J. Alloy. Compd. **698**, 532 (2017)
15. Lima, M.K., et al.: Sol-Gel Sci. Technol. **72**, 301 (2014)
16. Salem, J.K. Hammad, T.M. Harrison, R.R.: J. Mater. Sci. Mater. Electron. **24**, 1670 (2013)
17. He, R., Tang, B., Ton-That, C., Phillips, M., Tsuzuki, T.: J. Nanopart. Res. **151**, 8 (2013)
18. Li, P., Wang, S., Li, J., Wei, Y.: J. Lumin. **132**, 220 (2012)
19. Diouri, J., Lascaray, J.P. El Amrani, M.: Phys. Rev. B **31**, 7995 (1985)
20. Maensiri, S., Laokul, P., Phokha, S.: J. Magn. Magn. Mater. **305,** 381 (2006)
21. Singhal, S., Kaur, J., Namgyal, T., Sharma, R.: Phys. B Condens. Matter **407**, 1223 (2012)
22. Boris, S., Brigittle, B., Andrei, M., Svetlana, P., Atamytiev, P.l: J. Eur. Ceram. Soc. JECS **29**, 1963 (2009)
23. Kumar, S., Singh, S.: Band gap engineering of cobalt-doped ZnO nanostructures for enhanced photocatalytic activity. J. Mater. Sci. Mater. Electron. **31**(9), 7433–7442 (2020)
24. Bera, S., Pal, A.K.: Effect of cobalt doping on the structural, optical, and electrical properties of ZnO nanostructures. J. Appl. Phys. **131**(6), 064302 (2022)

Aperiodic Distributed Bragg Reflectors Based on GaGeSe Chalcogenide Glass: Synthesis, Characterisation, and Numerical Investigation

Arun Pappachan[1], P. M. Sumayyabi[1], Soumya Suresh[1], Sheenu Thomas[1], and T. Priya Rose[1,2](✉)

[1] International School of Photonics, Cochin University of Science and Technology, Cochin, India
priyarose@cusat.ac.in
[2] Inter University Centre for Nanomaterials, Devices, Cochin University of Science and Technology, Cochin, India

Abstract. An aperiodic distributed Bragg reflector (A-DBR) based on $Ga_5Ge_{40}Se_{55}$ chalcogenide alloy and SiO_2 multilayers was designed for photonic applications in the visible region. The $Ga_5Ge_{40}Se_{55}$ alloy was synthesised via the conventional melt-quenching technique, and the optical constants of its thin films were precisely extracted using spectroscopic ellipsometry. These experimentally obtained dispersion data were employed in the numerical modelling of an 8-layer A-DBR stack comprising alternating $Ga_5Ge_{40}Se_{55}$ and SiO_2 layers. The simulated reflectance spectra reveal high reflectivity and excellent colour purity, making the structure suitable for a narrow-band, polarisation-independent colour filter. Unlike conventional periodic DBRs, the aperiodic design enables spectral flexibility and reduced side lobes, thereby enhancing the device's filtering selectivity and performance. The proposed structure offers significant potential in integrated photonic circuits, optical sensing, and display technologies, addressing limitations of traditional pigment- or dye-based filters such as instability, angular dependence, and limited tuning capability.

Keywords: Chalcogenide glass · A-DBR · Colour filters

1 Introduction

Conventional colour filters [1] are limited in application due to colour degradation, poor scalability and colour purity. Conventional structural colour filter techniques often need more complex nanofabrication and are sensitive to incident angle and polarisation effects. Aperiodic Bragg reflectors (A-DBR) [2] are an alternative to conventional DBR structures for this problem. A-DBR is an optical structure that uses a non-periodic sequence of layers to achieve high reflectivity over a broad range of wavelengths or angles. Unlike traditional periodic DBRs, which rely on a regular repeating pattern of alternating high and low refractive index materials, ADBRs use an aperiodic sequence to enhance performance in specific applications. By eliminating periodicity, ADBRs

A. Antony et al. (Eds.): icONMAT 2025, SPPHY 328, pp. 190–201, 2025.
https://doi.org/10.1007/978-981-95-2697-0_24

can become more resilient and consistently function better by being less susceptible to manufacturing process flaws and layer thickness changes. Traditional DBR usually produces broader reflection spectra and needs more layers (20 to 40), which can lead to poor colour purity due to off-resonant reflections and colour mixing. However, A-DBR allows precise tailoring of each layer thickness and enables a narrow reflection band with good colour purity. Chalcogenide glasses are amorphous materials consisting of one or more elements from the chalcogen group elements (S, Se or Te) doped with other elements such as Ge, Sb, As, Ga, etc. These glasses are made using the conventional melt-quenching method. Chalcogenide glasses [3] are a better option as an alternative layer for A-DBR because they offer high refractive index contrast with SiO_2 and low loss compared to conventional metal-dielectric structures, and tunability and reconfigurability by utilising optical, thermal and electrical stimuli, essential for developing an ideal narrow photonic band gap with the least number of alternative layers.

2 Materials and Methods

2.1 Preparation of Bulk Glass and Thin Film

An appropriate amount of 5 N purity of Ga, Ge, and Se is weighed and loaded in pre-cleaned ampoules, which are then vacuum sealed. The sealed ampoules are kept in a rocking and rotating furnace for 24 h at 1050 °C to ensure homogenisation of the melt. Amorphous bulk glass of the alloy $Ga_5Ge_{40}Se_{55}$ formed by the rapid melt quenching process in ice-cooled water. Thin films of different thicknesses have been prepared from bulk glasses by the thermal evaporation technique, resistively heating a tungsten boat containing bulk glass at a pressure of 10^{-5} mbar by using an INDVAC 12A4-D thermal evaporation unit. The thin film was deposited over a pre-cleaned glass substrate.

2.2 XRD Analysis

The XRD (X-Ray Diffraction) technique was used to analyse the amorphous nature of the as-prepared $Ga_5Ge_{40}Se_{55}$ bulk glass and thin film, using a Bruker AXS D8 Advance X-ray powder diffractometer (Cu-kα, $\lambda = 1.5406$ Å).

2.3 EDS, UV–VIS-NIR Spectroscopy and Spectroscopic Ellipsometry (SE)

SEM–EDX (Scanning Electron Microscope–Energy Dispersive X-ray, Bruker-QUANTAX 200) analysis was used to measure the composition of the prepared bulk sample and a JASCO V-770 spectrophotometer for optical spectroscopy over the range 300 nm to 2500 nm. SE studies were done over the range 380 nm to 890 nm using a JA Woollam ellipsometer by analysing the reflected S and P polarised light from the thin film surface. Glass with transparent thin film with back side reflection was used as the model were the thin film is modelled as the general oscillating model with Tauc Lorentz model as the sub model. Substrates were modelled as the Cauchy.

2.4 Differential Scanning Calorimetry (DSC)

DSC measurement done at a heating rate of 5 ° C/min over a temperature range of 90 °C to 450 °C by using a DSC Q 100 TA Instruments Inc.

3 Results and Discussion

Absence of sharp peaks in the XRD spectra of the bulk sample confirms the amorphous nature of the prepared bulk shown in Fig. 1. The presence of specific broad humps, notably observed at approximately 15°, 30°, and 50°, indicates a broad, amorphous halo. This behaviour can be associated with the structure of the Ga-Ge-Se (GGS) glass composition, which is made up of cross-linked three-dimensional tetrahedral structural units of $GeSe_2$ and pyramidal units of Ga_2Se_3 with excess Selenium (Se) dispersed among these units [4].

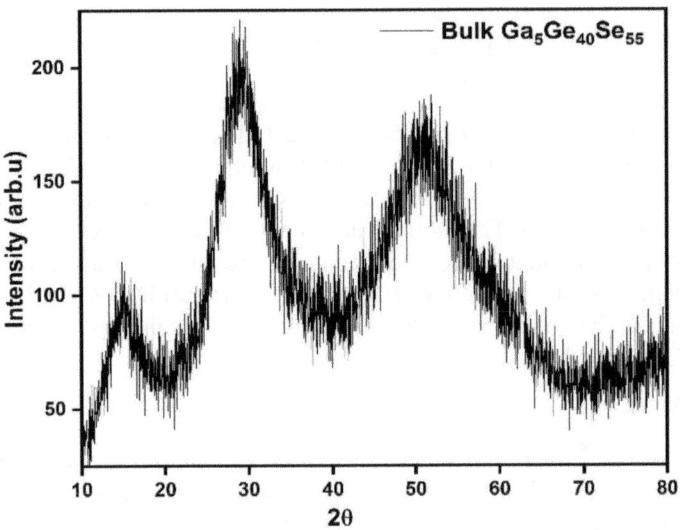

Fig. 1. XRD spectra of as prepared bulk sample of $Ga_5Ge_{40}Se_{55}$

The composition of the formed glass is confirmed by SEM-EDS plotted in Fig. 2 and tabulated in table 1.

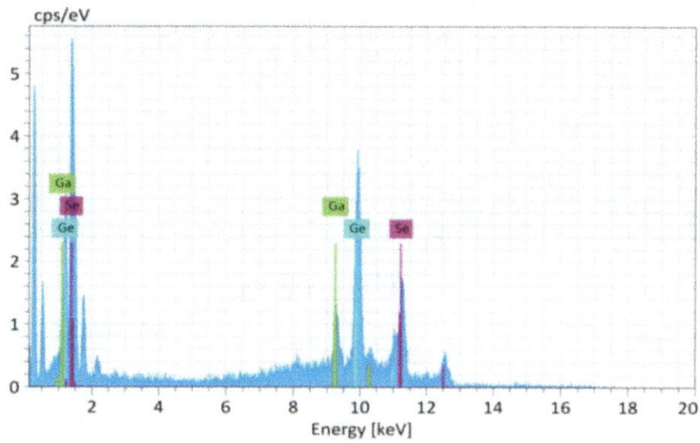

Fig. 2. EDS spectra of as prepared bulk sample of $Ga_5Ge_{40}Se_{55}$

Table 1. Results obtained from the SEM-EDS

Element	Atomic %
Gallium	5.64
Germanium	40.81
Selenium	53.56

The glass transition temperature (Tg) was determined from DSC measurement. The Observed Tg value for the $Ga_5Ge_{40}Se_{55}$ glass is 280 °C and plotted in Fig. 3.

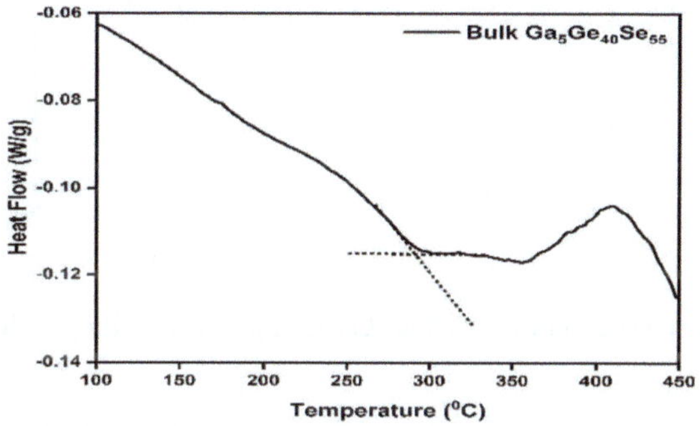

Fig. 3. DSC spectra of as prepared bulk sample of $Ga_5Ge_{40}Se_{55}$ at a heating rate of 5 °C/min

The thermal evaporation technique was used to fabricate the thin films from the bulk glass. A broad spectrum of XRD confirms the amorphous nature of the thin film grown on the pre-cleaned glass substrate, shown in the Fig. 4 given below.

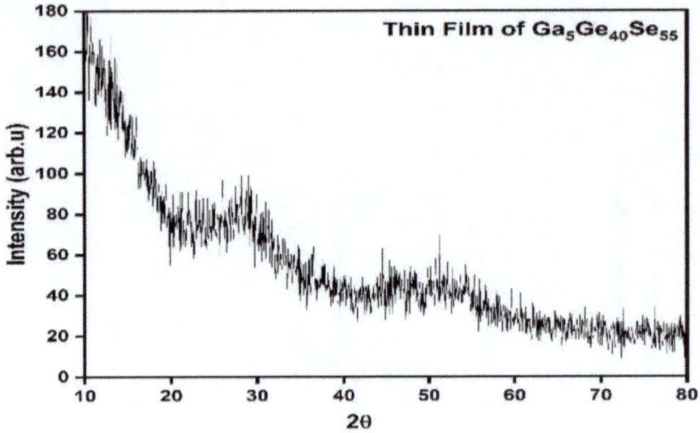

Fig. 4. XRD spectra of as prepared thin film of $Ga_5Ge_{40}Se_{55}$ coated on glass substrate

We have prepared three thin films of different thicknesses. The interference fringes in the UV–VIS-NIR transmission spectra of the thin film shown in the Fig. 5 arise due to the reflection of light waves between the substrate surface and film-air interface. It is obvious from the figure that the number of interference fringes increases as the thickness of the films increases and is more closely spaced.

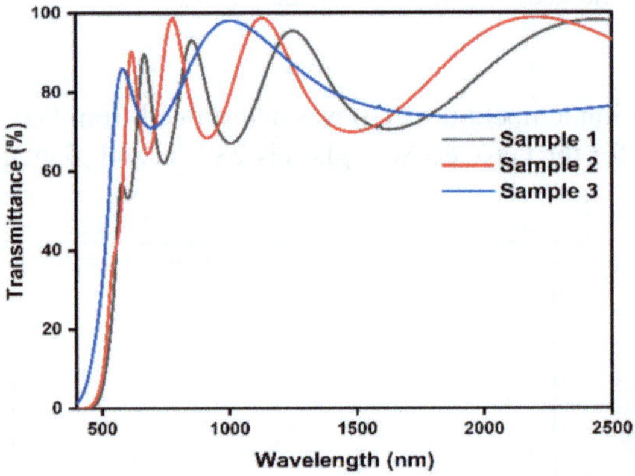

Fig. 5. Transmission spectra of amorphous $Ga_5Ge_{40}Se_{55}$ thin films of different thicknesses

The swanepoel method was employed to find out the optical constants of the thin film, such as refractive index (n), extinction coefficient (k), thickness (d) of the thin film and Tauc plot used to extract the band gap (Eg). The extracted parameters are plotted in Fig. 6 (a) and (b).

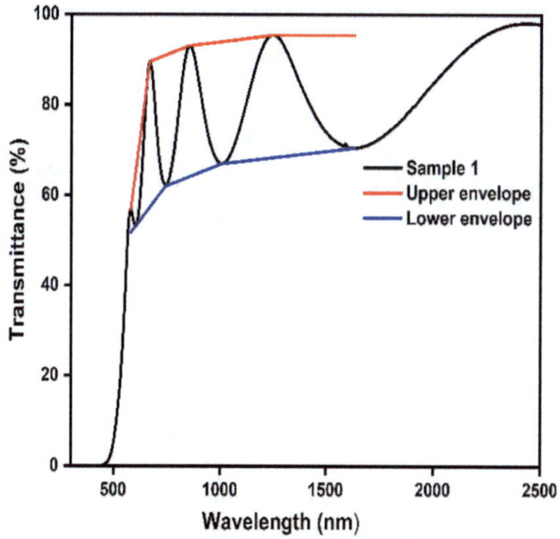

Fig. 6. (a) Swanepoel plot of Sample 1 thin film

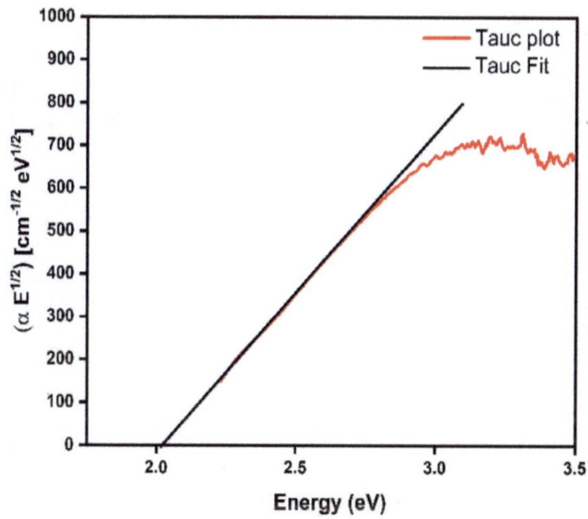

Fig. 6. (b) Tauc plot of sample 2 for determining bandgap

Spectroscopic ellipsometry studies were conducted to compare the optical constants obtained from the envelope method. The Tauc-Lorentz model was used to fit the experimental psi and Delta. The Fig. 7 shows that the theoretical fit agrees with the experimental data.

The extracted n and k are plotted in Fig. 8.

The refractive index plot shows a normal dispersion behaviour, i.e. the value of n decreases with increasing wavelength. Refractive index values decrease as the thickness of the thin film decreases from sample 1 to 3. The increase in refractive index 'n' with the

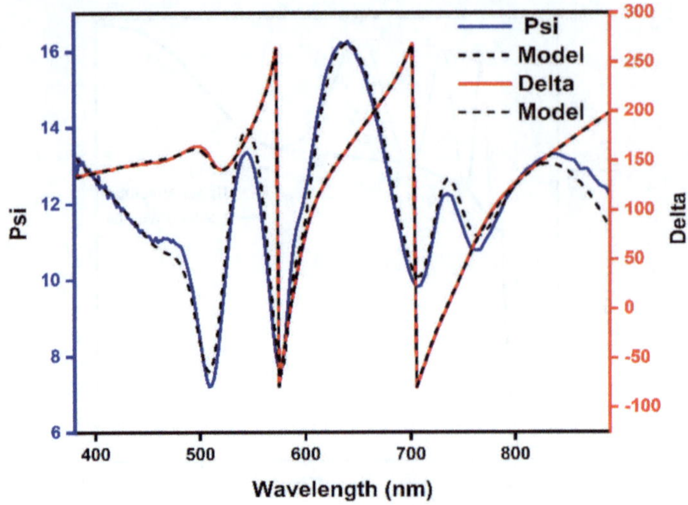

Fig. 7. Spectroscopic ellipsometry parameters of sample 2, Psi and Delta fitted with Tauc Lorentz model

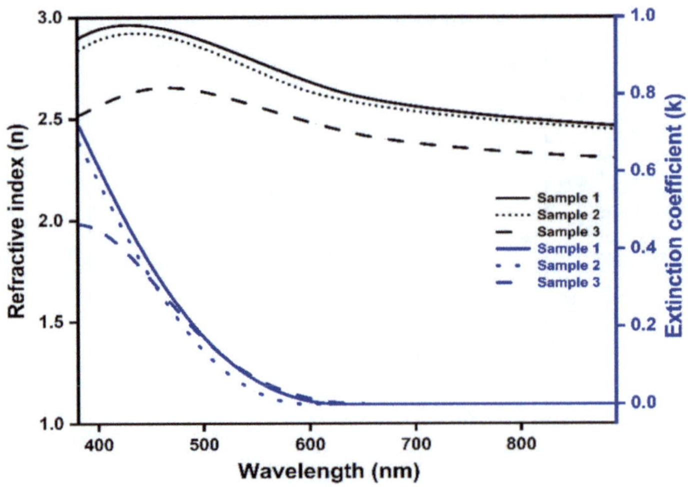

Fig. 8. Extracted n and k of all sample from ellipsometry by using Tauc Lorentz model

thickness of the thin film could be attributed to the enhancement of the packing density of films [5]. The extinction coefficient of all three samples seems to be ultra-lossless beyond 600 nm and suffers significant losses in the UV region to 600 nm. We compared n and k extracted from the swanepoel and the ellipsometry method and plotted in Fig. 9. We could see a slight alteration of the refractive index spectrum near the UV region and up to 550 nm. This is because the thin film absorption lies in this region, and the swanepoel model is not capable of accounting for the absorption, but the Tauc Lorentz model can. To validate the reliability of extracted optical constants from ellipsometry, we have done theoretical modelling by employing the transfer matrix method [6], whereas the extracted parameters from the spectroscopic ellipsometry are given as input to the theoretical modelling and the corresponding plot is shown in Fig. 10.

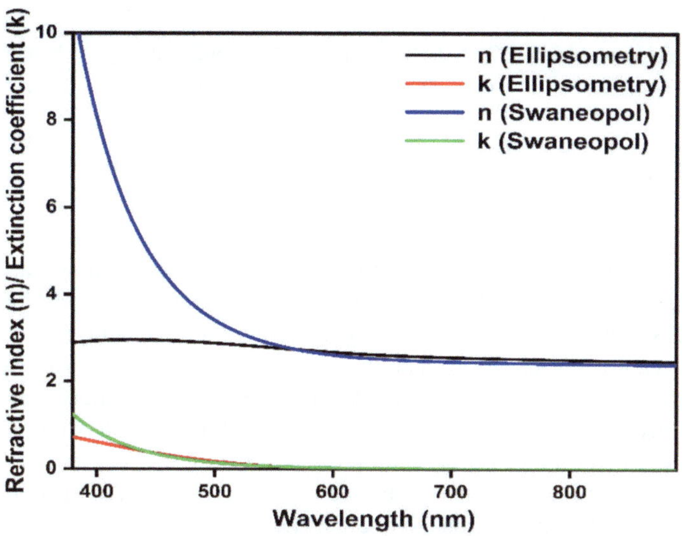

Fig. 9. Comparison plot of n and k of sample 2 obtained from swanepoel and ellipsometry method

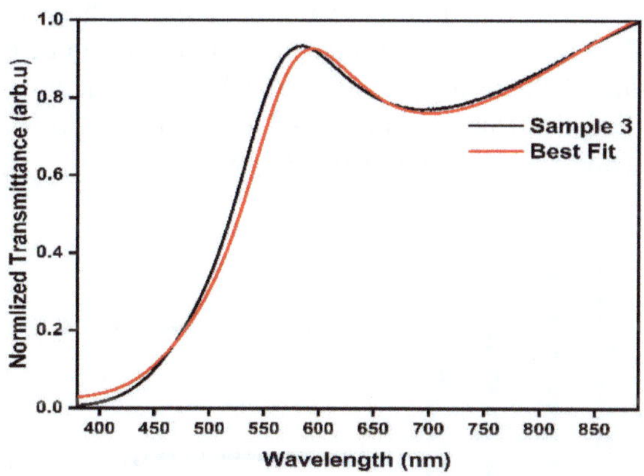

Fig. 10. Experimental transmission of sample 3 is theoretically fitted with transfer matrix method

A comparison of the optical parameters obtained from the different methods is tabulated below in Table 2.

The table summarizes that the results obtained from the different methods are comparable. There is a slight increase in the band gap as the thickness decreases, which can be attributed to the enhancement in morphological changes in films, like a change in

Table 2. Comparing the optical constants obtained from different extraction methods

Parameters	Tauc Lorentz			Swaneopol			Transfer matrix method		
	1	2	3	1	2	3	1	2	3
Mean Squared Error (MSE)	7	9	10	Not applicable			Not applicable		
Thickness (d) (nm)	510	470	228	531	482	NA	520	480	225
Roughness (R) (nm)	5.94	4.70	14	Not applicable			Not applicable		
Band gap (E_g) (eV)	1.97	2.07	2.18	1.98	2.02	2.04	Not applicable		

atomic distance. The preparation of thin films at room temperature may lead to structural defects, which thereby form allowed states in the forbidden region near the conduction band, and these allowed states to merge with the conduction band, rendering a reduction in band gap in thicker films. The high concentrations of localised states formed due to unsaturated bonds also play a role in decreasing the optical band gap of the deposited films [5]. We have designed and simulated an A-DBR structure for colour filter applications by using Comsol Multiphysics software [8]. The extracted n and k of the $Ga_5Ge_{40}Se_{55}/SiO_2$ from ellipsometry were given as the input of the numerical simulation as the interpolation function to simulate an A-DBR structure containing a stack of 8 layers. Scattering boundary conditions and a perfectly matched layer used to account for scattering, truncate the computational domain and prevent unwanted reflections at the simulation domain's boundaries by absorbing outgoing waves. Periodic boundary conditions, along with the floquot periodicity, were used to address the periodicity of the structure. Port boundary conditions were used to excite the structure and calculate the S-parameter. The reflectance/transmittance/absorptance can be calculated from the S parameter. The Fig. 11 shows that the simulated A-DBR structure consists of an alternative layer of $Ga_5Ge_{40}Se_{55}/SiO_2$ having a thickness of 170,185,195,165/50, 75, 60, 75, respectively and having a total thickness of 975 nm.

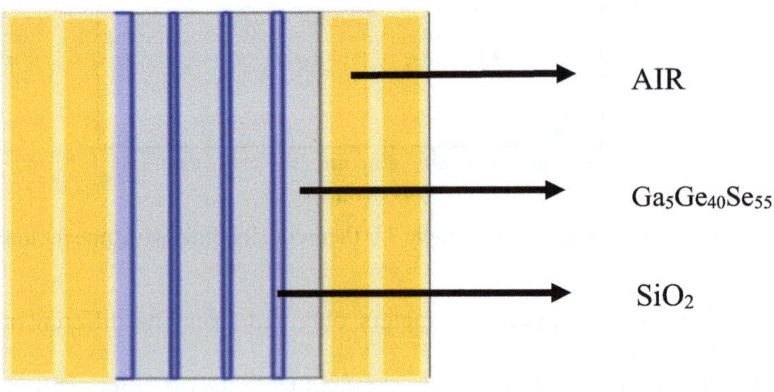

Fig. 11. Simulated A-DBR structure by comsol multiphysics software

The simulated structure shows a peak reflectance of 80% at 590 nm in the Fig. 12. A comparison plot of reflectance, transmittance and absorptance is also plotted.

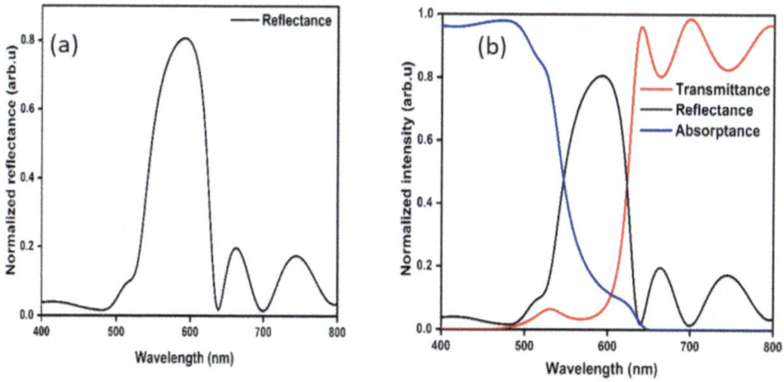

Fig. 12. (a) Reflectance spectra of simulated A-DBR structure (b) Comparison of reflectance, Transmittance and absorptance of simulated A-DBR structure

We find out that the colour purity of the simulated reflectance shows a purity of 86% and can be used a colour filter for photonic applications. The CIE chromaticity diagram for the calculated reflection spectra is plotted in Fig. 13 given below. The CIE plot validates the high-purity yellowish colour for the simulated reflectance spectra.

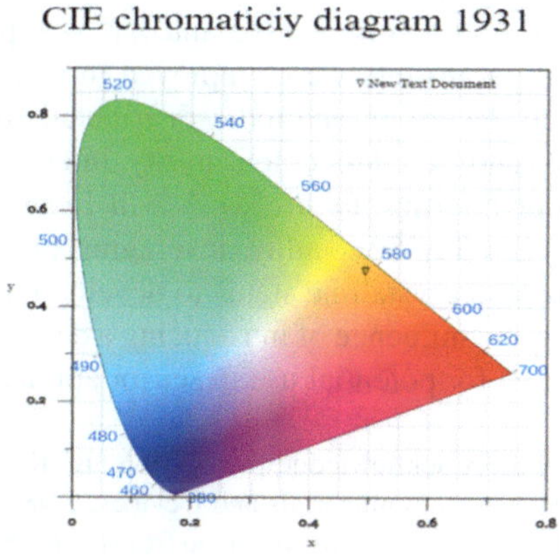

Fig. 13. CIE chromaticity diagram of simulated reflectance spectra

We have also studied the polarisation sensitivity of the simulated A-DBR and found it to be polarisation insensitive for both Transverse electric and magnetic waves, and plotted in Fig. 14 (a) and (b) below.

The incident angle-dependent studies on reflection were done on the angles from 0° to 85° with a step size of 5°. Only for 50° and 60°, we got almost the same reflectance

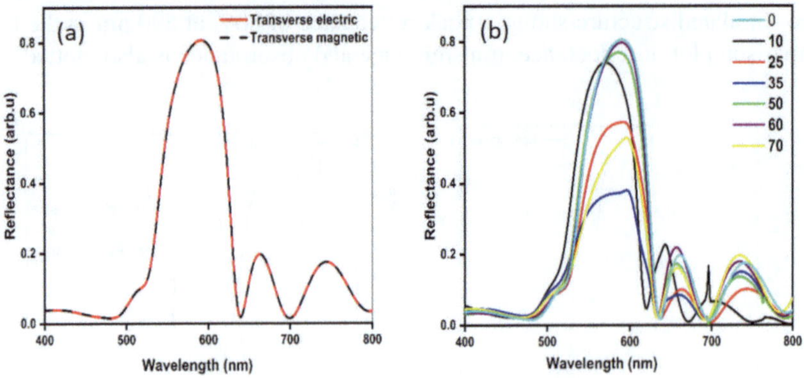

Fig. 14. (a) Simulated reflectance spectra for TE and TM polarisation (b) Simulated incident angle dependent reflectance

as compared to the normal incidence. A slightly blue-shifted but comparable reflectance spectrum was obtained for 10°. A strategic optimisation of thickness may lead to angle-independent reflection spectrum.

4 Conclusion

Bulk and thin films of amorphous $Ga_5Ge_{40}Se_{55}$ chalcogenide alloy were synthesised and thoroughly characterised. The optical constants of the thin films were extracted using multiple techniques, including the Swaneopol method, the transfer matrix method, and spectroscopic ellipsometry, and a comparative analysis was performed. These experimentally obtained optical parameters were employed in numerical simulations using COMSOL Multiphysics to design an aperiodic distributed Bragg reflector (A-DBR) exhibiting high reflectance and excellent colour purity, intended for advanced photonic applications such as colour filtering. Future work will focus on exploring the phase-change behaviour of the $Ga_5Ge_{40}Se_{55}$ alloy, leveraging the contrast in optical constants between amorphous and crystalline states to develop reconfigurable and tunable colour filters. Additionally, the influence of introducing a defect layer within the A-DBR structure will be investigated for potential use in sensor applications.

Acknowledgements. The authors acknowledge the SAIF-STIC Kochi, Department of Physics and the Inter University Centre for Nanomaterials and Devices, CUSAT, for various characterisation facilities. The author acknowledges funding received from the RUSA-CUSATECH foundation and Cusat for the fellowship to carry out the study.

References

1. Gildas, F., Dan, Y.: J Nanophotonics. **13**, 1 (2019)
2. Jana, S., et al.: Nano Lett. **24**, 3922 (2024)
3. Thomas, S., Philip, J.: Phys. Status Solidi B. **200**, 359 (1997)

4. Soumya, S., Thomas, S.: AIP Conf Proc. **2244**, 100007 (2020)
5. pn, M., Nampoori, V.P.N., Thomas, S.: Mater Res Express. **6** (2019)
6. Yeh, P., Hendry, M.: Opt. waves layer. media. **43**, 77–78 (1990) https://www.comsol.com

Investigation of the Structural and Magnetic Properties of Bi_2Se_3 Nanosheets

K. V. Shilna and K. J. Thomas(✉)

Department of Physics, Central University of Kerala, Kasaragod, Kerala 671316, India
kj_thomas@cukerala.ac.in

Abstract. Bismuth selenide (Bi_2Se_3) is a widely recognized three-dimensional topological insulator with promising applications in the field of thermoelectric, optoelectronics, and spintronics. Here, we have synthesized Bi_2Se_3 nanosheets by solvothermal technique and studied their structural, morphological, and magnetic properties with varying solvothermal reaction times. The X-ray diffraction (XRD) studies showed a phase-pure trigonal Bi_2Se_3 with (015) plane-oriented growth. Fourier-transform infrared spectroscopy (FTIR) analysis confirmed the characteristic Bi–Se bonds and some residual organic species from precursors. The nanosheet-like morphology was demonstrated through scanning electron microscopy (SEM) analysis. Magnetic measurements exhibit dominant diamagnetic behavior, mixed with a subtle ferromagnetic response at 60 K and 300 K. The observed weak ferromagnetism may be attributed to intrinsic point defects or unintentional doping during synthesis.

Keywords: Bismuth selenide · Solvothermal · Weak ferromagnetism

1 Introduction

Bismuth selenide (Bi_2Se_3), a well-known three-dimensional topological insulator, has been widely studied due to its unique surface state properties and strong spin-orbit coupling, making it a promising candidate for spintronic devices [1, 2]. It has potential applications in thermoelectrics, quantum computing, and photodetectors [3, 4]. Conventional approaches, such as chemical vapor transport [5], Bridgman [6] and hydrothermal/solvothermal reactions [3, 4] are employed to synthesize Bi_2Se_3. Among these, some of the methods require high temperatures or expensive equipment; on the other hand, here we employed a solvothermal approach, which is a low-temperature solution-based method, using ethylene glycol as both solvent and reducing agent. Ethylene glycol solution enhances nucleation and growth of nanostructures because of its high boiling point and binding capabilities.

Generally, Bi_2Se_3 is a diamagnetic material; recent studies have revealed that the emergence of weak ferromagnetism may be induced by defects, doping, or proximity coupling [7–9]. Here, we investigate solvothermal synthesis of Bi_2Se_3 nanosheets and their structural and magnetic properties. Magnetic measurements at room temperature show the coexistence of strong diamagnetism mixed with a weak ferromagnetism.

A. Antony et al. (Eds.): icONMAT 2025, SPPHY 328, pp. 202–208, 2025.
https://doi.org/10.1007/978-981-95-2697-0_25

2 Experimental Techniques

For the solvothermal synthesis of Bi_2Se_3 nanoparticles, 1 mmol bismuth nitrate pentahydrate ($Bi(NO_3)_3.5H_2O$) and 1.5 mmol selenium (Se) powder were dissolved in 80 ml of ethylene glycol (EG) solution. The prepared mixture was stirred for 45 mins and transferred into a Teflon-lined autoclave and heated to 180 °C and held for a reaction time of 24 h. After the solvothermal synthesis procedure, the autoclave was cooled to room temperature, and the black precipitate was collected. The precipitate obtained was filtered out, washed several times with distilled water, and dried at 200 °C for 1 h. The final product is crushed into a fine powder. The solvothermal experiment was repeated for different reaction times (6, 12, 18, 48 h) at a constant reaction temperature of 180 °C. A schematic representation of the solvothermal reaction of Bi_2Se_3 nanosheets is shown in Fig. 1.

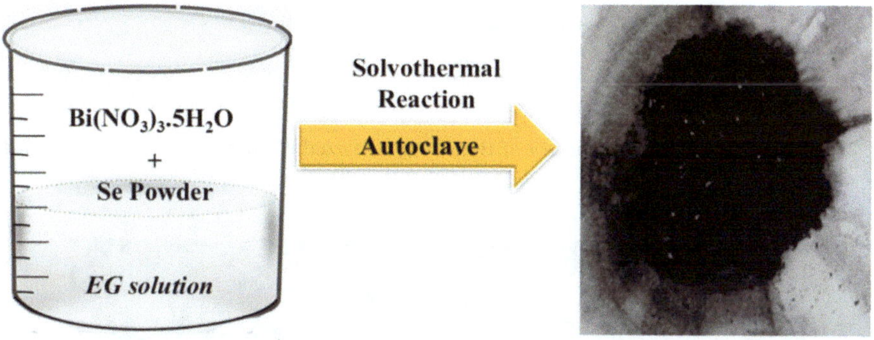

Fig. 1. Schematic representation of the solvothermal reaction of Bi_2Se_3 nanosheets and the actual black precipitate of Bi_2Se_3.

Structural studies were performed on synthesized powder samples using an X-ray diffractometer (XRD) (Model: Rigaku, Miniflex, Japan) with 0.154 nm CuKα source radiation and a Fourier Transform Infrared Spectrometer (FTIR) (Model: Perkin Elmer Spectrum 2 infrared spectrometer). Morphological studies were performed by a Field Emission Scanning Electron Microscope (FE-SEM) (Model: Zeiss, Ultra 55). Magnetic measurements were carried out using a vibrating sample magnetometer (VSM) module of a physical property measurement system (Model: Versa Lab, Quantum Design, USA).

3 Results and Discussion

The XRD patterns of the Bi_2Se_3 nanosheets synthesized at different solvothermal reaction times at a constant temperature of 180 °C are presented in Fig. 2(a). Samples show no impurity peaks, which indicates phase purity. Also, the well-resolved, sharp, and intense peaks for all the solvothermal reaction times confirm the polycrystalline nature of the samples. The measured XRD spectra are compared with the standard JCPDS file No. 01–078-3528, and all the observed peaks are indexed to the trigonal phase of Bi_2Se_3 with the R-3 m space group.

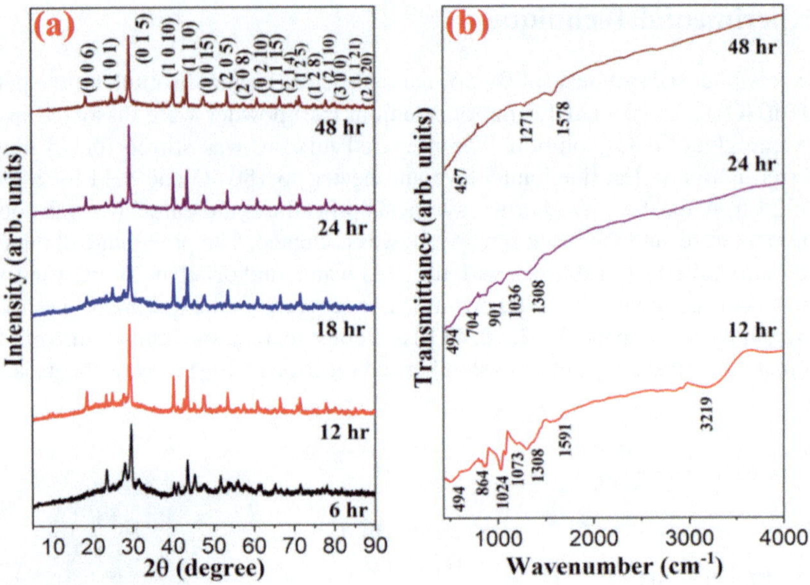

Fig. 2. (a) XRD and (b) FTIR spectra of Bi_2Se_3 synthesized at different solvothermal reaction time durations at 180 °C.

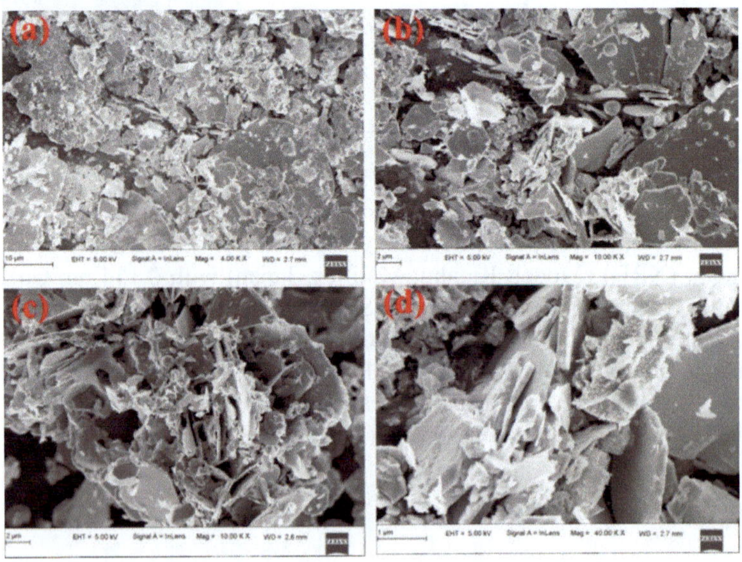

Fig. 3. FE-SEM images of Bi_2Se_3 nanosheets synthesized at 180 °C for a reaction time of 24 h. with different magnifications.

Average crystallite size (D) of the samples is calculated using the Debye-Scherrer formula (Eq. (1)):

$$D = \frac{0.9\lambda}{\beta cos\theta} \quad (1)$$

where λ is the wavelength of the X-ray used (1.5406 Å), β is the full width at half maximum (FWHM), and θ is the Bragg's angle.

The lattice constants *a* and *c* are calculated using the following Eq. (2):

$$\frac{1}{d^2} = \frac{4}{3}\left(\frac{h^2 + hk + k^2}{a^2}\right) + \frac{l^2}{c^2} \quad (2)$$

The calculated crystallite size and lattice parameters are shown in Table 1. The estimated crystallite size was 16–28 nm. Lattice parameters are consistent with previously reported data [10].

Table 1. The calculated crystallite size and lattice parameters of Bi_2Se_3 nanosheets. (Standard lattice constants from the JCPDS file are $a = 4.1355$ Å, $c = 28.6150$ Å)

Reaction Time	**Lattice Parameters** ***a*** **(Å)** ***c*** **(Å)**	**D (nm)**
6 h	4.139 28.707	16
12 h	4.139 28.799	27
18 h	4.139 28.799	20
24 h	4.139 28.815	28
48 h	4.139 28.615	24

Figure 2(b) shows the FTIR spectra of Bi_2Se_3 samples with peaks located around 457, 494, 704, 864, 901, 1024, 1073, 1308, 1591, and 3219 cm^{-1}. The bands near 440 and 500 cm^{-1} are characteristic of Bi_2Se_3 vibrations [11]. The characteristic bond stretching vibrations of Bi_2Se_3 occur in the 600–900 cm^{-1} range [11–13]. Hence, observation of Bi–Se vibrational modes confirms the formation of Bi_2Se_3 and is consistent with previously reported data [11–13]. The peaks around 1024, 1073, 1308, 1591, and 3219 cm^{-1} correspond to C-O stretching, O–H bending, and O–H stretching, which may be attributed to residual organic species originating from the precursors [14, 15].

Figure 3 (a)-(d) shows SEM images of Bi_2Se_3 synthesized at 180 °C for a reaction time of 24 h. with different magnifications. SEM images reveal a nanosheet-like morphology with a thickness in the range of 20–30 nm.

The magnetic measurements of the Bi_2Se_3 samples synthesized at 180 °C for a reaction time of 12, 24, and 48 h. are presented in Fig. 4 (a) - (c). The M–H curves demonstrate that undoped Bi_2Se_3 exhibits predominant diamagnetic behavior with a weak ferromagnetic signal at 60 K and 300 K. The observed diamagnetism is mainly ascribed to the closed-shell electronic configurations of Bi^{3+} and Se^{2-} ions. Interestingly, a weak ferromagnetic behavior is observed, which may suggest the presence of defects [16, 17], possibly due to vacancies or antisite defects, or may be due to traces of magnetic impurities [7, 18]. Antisite defects, such as Bi atoms occupying Se sites or vice versa, Se vacancies, and traces of magnetic impurities introduced during synthesis, may act as potential sources of localized magnetic moments [7, 18–21].

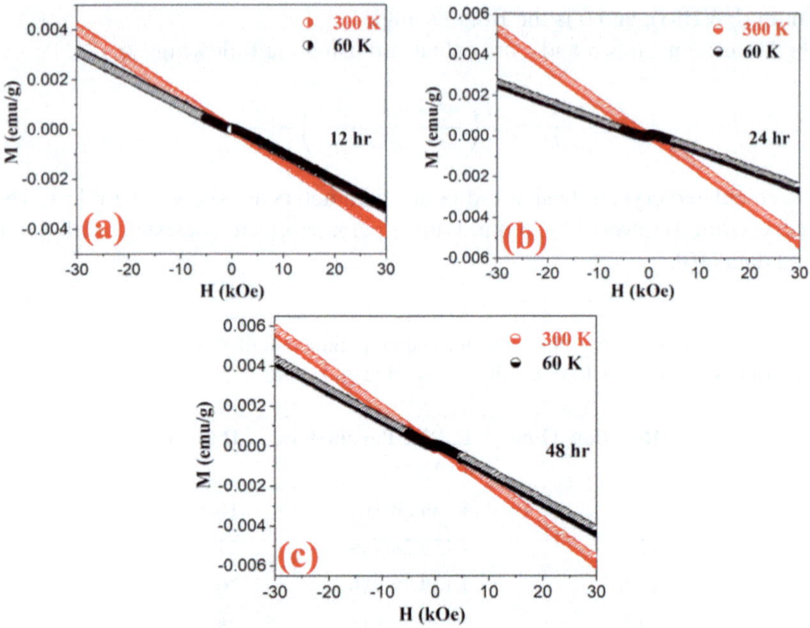

Fig. 4. M-H curves of Bi_2Se_3 synthesized at different solvothermal reaction times of (a) 12, (b) 24, and (c) 48 h. at 180 °C.

4 Conclusions

We have successfully synthesized Bi_2Se_3 nanosheets by a simple solvothermal synthesis technique and characterized their structural, morphological, and magnetic properties. XRD and FTIR spectroscopy confirm the structural purity and formation of Bi_2Se_3. XRD analysis confirmed the formation of pure trigonal Bi_2Se_3. FTIR spectra validated the presence of Bi–Se bonding, while SEM revealed sheet-like morphology with nanoscale dimensions. Magnetic measurements showed predominantly the inherent diamagnetic behavior, but also revealed a subtle ferromagnetic nature. The presence of such weak ferromagnetism is attributed to intrinsic defects, such as Se vacancies or antisite disorder. This study points at the future scope of defect engineering in tailoring the magnetic properties of Bi_2Se_3 for advanced spintronic and topological applications.

Acknowledgements. The authors thank INUP, IIT Bombay. A part of the reported work (SEM characterization) was carried out at the IITBNF, IITB under INUP which is sponsored by Deity, MCIT, Government of India.

References

1. Zhang, H., Liu, C.X., Qi, X.L., Dai, X., Fang, Z., Zhang, S.C.: Topological insulators in Bi_2Se_3, Bi_2Te_3 and Sb_2Te_3 with a single Dirac cone on the surface. Nat. Phys. **5**, 438–442 (2009)
2. Jamali, M., et al.: Giant spin pumping and inverse spin hall effect in the presence of surface and bulk spin-orbit coupling of topological insulator Bi_2Se_3. Nano Lett. **15**, 7126–7132 (2015)
3. Cui, H., et al.: Synthesis of Bi_2Se_3 thermoelectric nanosheets and nanotubes through hydrothermal co-reduction method. J. Solid State Chem. **177**, 4001–4006 (2004)
4. Lin, G., Veeramalai, C.P., Wang, J., Zhou, H., Wang, Y., Li, C.: Bi_2Se_3 nanosheets prepared by solvothermal method for high performance NIR photodetector. Colloids Surf. Physicochem, Eng, Asp. **694**, 134090 (2024)
5. Sun, Z., Liufu, S., Liu, R., Chen, X., Chen, L.: A general strategy to bismuth chalcogenide films by chemical vapor transport. J. Mater. Chem. **21**, 2351–2355 (2011)
6. Wu, K.K., Ramachandran, B., Kuo, Y.K., Sankar, R., Chou, F.C.: Influence of induced defects on transport properties of the bridgman-grown Bi_2Se_3-based single crystals. J. Alloys Compd. **682**, 225–231 (2016)
7. Das, S.K., Sahoo, U.P., Das, B., Ghorai, G., Samal, P., Sahoo, P.K.: Defect induced unconventional ferromagnetism in au intercalated Bi_2Se_3 nanocrystals for spintronic applications. Adv. Quantum Technol. **6**, 1–12 (2023)
8. Tarasenko, R., et al.: Magnetic and structural properties of Mn-doped Bi_2Se_3 topological insulators. Phys. B Condens. Matter. **481**, 262–267 (2016)
9. Wang, H., et al.: Proximity-induced magnetic order in topological insulator on ferromagnetic semiconductor. Sci. China Inf. Sci. **66**(222403), 1–8 (2023)
10. Zhou, F., Zhao, Y., Zhou, W., Tang, D.: Temperature-dependent Raman scattering of large size hexagonal Bi_2Se_3 single-crystal nanoplates. Appl. Sci. **8**(1794), 1–10 (2018)
11. Abd El-Fattah, Z.M., Ashoush, M.A.: Structural characterization of pure and magnetic-doped Bi_2Se_3 nanoparticles. J. Mater. Sci. Mater. Electron. **29**, 2593–2599 (2018)
12. Batool, Z., et al.: Simple synthesis of multifunctional bismuth selenide nanoparticles; structural, optical characterizations and their effective antibacterial activity. Appl. Phys. A Mater. Sci. Process. **127**, 1–9 (2021)
13. Vijila, J.J.J., Mohanraj, K., Henry, J., Sivakumar, G.: Microwave-assisted Bi_2Se_3nanoparticles using various organic solvents. Spectrochim, Acta - Mol, Biomol, Spectrosc. **153**, 457–464 (2016)
14. Xie, X., Shang, P., Liu, Z., Lv, Y., Li, Y., Shen, W.: Synthesis of nanorod-shaped cobalt hydroxycarbonate and oxide with the mediation of ethylene glycol. J. Phys. Chem. C. **114**, 2116–2123 (2010)
15. Al-Bermany, E., Chen, B.: Preparation and characterisation of poly(ethylene glycol)-adsorbed graphene oxide nanosheets. Polym. Int. **70**, 341–351 (2021)
16. Shilna, K.V., Sahoo, S.C., Thomas, K.J.: Novel ferromagnetism and negative magnetoresistance in $BaBiO_3$ nanoparticles. Appl. Mater. Today. **27**(101427), 1–8 (2022)
17. Shilna, K.V., Thomas, K.J.: Tuning the magnetic behavior of $BaBiO_3$ nanoparticles by Pb doping. J. Supercond. Nov. Magn. **9**, 1–13 (2025)
18. Nahas, S., Sanyal, B., Black-Schaffer, A.M.: Spontaneous ferromagnetism and finite surface energy gap in the topological insulator Bi_2Se_3 from surface BiSe antisite defects. Phys. Rev. B. **102**(140407), 1–6 (2020)
19. Kander, N.S., Islam, S., Guchhait, S., Das, A.K.: The effect of Fe-doping on structural, elemental, magnetic, and weak anti-localization properties of Bi_2Se_3 topological insulator. Appl. Phys. A Mater. Sci. Process. **129**, 1–13 (2023)

20. Wei, X.Y., Zhang, J.Y., Zhao, B., Zhu, Y., Yang, Z.Q.: Ferromagnetism in Fe-doped Bi_2Se_3 topological insulators with Se vacancies. Phys, Lett, Gen, At, Solid State Phys. **379**, 417–420 (2015)
21. Chen, T., et al.: High-mobility Sm-Doped Bi_2Se_3 ferromagnetic topological insulators and robust exchange coupling. Adv. Mater. **27**, 4823–4829 (2015)

In-situ Assembly of Bi_2S_3 on TiO_2 as Photoanode in Sensitized Solar Cells

Arya Mohan, Arya Krishna, and K. G. Deepa(✉)

Laboratory for Energy and Environmental Devices, Department of Physics, University of Kerala, Thiruvananthapuram, Kerala, India 695581
deepachaithanya@gmail.com

Abstract. In this work, an easy and cost-effective deposition technique of successive ionic layer adsorption and reaction (SILAR) is used for in-situ sensitization of bismuth sulphide (Bi_2S_3). The Bi_2S_3 sensitized solar cells are grown on anatase phase titanium dioxide (TiO_2). The number of SILAR cycles is varied from 5 to 25 with an interval of 5. The SILAR grown Bi_2S_3 are characterized using XRD, Raman spectroscopy, SEM, UV-Vis absorption spectroscopy and cyclic voltammetry. The band gap of the sample decreased from 1.8 to 1.44 eV with increase in the deposition cycle indicating an increase in the particle size. The HOMO-LUMO levels estimated through cyclic voltammetry predicts the likelihood of carrier transfer from Bi_2S_3 to TiO_2 photoanode layer. Out of all the devices fabricated using various SILAR cycles, the best results are obtained for the film employing five SILAR cycles. By incorporating an additional TiO_2 buffer layer, which serves to mitigate electron leakage, further improvement in the solar cell parameters is achieved. The device demonstrated an open circuit voltage of 400 mV, a short circuit current of 0.28 mA/cm^2, and a power conversion efficiency of 0.045%.

Keywords: SILAR · solar cells · cyclic voltammetry · bandgap · sensitization

1 Introduction

Nanoparticles and quantum dots (QD) are used as photosensitizers in place of organic dyes because they have versatile optical and electrical properties including a tunable band gap which is dependent on the size of the particle, a larger extinction coefficient, higher stability toward water and oxygen, and multiple exciton generation (MEG) with single-photon absorption [1]. It is anticipated that QD-based solar cells will have a maximum theoretical efficiency of 44% [2].

The global demand for clean energy could be greatly met by quantum dot sensitized solar cells (QDSSCs), a next-generation solar cell with simple and inexpensive fabrication methods. QDSSCs, typically have an electrolyte positioned between the counter and working electrodes. In a QDSSC, the counter electrode is typically made of platinum, conducting polymers, metal sulphides, or carbon derivatives, while the photoanode is typically made of TCO glass, metal oxide particles, or thin film, followed by a layer of

A. Antony et al. (Eds.): icONMAT 2025, SPPHY 328, pp. 209–218, 2025.
https://doi.org/10.1007/978-981-95-2697-0_26

quantum dots. The device typically operates by sandwiching an electrolyte, between the working and counter electrodes. TiO_2 particles make an excellent choice for the metal oxide in the device's working electrode in an effective QDSSC. The crystal phase, surface area, pore size, and desired morphology, are the primary characteristics that make it one of the best materials to utilize as a metal oxide in QDSSC [3].

QDSSCs utilizing chalcogenide quantum dots like PbS, CdSe, and CdTe have seen significant advancements over the past decade. However, these QDSSCs contain highly toxic elements, namely Pb and Cd, which are subject to strict regulations for electronic consumer products in many countries. This is a major drawback for the adoption of the QDSSC at the consumer level, accordingly less toxic elements in the fabrication of solar cells are desirable.

Bi_2S_3 is a non-toxic metal chalcogenide which has the potential to replace toxic materials in QDSSCs. Bi_2S_3 owns a layered crystal structure in an orthorhombic unit cell in the Pbnm space group [4]. A large absorption coefficient of 10^4 cm^{-1} in the visible region, a band gap of 1.3 eV which is close to the optimal band gap for single-junction solar cells according to the Shockley Queisser value, and a conduction band (CB) positioned at more negative than those of TiO_2 that favour the injection of photogenerated electrons from the chalcogenide into the oxide are some of the essential characteristics of Bi_2S_3 that make it a good sensitizer [5]. Furthermore, Bi_2S_3 can be produced using simple, inexpensive, and energy-efficient techniques, which is advantageous for its scalable application. Bismuth chalcogenide-based PV cells, including those based on bismuth iodide and bismuth sulphide, achieved PCEs of up to 4.37% and 7.47%, respectively [6, 7].

Herein, Bi_2S_3 is used as the photosensitizer. The Bi_2S_3 quantum dots are sensitized on TiO_2 photoanode via SILAR technique. Structural, morphological, optical, and cyclic voltammetric analysis are conducted on the synthesized photoanodes. The Bi_2S_3 sensitized solar cells are fabricated with varying SILAR cycles and the IV characteristics are studied.

2 Materials and Methods

All chemicals used in the experiments are of analytical grade purchased from Sigma-Aldrich and are used without further purification.

2.1 Synthesis of TiO_2 Thin Films

Fluorine-doped Tin Oxide (FTO) glass substrates kept in chromic acid solution are ultrasonically cleaned in soap solution, distilled water, acetone, and isopropyl alcohol, respectively for 15 min each. Finally, the FTOs are treated in a UV-ozone cleaner for 10 min. The well-cleaned FTO substrate is coated with TiO_2 paste. To obtain a homogeneous TiO_2 layer, 2 g of TiO_2 (20 nm particles) purchased from Degussa is mixed well with 12 mL of ethanol through a vigorous stirring of the mixture in a magnetic stirrer for 1 h. The as-prepared TiO_2 paste is coated on the FTO via doctor blade method. Then the photoelectrode is kept in a muffle furnace for air annealing at 4000 °C for 1 h.

For the preparation of double layer TiO_2, second layer of TiO_2 (sigma) paste is doctor-bladed on top of the coating and the annealing process is repeated. Annealing helps to obtain good electrical contact between the nanoparticles.

2.2 Sensitization of Bi_2S_3 Thin Films and Fabrication of Solar Cells

The deposition of the Bi_2S_3 quantum dot on the photoelectrode is done through the SILAR (Successive ionic layer and adsorption reaction) method. Here, the photoelectrode is first dipped in 0.025 M of bismuth nitrate $Bi_2(NO_3).5H_2O$ solution. To dissolve the $Bi_2(NO_3).5H_2O$ completely in the solution, 5 mL of acetic acid is also added while preparing the solution. 0.025 M of sodium sulphide (Na_2S) solution is used as the anionic precursor. The photoelectrode is first dipped in the cationic solution for 60 s in which, the Bi^{3+} ions are deposited on the substrate surface. The excess amount of Bi^{3+} ions is removed from the surface when the substrate is dipped in distilled water for 15 s. The substrate is then dipped in the anionic solution for 60 s for the deposition of S^{2-} ions. The unabsorbed ions are removed from the surface on dipping again in distilled water for 15 s. Hence, one cycle of SILAR is completed for the deposition of Bi_2S_3 film. Five different samples were obtained by varying the number of SILAR cycles to 5,10,15,20, and 25 (denoted as S5, S10, S15, S20, and S25 respectively). Electrolyte is prepared by using 2 M sodium sulphide (Na_2S), 2 M sulphur (S) and 0.2 M potassium chloride (KCl) in 20 mL distilled water. A yellow-colored uniform solution rich in sulphur ions, is obtained after vigorous stirring in a magnetic stirrer for 1 h. Platinum coated glass is used as the counter electrode.

2.3 Characterization Techniques

Bruker D8 Advance X-ray diffractometer (40 kV, 40 mA) with Cu Kα radiation (λ- 1.5406 Å) in 2theta range of 10–90^0 is used to characterize the crystalline properties of the films. LabRAM HR Evolution Raman spectrometer (Horiba) is used to record the Raman spectra of the films in the spectral range 50–1000 cm^{-1}. The morphological features are analysed using Jeol 6390 LV scanning electron microscopy (SEM). JASCO V-750 UV–Vis Spectrophotometer is used to measure the optical properties of Bi_2S_3/TiO_2 films in a wavelength range of 300–1600 nm. The CV measurements are carried out in a Metrohm auto lab system using 1 M KOH solution as the electrolyte. Bi_2S_3 adsorbed TiO_2 on FTO serves as the working electrode whereas the FTO and Ag/AgCl serve as the counter and reference electrode, respectively. The scan rate is set to 20 mV/s in a potential range -1 V to 1 V. Photo Emission Tech solar simulator with an AM 1.5 illumination is used as the light source for recording solar cell IV characteristics.

3 Results and Discussions

The X-ray diffraction pattern of the Bi_2S_3 sensitized TiO_2 thin film is shown in Fig. 1. Here the TiO_2 peak positions can be indexed to the pure tetragonal anatase structure with JCPDS No. 00–002-0387. The peak positions at $2\theta = 25.35$, 37.93, 48.10, and 55.29° correspond to the (101), (004), (200), and (211) planes, respectively. Similarly, the peak positions at 2θ = 28.97, 45.79, 52.88, 54.23, 62.73, and 71.40° corresponding to (320), (440), (312), (061), (171) and (660) planes match exactly with the orthorhombic structure of Bi_2S_3 with the JCPDS No.00–002-0391 having Pbnm space group. The X-ray diffraction pattern confirms the successful deposition of Bi_2S_3 onto the TiO_2 thin films. Also, the absence of any additional binary phases makes the SILAR deposition process even more convenient. Hence, the XRD analysis confirms the formation of pure anatase TiO_2 with orthorhombic Bi_2S_3 Fig. 2.

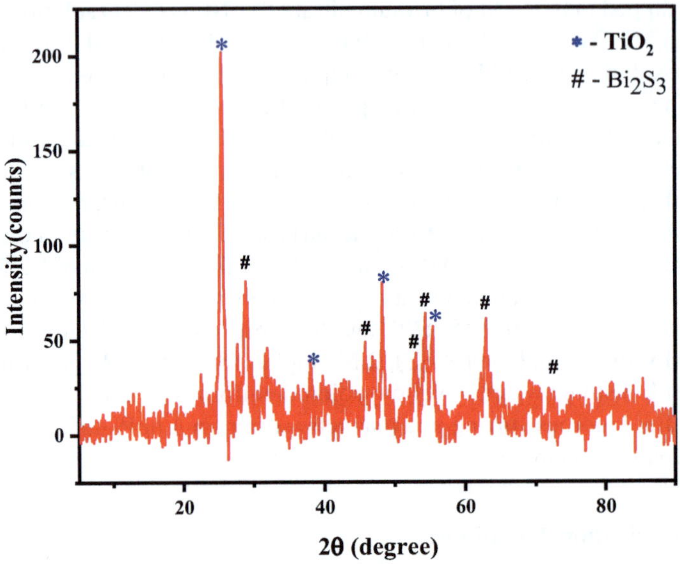

Fig. 1. X-ray diffraction pattern of Bi_2S_3 sensitized TiO_2 (S20) thin film.

Raman spectrum provides detailed information about molecular vibrations, which can be directly related to the chemical structure and composition of a sample. Fig. 3a shows Raman spectra of all the Bi_2S_3 sensitized TiO_2 thin films with different SILAR cycles. The peaks observed at 148(Eg), 191(Eg), 396(B1g), 511 (A1g/B1g) and 633 (Eg) cm^{-1} originate from anatase TiO_2 [8]. 96 cm^{-1} is the more active mode of Bi_2S_3 [9]. The peaks around 191 cm^{-1} and 258 cm^{-1} correspond to Ag and B1g mode of Bi_2S_3 [10].

Figure 3(a) shows the SEM images of pure TiO_2 particles covered throughout the glass substrate. The SEM images show uniform nanosized particles with spherical morphology having porous structure. Figure 3(b) shows the Bi_2S_3 nanoparticles deposited at 20 SILAR cycles on TiO_2. The surface retained the same morphology even after the deposition of Bi_2S_3. With the deposition of the Bi_2S_3 on to the TiO_2 photoanode layer, the surface becomes well-covered with the Bi_2S_3 nanoparticles which is evident from the higher particle density.

The optical properties of Bi_2S_3 thin films with different SILAR cycles are studied using the UV-Vis absorption spectra recorded in the spectral range of 300–1600 nm. From the spectral analysis, the Bi_2S_3-sensitized TiO_2 photoanode had an absorption band that

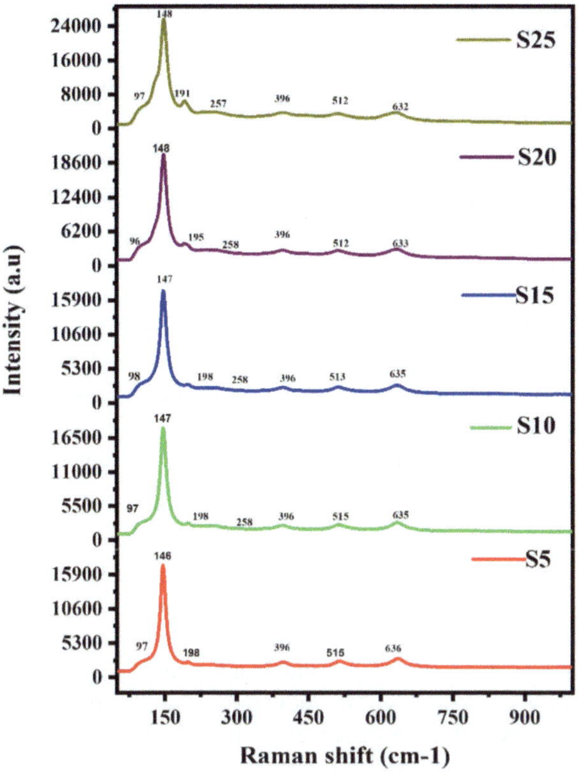

Fig. 2. Raman spectra of Bi_2S_3 sensitized TiO_2 thin film.

Fig. 3. SEM images of (a) bare TiO_2 (b) Bi_2S_3 /TiO_2 film of S20.

extended from the ultraviolet to the visible region, indicating that the sensitization of TiO_2 with Bi_2S_3 nanoparticles improved its light absorption capabilities across a wider range of wavelengths. TiO_2 primarily absorbs light in the UV region; however, when Bi_2S_3 is coated on it, the absorption region extends into the visible light spectrum as well. This enables the sensitized TiO_2 photoanode to capture a broader spectrum of light wavelengths, potentially improving its efficiency and performance in solar cell applications.

The bandgap of films is calculated from the Tauc plot using the following formula:

$$(\alpha h\upsilon)^{n} = A(h\upsilon - E_g) \tag{1}$$

where α is the optical absorption coefficient, υ is the incident photon frequency, h is the Plank's constant, A is a constant, and E_g is the bandgap of the material. Here, n is a constant which take on values of 2 or ½, depending on the kind of band transition present in the sample. For direct transition, n = ½ and for indirect transitions, n = 2 [11]. On the contrary, the bandgap of the bare TiO_2 film is deducted from the diffuse reflectance spectroscopy. The Kubelka-Munk function F(R) can be used to convert the measured reflectance spectra to the corresponding absorption spectra. Using the following equation, the band gap can be extracted from the Tauc plot of the Kubelka-Munk function, by extending the linear portion of the $F(R)^2$ Vs hυ graph to the x-axis [12]:

$$(F(R) \times h\upsilon)^{n} = A(h\upsilon - E_g) \tag{2}$$

Figure 4(a) shows the bandgap of bare TiO_2 film estimated from the Tauc plot derived from using the Kubelka Munk function and is found to be 3.2 eV. The bandgaps of Bi_2S_3 thin films determined from the Tauc plot of (αhυ) 2 Vs photon energy is shown in Fig. 5. Bandgap for SILAR cycles 5, 10, 15, 20 and 25 are found to be 1.8 eV, 1.64 eV, 1.56 eV, 1.44 eV, and 1.44 eV, respectively. The reduction in the band gap with increase in the SILAR cycle clearly indicates the increase in particle size of Bi_2S_3.

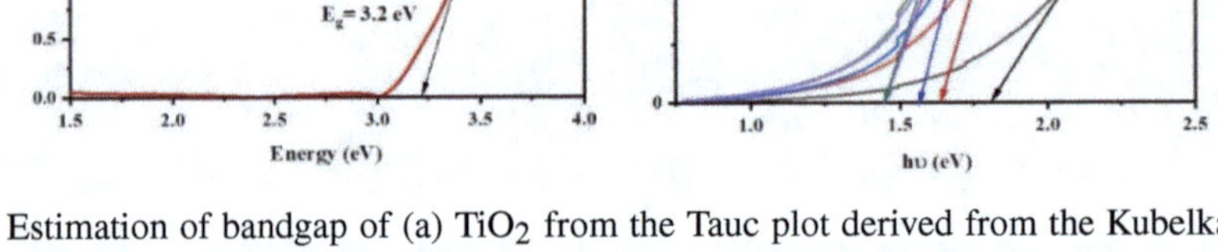

Fig. 4. Estimation of bandgap of (a) TiO_2 from the Tauc plot derived from the Kubelka Munk function (b) Tauc plots of all SILAR cycles of Bi_2S_3 sensitized TiO_2 thin films.

Cyclic voltammetry is a powerful and popular electrochemical tool commonly used to investigate molecular species' reduction and oxidation processes. The peak at positive current corresponds to oxidation and the peak at negative current corresponds to reduction. The carrier transport between TiO_2 and Bi_2S_3 could be understood by determining their HOMO-LUMO levels. Here, the oxidation and reduction peaks observed in the TiO_2 and TiO_2/Bi_2S_3 system are used for estimating the individual HOMO-LUMO

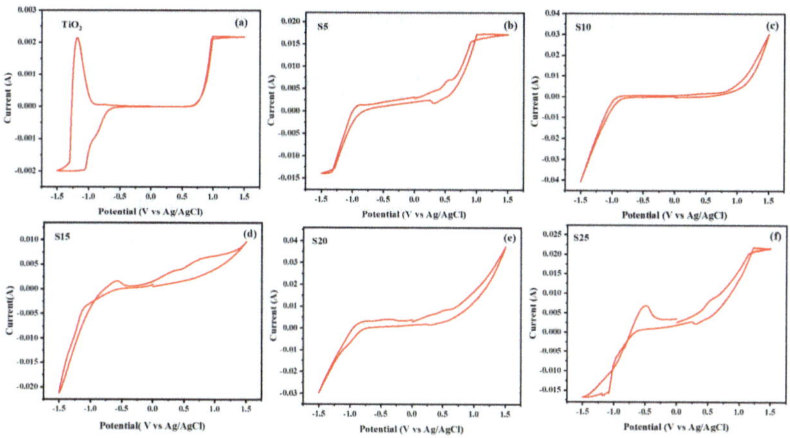

Fig. 5. Cyclic Voltammogram of TiO_2 and Bi_2S_3 sensitized TiO_2 layers of all five SILAR cycles.

levels. Once the oxidation/reduction peaks are obtained, HOMO-LUMO levels can be determined from the following equations [13]:

$$E_{HOMO} = -(\mathbf{Eox + 4.71})\ \mathbf{eV} \quad (3)$$

$$E_{LUMO} = -(\mathbf{Ered + 4.71})\ \mathbf{eV} \quad (4)$$

where 4.71 eV is the difference between vacuum level potential of the normal hydrogen electrode and the potential of the Ag/AgCl electrode.

Figure 5 shows the cyclic voltammogram of TiO_2 and TiO_2/Bi_2S_3 layers. In the case of TiO_2, the reduction peak is obtained at −0.85 V and there is no oxidation peak. The LUMO level of TiO_2 is calculated to be at −3.86 eV using Eq. (4). The optical band gap of 3.2 eV is subtracted from the LUMO level to estimate the position of HOMO at −7.06 eV. Oxidation peaks are observed at 0.311 V, 0.366 V, 0.4373 V, 0.574 V, 0.733 V for SILAR cycles 5, 10, 15, 20 and 25 respectively. No reduction peaks are observed in these graphs

From the observed oxidation and reduction peaks, the HOMO-LUMO levels are determined using Eq. (3) and (4). The results are tabulated in Table 1.

With increase in particle size, HOMO levels of Bi_2S_3 are going down and LUMO going up. In the solar cell structure, Bi_2S_3 absorbs electrons and transfer to TiO_2. For the transfer to be possible, the LUMO level of Bi_2S_3 should be above that of TiO_2. As observed in Table 1, the LUMO levels of samples S5 to S15 occurs above that of TiO_2 enabling the electron transport from Bi_2S_3 to TiO_2. The LUMO levels of samples, S20 and S25 occurs at nearly the same value and above that of TiO_2 respectively. Electron transfer probability is comparatively lesser in these samples.

Bi_2S_3 QDSSC is fabricated and the cell is illuminated under AM1.5 condition. The measurements are taken for the photoanode made up with various number of SILAR cycles. The I–V characteristics are shown in Fig. 6, and the solar cell parameters are listed in Table 2.

Table 1. Estimation of HOMO-LUMO levels.

Sample	E_{LUMO} (eV)	E_{HOMO} (eV)	Bandgap (eV)
TiO_2	−3.86	−7.06	3.2
S5	−3.22	−5.021	1.8
S10	−3.436	−5.076	1.64
S15	−3.59	−5.147	1.56
S20	−3.844	−5.284	1.44
S25	−4.0	−5.443	1.44

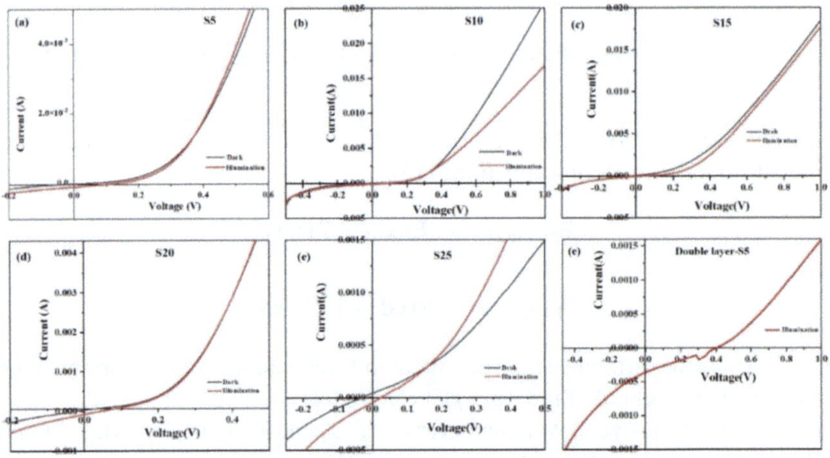

Fig. 6. IV characteristics of Bi_2S_3/TiO_2 with various SILAR cycles.

Table 2. The solar cell parameters obtained from the I–V characteristics.

Condition	J_{sc} (mA/cm^2)	V_{oc}(V)	Fill Factor	Efficiency (%)
SILAR cycle 5	4.8×10^{-2}	0.112	0.4169	0.0028
SILAR cycle 10	3.69×10^{-1}	0.074	0.5467	0.0016
SILAR cycle 15	4.51×10^{-2}	0.0227	0.4989	0.0005
SILAR cycle 20	5.24×10^{-2}	0.0863	0.6092	0.0027
SILAR cycle 25	2.76×10^{-2}	0.0322	0.1874	0.00016
Double layer-cycle 5	2.8×10^{-1}	0.4	0.3765	0.045

Maximum efficiency of 0.0028% is obtained for TiO_2/Bi_2S_3 solar cells having a SILAR cycle of 5 and the efficiency decreases with increase in the SILAR cycles. Though more Bi_2S_3 gets absorbed with increase in the SILAR cycle, the band alignment between

TiO_2 and Bi_2S_3 gets dislocated as observed in cyclic voltammetry, thus hindering the transfer of carriers. The efficiency was the lowest for cycle 25, where the LUMO of Bi_2S_3 is located below that of TiO_2 which prevents the motion of electrons to TiO_2. To improve the solar cell parameters, another layer of TiO_2 is added as buffer layer. This layer prevents the electron leakage from TiO_2 to electrolyte. Sample with a SILAR cycle of 5, which has shown best efficiency among the last set is chosen for this device. As shown in Fig. 6(e), the efficiency of the solar cell, improved by a factor of 16. V_{oc} increased to 400 mV and J_{sc} to 0.28 mA/cm^2.

4 Conclusion

The Bi_2S_3 nanoparticles are deposited in-situ on mesoporous TiO_2 thin film using low-cost fabrication method of successive ionic layer adsorption and reaction for various cycles. The structural analysis confirmed the formation of orthorhombic Bi_2S_3 on top of anatase TiO_2 thin films. Bi_2S_3 sensitized solar cell is fabricated with Bi_2S_3 samples having various SILAR cycles. With larger number of cycles, a misalignment between the band positions of Bi_2S_3 and TiO_2 has occurred hindering the movement of charge carriers. Solar cell fabricated with 5 cycles of SILAR yielded the best result. Further improvement in the solar cell parameters is obtained with an additional TiO_2 buffer layer which protects the leakage of electrons. The device exhibited an open circuit voltage of 400 mV, short circuit current of 0.28 mA/cm^2 and a power conversion efficiency of 0.045%.

Acknowledgements. Arya Mohan acknowledge KSCSTE, Government of Kerala, India for providing research fellowship (Ref No.317, 2021).

References

1. Surana, K., Mehra, R.M., Bhattacharya, B., Rhee, H.-W., Polu, A.R., Singh, P.K.: A comprehensive study of chalcogenide quantum dot sensitized solar cells with a new solar cell exceeding 1 V output. Renew. Sust. Energ. Rev. **52**, 1083–1092 (2015)
2. Hanna, M.C., Nozik, A.J.: Solar conversion efficiency of photovoltaic and photo electrolysis cells with carrier multiplication absorbers. J. Appl. Phys. **100**(7), 074510 (2006)
3. Latif, H., et al.: A novel, PbS quantum dot-Sensitized solar cell structure with TiO2-fMWCNTS nano-composite filled meso-porous anatase TiO2 photoanode. Sol. Energy. **204**, 617–623 (2020)
4. Senevirathne, K.M.B.B., et al.: Non-toxic Bi2S3 quantum dot-sensitized TiO2 electrodes for cost-efficient solar power conversion. Electrochim. Acta. **507**, 145121 (2024)
5. Solís, M., Rincón, M.E., Calva, J.C., Alvarado, G.: Bismuth sulfide sensitized TiO2 arrays for photovoltaic applications. Electrochim. Acta. **112**, 159–163 (2013)
6. Guan, X.N., et al.: A 4.37% efficiency on bismuth triiodide sensitized solar cells with aqueous electrolyte. Mater. Sci. Eng. B. **294**, 116557 (2023)
7. Xu, B., Wang, G., Fu, H.: Enhanced photoelectric conversion efficiency of dye-sensitized solar cells by the incorporation of flower-like Bi2S3:Eu3+ sub-microspheres. Sci. Rep. **6**(1), 23327 (2016)

8. Ahmad, A., Tezcan, F., Yerlikaya, G., Zia-ur-Rehman, Paksoy, H., Kardaş, G.: Three dimensional rosette-rod TiO2/Bi2S3 heterojunction for enhanced photoelectrochemical water splitting. J. Alloys Compd. **868**, 159133 (2021)
9. Monteiro, T.O., do M., S., Damos, F.S., Rita, F.-F.O.: Highly sensitive photoelectrochemical immunosensor based on anatase/rutile TiO2 and Bi2S3 for the zero-biased detection of PSA. J. Solid State Electrochem. **24**(8), 1801–1809 (2020)
10. Zumeta-Dubé, I., Ortiz-Quiñonez, J.-L., Díaz, D., Trallero-Giner, C., Ruiz-Ruiz, V.-F.: First order Raman scattering in bulk Bi_2S_3 and quantum dots: Reconsidering controversial interpretations. J. Phys. Chem. C. **118**(51), 30244–30252 (2014)
11. Rasool, S., et al.: Optical properties of thermally evaporated In 2 S 3 thin films measured using photoacoustic spectroscopy. Mater. Sci. Semicond. Process. **72**, 4–8 (2017)
12. Landi, S., Segundo, I.R., Freitas, E., Vasilevskiy, M., Carneiro, J., Tavares, C.J.: Use and misuse of the Kubelka-Munk function to obtain the band gap energy from diffuse reflectance measurements. Solid State Commun. **341**, 114573 (2022)
13. Deepa, K.G., Lekha, P., Sindhu, S.: Efficiency enhancement in DSSC using metal nanoparticles: A size dependent study. Sol. Energy. **86**(1), 326–330 (2012)

Synthesis and Characterization of MoS_2 Thin Films via Spray Pyrolysis for Solar Cell Application

Frenson P. Jose[1], Reeba Dais[2], Carmen Martinez-Tomas[3], Oleksii Klymov[3], and A. S. Asha[1,4,5](✉)

[1] Nanomaterial for Emerging Solid-state Technology (NEST) Research Laboratory, Department of Physics, Cochin University of Science and Technology, Kochi 682022, Kerala, India
asa@cusat.ac.in

[2] DST -Inspire fellow, Union Christian College, Aluva, India

[3] Department of Applied Physics and Electromagnetism, University of Valencia, c/Dr. Moliner n° 50, Burjassot 46100, Spain

[4] Centre of Excellence in Advanced Materials, Cochin University of Science and Technology, Kochi 682022, Kerala, India

[5] Inter University Centre for Nanomaterials and Devices, Cochin University of Science and Technology, Kochi 682022, Kerala, India

Abstract. Quantum Dot Sensitized Solar Cells (QDSSCs) represent a promising third-generation photovoltaic technology. While platinum (Pt) has been the benchmark counter electrode (CE) due to its excellent catalytic activity, its high cost necessitates the exploration of affordable alternatives. Molybdenum disulfide (MoS_2), has emerged as a strong candidate owing to its catalytic activity and stability for various redox couples. In this work, we report the synthesis of MoS_2 thin films on both Fluorine-doped Tin Oxide coated glass substrates (FTO) and soda lime glass substrates using the spray pyrolysis method, systematically varying the substrate temperature and duration. SEM images revealed a highly porous and interconnected network morphology with distinct hexagonal-like MoS_2 nanocrystals. XRD and Raman analysis confirmed the formation of a mixed phase, primarily composed of: MoO_3, 2H MoS_2, 1 T MoS_2 for the thin films deposited on FTO and glass substrates. The formation of the 1 T-MoS_2 phase, coupled with its advantageous porous morphology, underscores its potential as an efficient and cost-effective counter electrode for QDSSCs.

Keywords: Quantum Dot Sensitized Solar Cells · Molybdenum Disulfide · Counter Electrode · Spray Pyrolysis

1 Introduction

Quantum Dot Sensitized Solar Cells (QDSSCs) are a promising third-generation photovoltaic technology, surpassing silicon solar cells in photo-stability, durability, and recyclability. Unlike other solar cells, QDSSCs generate electricity both indoors and

A. Antony et al. (Eds.): icONMAT 2025, SPPHY 328, pp. 219–224, 2025.
https://doi.org/10.1007/978-981-95-2697-0_27

outdoors, converting artificial and natural light into electricity for powering various electronic gadgets [1]. Platinum (Pt), due to its high conductivity and catalytic activity, acts as a good counter electrode for QDSSC. But the high cost of Pt makes researchers investigate more affordable alternatives, including carbonaceous materials (graphene, carbon black, carbon nanotubes) and transition metal compounds like sulfides, nitrides, and carbides. Molybdenum disulfide (MoS_2), due to its strong catalytic activity and thermal stability, and natural abundance, is one among these alternatives. MoS_2 is of particular inter due, amongst other properties, to its direct band gap of 1.8 eV in monolayer and layer dependence of band structure tackling the gapless problems of graphene [2–5]. MoS_2 demonstrates excellent catalytic performance and is being developed as a counter electrode for QDSSCs. MoS_2 is stable for both Sn^{2-}/S^{2-} and I^-/I^{3-} redox couples making it suitable for QDSSCs and offers a significant advantage over other materials typically limited to a single redox system [6, 7]. An FTO/MoS_2 counter electrode significantly boosted QDSSC power conversion efficiency (PCE) to 3.69%, a 70% increase compared to FTO/Pt CE (2.16%). Furthermore, this MoS_2-based counter electrode (CE) achieved a comparable cell efficiency of 7.16% in sensitized solar cells, nearly matching that of FTO/Pt CE (7.48%). This enhanced performance is attributed to FTO/MoS_2 superior resistance to charge and mass transfer, increased electrochemical active sites, and optimized nanopore morphology [8].

A variety of methods, such as chemical bath deposition, electrodeposition, sputtering, mechanochemical method, spin coating technique, dip technique, high-power impulse magnetron sputtering (HIPIMS), pulsed laser deposition, and chemical vapor deposition (CVD), can be used to deposit MoS_2 thin films [5–9]. The present work describes the synthesis of MoS_2 film on the fluorine-doped tin oxide coated sodalime glass substrates (FTO) and sodalime glass substrates by chemical spray pyrolysis.

2 Experimental and Characterization

FTO was ultrasonically cleaned with ethanol for 10 min, followed by two rinses with deionized (DI) water and drying. Soda lime glass substrates were, ultrasonically cleaned with acetone and 2-propanol for 15 min, followed by two rinses with deionized (DI) water and drying. For the synthesis of MoS_2 thin films, an aqueous precursor solution was prepared using 0.1960 g of ammonium molybdate tetrahydrate ($(NH_4)_6Mo_7O_{24}.4H_2O$) and 0.3044 g of thiourea ($CS(NH_2)_2$) dissolved separately in 50 ml of DI water and subsequently mixed at room temperature to achieve a final volume of 100 ml.

The substrates were placed on the hot plate of the chemical spray pyrolysis system. The prepared precursor solution was sprayed onto the substrates for different durations (2 h, 2.5 h, 3 h). The experiment was repeated for two different substrate temperatures, 300 °C and 350 °C. The samples deposited on FTO at 300 °C for a spray time of 2 h and 2.5 h were named as MS5 and MS6, respectively, and those deposited for a duration of 3 h at a substrate temperature of 350 °C on FTO and soda lime glass were named as MS8 and MS10 respectively.

High resolution X-ray diffraction (HRXRD) patterns of the thin film at 2θ-ω configuration were carried out using a Panalytical X-Pert MRD diffractometer. Raman spectroscopy was done using a HORIBA MTB. XPlora spectrometer equipped with 532 and

785 nm lasers. A scanning electron microscope (SEM), HITACHI S4800 equipped with an X-ray detector, was employed to examine the morphology.

3 Result and Discussion

Figure 1 (a) displays a top-view FESEM image of a MoS_2 sample, revealing a highly porous and interconnected network morphology distributed uniformly across the surface, suggesting a high surface area. Figure 1 (b) shows distinct, hexagonal-like flakes or nanocrystals of MoS_2. These flakes appear to be closely packed and randomly oriented, forming a relatively dense film at the nanoscale, consistent with the layered structure of MoS_2. Figure 1 (c) presents a cross-sectional SEM image, illustrating the overall thickness and internal structure of the MoS_2 film. The image reveals a porous and columnar growth, with large, irregular voids extending through the film.

Fig. 1. FESEM image of MoS_2 film deposited on FTO (a) image with low magnification (b) image with high magnification (c) cross sectional image

Figure 2 (a), (b), and (c) show the XRD patterns for samples MS5, MS6, and MS8, respectively, all of which were deposited on FTO-coated glass substrates. In all these thin films, along with the peak of FTO, a broad peak around 2θ value of 10° corresponds to the (002) plane of the 1 T-MoS_2 phase, indicating the formation of the metallic 1 T phase of MoS_2. The broad nature of this peak suggests that the MoS_2 films are either nanocrystalline or possess a highly disordered structure [6]. No distinct peaks corresponding to the 2H-MoS_2 phase were observed in these samples, indicating the prevalence of the 1 T phase. In contrast, Fig. 2 (d) exhibits a prominent broad peak around $2\theta = 25°$, corresponding to MoO_3. A relatively weak and broad peak consistent with 1 T-MoS_2 is still present around 2θ value 10°, but it is significantly less pronounced than in the thin films deposited on FTO. The presence of MoO_3 in the sample MS10 suggests that the deposition on soda lime glass at higher temperatures might lead to partial oxidation of the MoS_2 or the formation of molybdenum oxide as a dominant phase, possibly due to insufficient sulfurization or higher oxygen content in the glass surface compared to FTO [6]. These XRD results confirms the formation of MoS_2 films, primarily in the 1 T phase on FTO, while revealing the influence of the substrate and synthesis conditions on the final crystalline structure and phase composition.

Figure 3 shows the Raman spectra of the synthesized MoS_2 thin films. Figure 3 (a)–(d) exhibit characteristic Raman peaks associated with the 1 T-MoS_2 metallic phase, corroborating the findings from the XRD analysis. Specifically, the spectra show the

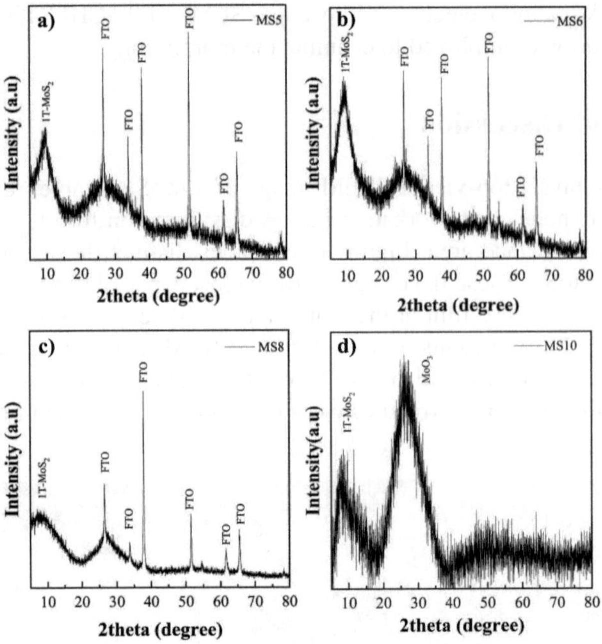

Fig. 2. X-ray diffraction (XRD) patterns of MoS_2 thin films synthesized under different conditions: (a) MS5 (2 h at 300 °C on FTO), (b) MS6 (2.5 h at 300 °C on FTO), (c) MS8 (3 h at 350 °C on FTO), and (d) MS10 (3 h at 350 °C on glass)

prominent in-plane vibration (E_{1g}) peak around 280–285 cm^{-1} and the out-of-plane vibration (A_{1g}) peak around 405–410 cm^{-1}. The relatively broad nature and suppressed intensity of the A_{1g} peak, combined with the clear presence of additional peaks, are indicative of the 1 T polytype rather than the semiconducting 2 H phase. Crucially, the spectra also display several distinct first-order Raman peaks viz., J1, J2, J3, J4, which are characteristic fingerprints of the 1 T-MoS_2 phase. The peaks at 112 cm^{-1} and 126 cm^{-1} correspond to MoO_3. The J_1 peak, typically found around 140–150 cm^{-1}, and the J_2 peak around 210–220 cm^{-1} are clearly resolved in all samples. The J_3 and J_4 peaks, located at approximately 330–340 cm^{-1} and 370–380 cm^{-1} respectively, are also discernible. Also, all the samples show the peak of MoO_3. This suggests that even when some oxidation occurs, the MoS_2 formed still retains its 1 T metallic structure. The intensity ratio between A_{1g} and $E^1{}_{2g}$ modes is less than one, suggesting a deviation from vertical morphology to horizontal growth orientation. The consistency in the Raman fingerprints across different deposition temperatures (300 °C and 350 °C) and durations, as well as on different substrates (FTO and glass), highlights the formation of the mixed phase.

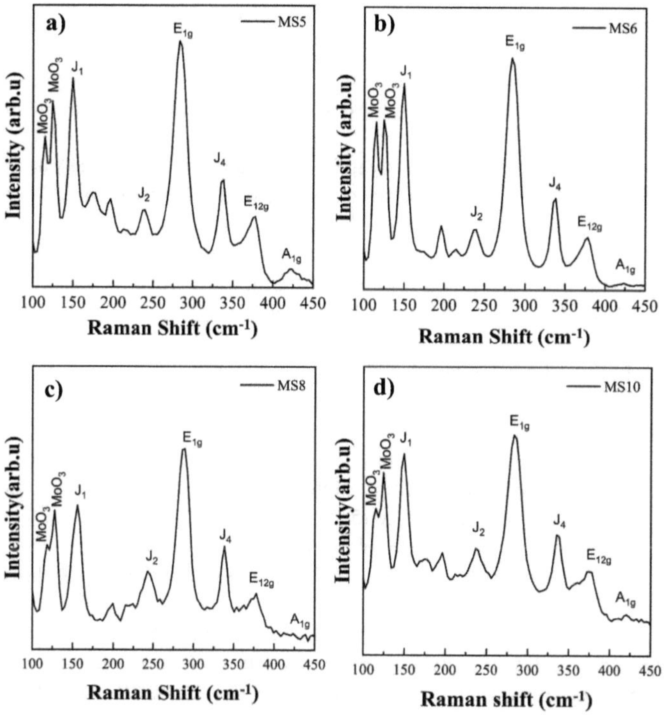

Fig. 3. Raman spectrum of MoS_2 thin films synthesized under different conditions: (a) MS5 (2 h at 300 °C on FTO), (b) MS6 (2.5 h at 300 °C on FTO), (c) MS8 (3 h at 350 °C on FTO), and (d) MS10 (3 h at 350 °C on glass)

4 Conclusion

In this study, MoS_2 thin films were deposited on FTO and soda lime glass substrates through the spray pyrolysis method. FESEM images revealed a porous and interconnected thin film morphology composed of distinct MoS_2 nanocrystals, indicating a high surface area beneficial for electrochemical reactions. XRD confirmed the predominant formation of the metallic 1 T-MoS_2 phase in thin films deposited on FTO substrates. Raman spectroscopy further corroborated the formation of the mixed phase across all synthesized samples. The consistent formation of the electrocatalytically active 1 T-MoS_2 phase, combined with its high surface area porous structure, highlights its significant potential as a cost-effective and efficient counter electrode material for Quantum Dot Sensitized Solar Cells (QDSSCs).

Acknowledgement. This work was partly supported by the projects, PID2020-114796RB-C2 funded by MCIN/AEI/10.13039/501100011033/; Project PDC2023-145827-C2 funded by MICIU/AEI/10.13039/501100011033/ and by European Union Next Generation EU/PRTR; and Generalitat Valenciana under Project No. PROMETEU/2021/066.

References

1. Sharma, D., Jha, R., Kumar, S.: Quantum dot sensitized solar cell: Recent advances and future perspectives in photoanode. Sol. Energy Mater. Sol. Cells. **155**, 294–322 (2016). https://doi.org/10.1016/J.SOLMAT.2016.05.06
2. Sabari Girisun, T.C., Durairaj, M., Vijaya, S., Anandan, S.: 1T and 2H phase molybdenum disulfide as a counter electrode for Pt free dye-sensitized solar cells. Mater. Sci. Eng. B. **287**, 116123 (2023). https://doi.org/10.1016/J.MSEB.2022.116123
3. Shahbazi, M., Khanlary, M.R., Taherkhani, A.: Enhancement of optical, morphological and electronic properties of MoS 2 thin film by annealing to improve the performance of silicon solar cells. J. Mater. Sci.: Mater. Electron. **34**, 57 (2023). https://doi.org/10.1007/s10854-022-09538-2
4. Li, X., Zhu, H.: Two-dimensional MoS 2 : Properties, preparation, and applications-NC-ND license. J. Mater. **1**, 33–44. (http://creativecommons.org/licenses/by-nc-nd/4.0/ (2015). https://doi.org/10.1016/j.jmat.2015.03.003
5. Abinaya, A., Jeyaprakash, B.G.: Structural, surface and mechanical characterization of spray-deposited molybdenum disulfide thin films. Mater. Sci. Semicond. Process. **31**, 582–587 (2015). https://doi.org/10.1016/j.mssp.2014.12.045
6. Tian, Z., Chen, Q., Zhong, Q.: Honeycomb spherical 1T-MoS2 as efficient counter electrodes for quantum dot sensitized solar cells. Chem. Eng. J. **396**, 125374 (2020). https://doi.org/10.1016/J.CEJ.2020.125374
7. Syrrokostas, G., Antonelou, A., Leftheriotis, G., Yannopoulos, S.N.: Electrochemical properties and long-term stability of molybdenum disulfide and platinum counter electrodes for solar cells: A comparative study. Electrochim. Acta. **267**, 110–121 (2018). https://doi.org/10.1016/J.ELECTACTA.2018.02.068
8. Quy, V.H.V., et al.: Electrodeposited MoS2 as electrocatalytic counter electrode for quantum dot- and dye-sensitized solar cells. Electrochim. Acta. **260**, 716–725 (2018). https://doi.org/10.1016/J.ELECTACTA.2017.12.023
9. Li, J., Liang-Yu, L., Gao, X.-F., Gan, L., Ren, T.-L.: Recent advances in MoS$_2$ preparation and its application. J. Mater. Chem. C. **4**, 2419–2433 (2016). https://doi.org/10.1039/C5TC04207H

Thickness Dependent Study of the Structural and Optical Properties of Bi_2Te_3 Thin Films Grown by RF Sputtering

Jeswin Mammen Raju, Rakesh Ramachandran, and K. J. Thomas(✉)

Department of Physics, Central University of Kerala, Kasaragod 671320, India
kj_thomas@cukerala.ac.in

Abstract. Topological insulators (TI) have attracted immense attention due to the unusual conducting surface states discovered in the recent past. Developing high-quality thin films of TI is crucial for the development of planar devices. Thin films of Bi_2Te_3, a well-known TI material, were grown by RF sputtering at room temperature for different thicknesses ranging from 70 nm to 320 nm. The structural and optical properties show organized growth in the c-axis from a thickness of 70 nm up to 200 nm. At a thickness of 320 nm, the sample exhibits bulk behaviour. The optical energy gap was found to be greater than its bulk value (0.19 eV), which could be attributed to quantum size effects, density of localized states and barrier height differences due to varying crystallite sizes.

Keywords: Topological Insulators · Thin Films · Quantum Confinement

1 Introduction

Chalcogenides based on bismuth (Bi_2Ch_3) have long been investigated as thermoelectric materials [1, 2]. In the last decade, a new state of quantum matter, namely topological insulators (TI) were predicted theoretically [3] and realized experimentally [4] in these compounds, leading to heightened research activity in this field. These materials have been classified as strong Z_2 topological insulators, characterized by an insulating bulk and gapless spin-helical surface states protected by time reversal symmetry. Spin-orbit coupling present in these systems drives a band inversion between the valence and conduction bands, leading to topologically non-trivial surface states. These surface states form a single Dirac cone, which is spin-momentum locked, wherein the spin orientation is perpendicular to the momentum. The surface states in combination with a large bulk energy gap make these materials ideal for spintronic and fault-tolerant quantum devices applications [5–10].

Bismuth chalcogenides belong to the V-VI semiconductor family and have a rhombohedral tetradymite crystal structure with space group R-3 m [11–14]. Here, we study Bi_2Te_3, which has a bulk energy gap of 0.19 eV and a layered structure. Each layer of Bi_2Te_3 is stacked in the order Te(1)-Bi-Te(2)-Bi-Te(1) forming a quintuple layer (QL). These QLs are connected by weak van der Waals force. While the interaction of Te(1)

A. Antony et al. (Eds.): icONMAT 2025, SPPHY 328, pp. 225–232, 2025.
https://doi.org/10.1007/978-981-95-2697-0_28

with Bi atom is covalent, that of Bi with Te(2) atom is strongly ionic. Unintentional doping and defects make the bulk conducting, which makes the growth of high-quality crystals that support topological surface states of great demand in principle. Thus, optimized growth of TI materials is essential, as surface states have a significant role in determining their nature of transport.

Though high-quality single crystals can be grown by the Bridgeman technique [15–17], thin films are preferred for industry applications. Several thin film deposition techniques like molecular beam epitaxy (MBE) [18, 19], sputtering [20], evaporation [21–23], chemical vapour deposition (CVD) [24–26] and pulsed laser deposition (PLD) [27] can be employed to grow high-quality films on different substrates. Due to the high cost and low yield, the use of MBE in industry is limited. In this study, we explore the growth of Bi_2Te_3 thin films by RF Sputtering on glass substrate by varying the thickness of the film.

2 Experimental Details

Before depositing the thin film, the substrates were initially washed with Extran® solution followed by ultrasonication in acetone, isopropanol and distilled water for 15 min before being loaded into the vacuum chamber. A high-purity Bi_2Te_3 target (99.99%) was used as the source material. Initially, the sputtering system was pumped down to a pressure less than 1×10^{-5} mbar; a working pressure of 1.5×10^{-2} mbar was maintained in the chamber, and the argon gas flow was kept at 10 sccm (standard cubic centimeter per minute). The substrate was kept at room temperature. After the growth, the samples were annealed in vacuum, where pressure is kept less than 3×10^{-2} mbar for 30 min at different temperatures like 150 °C, 200 °C, 250 °C, 300 °C, 350 °C. The thickness of the thin films was calibrated ex situ with a Bruker Dektak XT stylus profilometer. The structural analysis was performed using the X-ray diffractometer (XRD), Rigaku Miniflex 600 employing CuKα radiation ($\lambda = 1.54059$ Å). UV-Vis-NIR characterization was carried out using an Agilent Cary 5000 series spectrophotometer over a spectral range of 200 nm to 3300 nm.

3 Results and Discussion

3.1 Structural Studies

Figure 1 depicts the XRD pattern of the as-prepared Bi_2Te_3 films at different thicknesses. Films with a thickness below 120 nm are found to be amorphous. As thickness increases, the films become crystalline, with grains oriented along the (0 1 5) lattice plane, consistent with the rhombohedral structure of Bi_2Te_3 (JCPDS CAS No.15–0863). No additional diffraction peaks are observed, indicating the absence of secondary phases or impurities in the film samples.

The crystalline nature of the films improved through vacuum annealing. Figure 2 shows the XRD pattern of as-deposited Bi_2Te_3 thin films of different thickness after post-annealing at 150 °C, 200 °C, 250 °C, 300 °C, and 350 °C. The thin film with a thickness of 68 nm shows a preferred c-axis orientation upon annealing at 200 °C, and

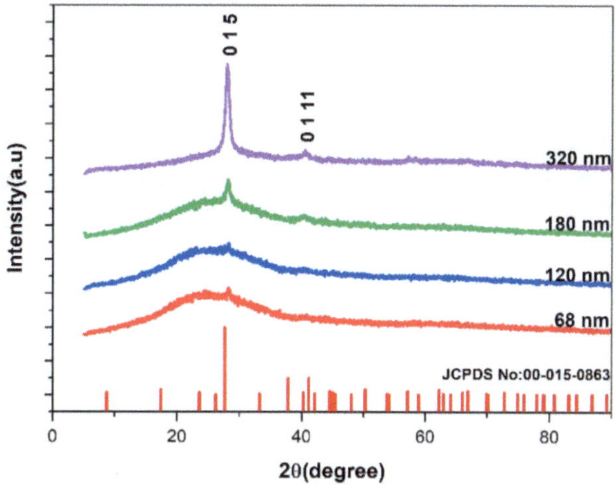

Fig. 1. X-ray diffraction (XRD) patterns for samples with different thicknesses: 320 nm, 180 nm, 120 nm, and 68 nm. Peaks corresponding to the (0 1 5) and (0 1 11) planes are indexed, showing increased intensity with increasing film thickness. A reference pattern is shown at the bottom with the label "JCPDS No: 00- 015-0863."

the intensities of the peaks (0 0 3), (0 0 6), and (0 0 15) increase progressively with annealing temperatures up to 350 °C. In addition, there is a peak at (1 0 10) lattice plane which increases with annealing temperature. This is the most intense peak, along with c-axis orientated peaks, for films with thicknesses of 120 nm and 180 nm (Fig. 2(b and c)). The 320 nm thick films were oriented along (0 1 5) plane (Fig. 2(d)). As the annealing temperature increases, crystallinity improves. The crystallite size increases from 10 nm to 25 nm on average for all the thicker films, indicating better crystallinity with increased annealing temperature. In crystalline materials, the dislocation density is the number of dislocations within a unit volume and is given by $\delta = 1/D^2$, where D is the crystallite size. As the annealing temperature increases, there is a decrease in the dislocation density; in other words, the crystallite size increases.

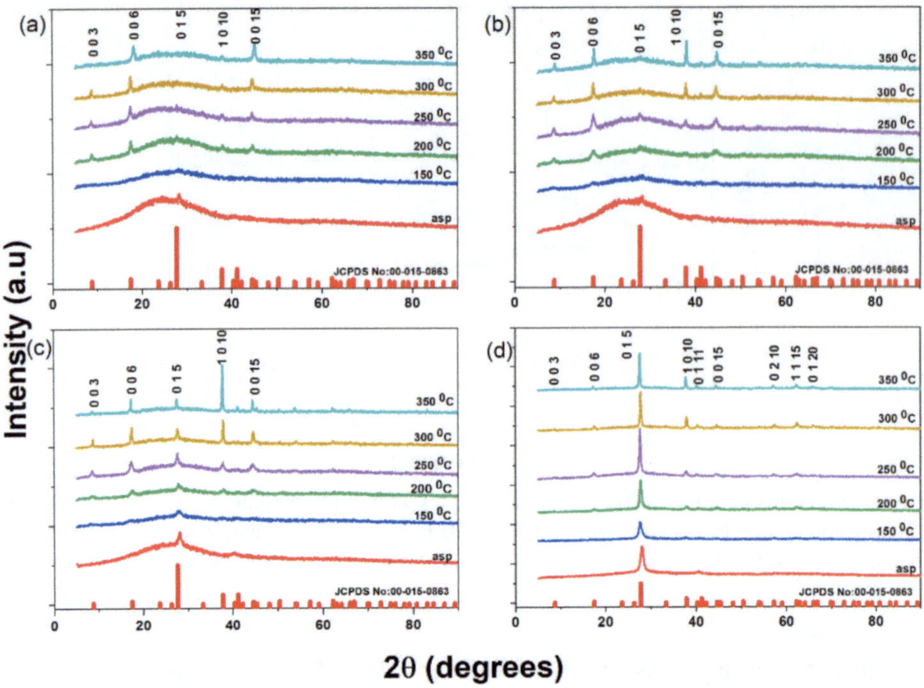

Fig. 2. XRD patterns of Bi_2Te_2 thin films of different thicknesses are shown in four panels labelled (a) 68 nm, (b) 120 nm, (c) 180 nm, and (d) 320 nm. Each panel displays multiple colored traces representing different annealing temperatures, as annotated, for each sample.

3.2 Optical Studies

As the thickness of the film increases, the transmittance decreases as shown in Fig. 3(a). This can be attributed to the increase in crystallite size and hence density of the films. The transmittance also increases with wavelength, indicating the films have significant absorption in the visible spectrum. On vacuum annealing, the transmittance decreases as the annealing temperature increases. There is a shift in the transmittance edge towards higher wavelength as the thickness of the film increases.

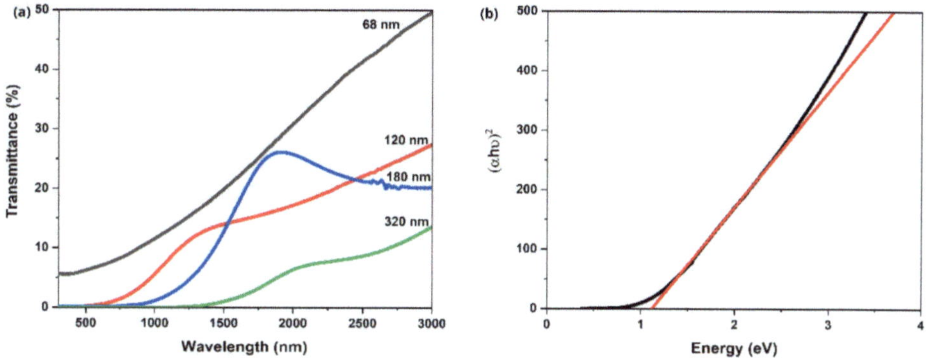

Fig. 3. (a) Transmittance spectra showing transmittance (%) versus wavelength (nm) for films with thicknesses of 68 nm, 120 nm, 180 nm, and 320 nm. (b) Tauc plot of $(\alpha h\nu)^2$ versus photon energy (hv) for the 180 nm film, showing the linear fit (red line) used to estimate the optical band gap.

The plot of $(\alpha h\nu)^2$ *vs hv* (Fig. 3(b)) was used to find the optical band gap, where α is the absorption coefficient. The linear part of the trace indicates a high absorption region. Extrapolating the linear part to the x-axis yields the optical energy gap. The band gap obtained for the samples investigated showed a value much higher than their bulk counterpart (0.19 eV), decreasing from 1.35 eV to 0.86 eV as the film thickness increased. This reduction is attributed to improved atomic ordering and reduced defects/disorder with increasing crystallite size. Band gap dependence on the thickness can be due to the quantum size effect, density of localized states and change in barrier height.

The influence of crystallite size on the optical band gap was discussed by Slater [28]. It was suggested that the barrier height varies with charge accumulation at the grain boundaries and is given by

$$E = E_0 + C\,(X - fD)^2 \quad (1)$$

where E_0 is the original barrier height, C is a term dependent on the charge carrier density, electronic charge, and the dielectric constant of material, X is the barrier width, D is the grain dimension, and f is the fraction that depends on the extent of charge accumulation and carrier concentration. It is reported in literature [29] that the grain size increases proportionally with film thickness. Replacing the grain size with film thickness in the above equation, E becomes proportional to $(X - ft)^2$. Our observations reveal that the energy band gap varies inversely with the square of the film thickness in the annealed samples (Fig. 4).

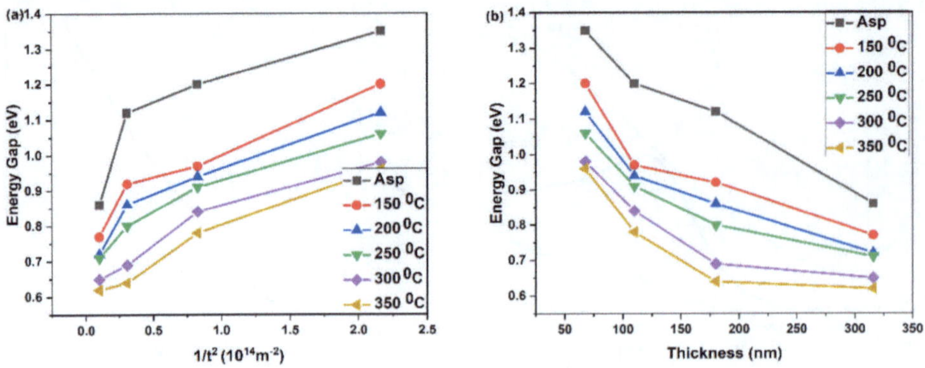

Fig. 4. Variation of band gap energy (Eg) of Bi_2Te_3 thin films (a) Eg (eV) plotted against 1/t^2 (10^(14) m^{-2}) for films Asp (as prepared) and annealed at different temperatures (150 °C, 200 °C, 250 °C, 300 °C, and 350 °C). A linear relationship is observed for 300 °C, and 350 °C annealed samples. (b) Band gap (Eg) plotted as a function of film thickness (nm), showing a decrease in Eg with increasing thickness for the same temperature set. Each temperature is represented by a distinct line and marker style.

In disordered semiconductors, the allowed energy band and the nearest neighbour coupling will be modified compared to their crystalline state. It is well understood that when atoms are arranged to form thin films, many dislocations/defects occur. These dislocations produce localized states in the film. The dislocation density increases as the thickness increases and saturates beyond a particular thickness. However, the dependence of dislocation density on thickness is complex and not established quantitatively. In semiconducting materials, the energy band gap increases due to the presence of a fairly large number of dislocations, provided the dislocations are separated by a distance exceeding their interatomic spacing [30]. This separation originates due to the alternate compression and dilation of the lattice around dislocations. In the present observation, E_g *vs* $1/t^2$ varies linearly for 300 °C and 350 °C annealed samples, effectively ruling out this possibility. However, the above effect cannot be ignored for the remaining samples.

In semi-metallic and semiconducting materials, Sandomirskii [31, 32] studied the quantum confinement effect for the first time. In thin films, a two-dimensional confinement model with hard-wall boundaries is often considered, where carrier motion is restricted in the direction perpendicular to the film surface (i.e., across the thickness), while it remains unrestricted along the other two directions. As the confining dimension decreases, the energy spectrum will become quantized, thus revealing the size dependency of the band gap. The quantum size effects dominate when the thin film thickness is of the order of the de Broglie wavelength and is less than the mean free path of the conduction electrons/holes. The quantization energy of the 1-D confinement will be given by

$$E = \frac{\hbar^2 \pi^2}{2mt^2} \tag{2}$$

where t is the thickness of the layer and, m is the effective mass of the carriers. As the thickness decreases, the band gap increases due to the upward shift of both the valence band maximum and the conduction band minimum relative to their bulk. The new energy gap is a function of the inverse square of thickness and in our present studies, is observed to be linear in the annealed case.

4 Conclusion

Our systematic study of Bi_2Te_3 thin films grown by RF magnetron sputtering reveals that film thickness significantly impacts both the structural quality and optical properties. X-ray diffraction analysis shows that the films, regardless of thickness, retain a polycrystalline, hexagonal (rhombohedral) phase, with a preferential c-axis orientation observed in thinner films- indicating layered growth. The crystalline quality improves with increasing thickness, as evidenced by larger crystallite sizes and reduced dislocation densities. Optical measurements reveal enhanced band gaps that decrease with increasing thickness, consistent with quantum confinement effects.

Acknowledgement. We gratefully acknowledge Professor Biju V (Department of Physics, University of Kerala, Thiruvanthapuram) for assistance with thickness measurements. We also thank the infrastructure support from the DST FIST (SR/FST/PSI-93/2019), Govt. of India.

References

1. Nolas, G.S., Sharp, J., Goldsmid, H.J.: Thermoelectrics: Basic Principles And New Materials Developments, vol. 45. Springer Science & Business Media (2013)
2. Tritt, T.M.: Holey and unholey semiconductors. Science. **283**(5403), 804–805 (1999)
3. Zhang, H., Liu, C.X., Qi, X.L., Dai, X., Fang, Z., Zhang, S.C.: Topological insulators in Bi_2Se_3, Bi_2Te_3 and Sb_2Te_3 with a single Dirac cone on the surface. Nat. Phys. **5**(6), 438–442 (2009)
4. Chen, Y.L., et al.: Experimental realization of a three-dimensional topological insulator, Bi_2Te_3. Science. **325**(5937), 178–181 (2009)
5. Katmis, F., et al.: A high-temperature ferromagnetic topological insulating phase by proximity coupling. Nature. **533**, 513–516 (2016)
6. Ferreira, G.J., Loss, D.: Magnetically defined qubits on topological insulators. Phys. Rev. Lett. **111**, 106802 (2013)
7. Kitaev, A.V.: Fault-tolerant quantum computation by anyons. Ann. Phys. **303**(1), 2–30 (2003)
8. Nayak, C., Simon, S.H., Stern, A., Freedman, M., Das Sarma, S.: Non-Abelian anyons and topological quantum computation. Rev. Mod. Phys. **80**, 1083 (2008)
9. Hasan, Zahid, M., Moore, J.E.: Three-dimensional topological insulators. Annu. Rev. Condens. Matter Phys. **2**(1), 55–78 (2011)
10. Dc, M., et al.: Room-temperature high spin-orbit torque due to quantum confinement in sputtered $Bi_xSe_{(1-x)}$ films. Nat. Mater. **17**, 800–807 (2018)
11. Jenkins, J.O., Rayne, J.A., Ure Jr., R.W.: Elastic moduli and phonon properties of Bi_2Te_3. Phys. Rev. B. **5**(8), 3171 (1972)
12. Kullmann, W., Geurts, J., Richter, W., Lehner, N., Rauh, H., Steigenberger, U., Geick, R.: Effect of hydrostatic and uniaxial pressure on structural properties and Raman active lattice vibrations in Bi_2Te_3. Phys. Status Solidi (b). **125**(1), 131–138 (1984)

13. Richter, W., Becker, C.R.: A Raman and far-infrared investigation of phonons in the rhombohedral V_2–VI_3 compounds Bi_2Te_3, Bi_2Se_3, Sb_2Te_3 and $Bi_2(Te_{1-x}Se_x)_3$ ($0 < x < 1$), $(Bi_{1-y}Sb_y)_2Te_3$ ($0 < y < 1$). Phys. Status Solidi (b). **84**(2), 619–628 (1977)
14. Wagner, V., Dolling, G., Powell, B.M., Landweher, G.: Lattice vibrations of Bi_2Te_3. Physica Status Solidi (b). **85**(1), 311–317 (1978)
15. Chen, Y.L., et al.: Experimental realization of a three-dimensional topological insulator, Bi_2Te_3. Science. **325**(5937), 178–181 (2009)
16. Qu, D.X., Hor, Y.S., Xiong, J., Cava, R.J., Ong, N.P.: Quantum oscillations and Hall anomaly of surface states in the topological insulator Bi_2Te_3. Science. **329**(5993), 821–824 (2010)
17. Satterthwaite, C.B., Ure Jr., R.W.: Electrical and thermal properties of Bi_2Te_3. Phys. Rev. **108**(5), 1164 (1957)
18. Mzerd, A., et al.: Crystal growth of Bi_2Te_3 on single-crystal substrate Sb_2Te_3 by molecular beam epitaxy. Int. J. Electron. **77**(3), 291–300 (1994)
19. Mzerd, A., Sayah, D., Brun, G., Tedenac, J.C., Boyer, A.: Crystal growth and sticking coefficient of Bi_2Te_3 thin films on Si (1 1 1) substrate. J. Mater. Sci. Lett. **14**(3), 194–197 (1995)
20. Francombe, M.H.: Crystal growth and orientation in sputtered films of Bismuth Telluride. Philos. Mag. **10**(108), 989–1010 (1964)
21. George, A.J., Pradeep, B.: Preparation and properties of co-evaporated bismuth telluride [Bi_2Te_3] thin films. Solid State Commun. **56**(1), 117–120 (1985)
22. Das, V.D., Soundararajan, N.: Size and temperature effects on the thermoelectric power and electrical resistivity of bismuth telluride thin films. Phys. Rev. B. **37**(9), 4552 (1988)
23. Zou, H., Rowe, D.M., Min, G.: Growth of p-and n-type bismuth telluride thin films by co-evaporation. J. Cryst. Growth. **222**(1–2), 82–87 (2001)
24. Venkatasubramanian, R., Colpitts, T., Watko, E., Lamvik, M., El-Masry, N.: MOCVD of Bi_2Te_3, Sb_2Te_3 and their superlattice structures for thin-film thermoelectric applications. J. Cryst. growth. **170**(1–4), 817–821 (1997)
25. Venkatasubramanian, R., Siivola, E., Colpitts, T., O'quinn, B.: Thin-film thermoelectric devices with high room-temperature figures of merit. Nature. **413**(6856), 597–602 (2001)
26. Venkatasubramanian, R., Colpitts, T., O'Quinn, B., Liu, S., El-Masry, N., Lamvik, M.: Low-temperature organometallic epitaxy and its application to superlattice structures in thermoelectrics. Appl. Phys. Lett. **75**(8), 1104–1106 (1999)
27. Liao, Z., et al.: Pulsed-laser epitaxy of topological insulator Bi_2Te_3 thin films. APL Mater. **7**(4), 041101 (2019)
28. Slater, J.C.: Barrier theory of the photoconductivity of lead sulfide. Phys. Rev. **103**(6), 1631 (1956)
29. Chopra, K.L.: Thin Film Phenomena MC. Graw Hill Co., New York, USA (1969)
30. Mataré, H.F.: Defect Electronics in Semiconductors. Wiley, New York (1971)
31. Das, V.D., Karunakaran, D.: Variations of energy gap, resistivity, and temperature coefficient of resistivity in annealed β-Ag_2Se thin films. Phys. Rev. B. **39**(15), 10872 (1989)
32. V. B.: Sandomirskii.: Quantum size effect in a semimetal film. J. Exptl. Theoret. Phys. (U.S.S.R.). **52**, 158–166 (1967)

Logic Gate Design for In-memory Computing with Sputtered Hafnium Oxide Based Memristor

Rose Mary Kuruvithadam[1(✉)], Antony Sharon[2], Aldrin Antony[2,3], M. K. Jayaraj[3], and S. Nalesh[1]

[1] Department of Electronics, Cochin University of Science and Technology, Kochi, Kerala 682022, India
{rosekuruvithadam,nalesh}@cusat.ac.in
[2] Centre of Excellence in Advanced Materials, Cochin University of Science and Technology, Kochi, Kerala 682022, India
[3] Department of Physics, Cochin University of Science and Technology, Kochi, Kerala 682022, India

Abstract. Artificial intelligence has advanced with the emergence of deep learning techniques. However, the performances of deep neural networks in Von Neumann architectures are limited by the memory bottleneck. In-memory computing using memristive devices, is a promising solution, with significant improvement in speed, efficiency, and scalability. This work demonstrates the application of RF-sputtered Hafnium Oxide (HfO_x) based memristive device for implementing logic gates targeted at In-memory computing applications. A model for Pt/HfO_x/Ag based memristor was developed, adopting the Voltage ThrEshold Adaptive Memristor (VTEAM) model, showing close match with experimental I-V data (rms error:0.0078). The Verilog-A model for this memristive device was developed and implemented in Cadence Virtuoso circuit simulator and realized basic logic gates using the Memristor Aided Logic design-MAGIC and analyzed its performances in terms of delay and energy. This work bridges the gap between experimental validation, theoretical modeling, and circuit implementation, illustrating the potential of HfO_x-based memristor for high performance, energy efficient computations.

Keywords: Memristor · In-memory computing · VTEAM model · MAGIC

1 Introduction

Traditional electrical circuit design is built around three essential passive components: capacitors, resistors, and inductors. These elements are theoretically defined as the relationships between two of the four fundamental circuit variables: current, voltage, charge, and magnetic flux. Capacitors connect voltage(v) to charge(q) as $dq = Cdv$, resistors link voltage(v) to current(i) as $dv = Rdi.$, and inductors relate current(i) into flux(φ) as $d\varphi = Ldi$. However, there was a critical gap in this framework, there should be a device that relates charge and flux. In 1971 Professor Leon Chua proposed Memristor - an abbreviation of memory resistor, as the fourth fundamental element that links

A. Antony et al. (Eds.): icONMAT 2025, SPPHY 328, pp. 233–240, 2025.
https://doi.org/10.1007/978-981-95-2697-0_29

charge with magnetic flux. Unlike its counterpart, a memristor has the unique property to retain the memory of past states, as its resistance termed memristance depends on the historical profile of charge or voltage applied across it. The memristor was originally a theoretical notion, but its practical significance was proved decades later, in 2008, when HP Labs researchers successfully constructed and demonstrated a physical memristor. This milestone validated its potential and highlighted its capabilities in energy-efficient memory technologies, particularly due to its non-volatile behavior. Alongside its low power consumption, the compatibility with CMOS technology allows memristors to easily integrate into existing hardware platforms. As a result, memristors emerged as the key component in energy-efficient memory technologies, bridging the gap between volatile RAM and long-term storage.

In-memory computing integrates memory and processing units to overcome the Von-Neumann bottleneck caused by the data transit between these units. Memristors are an ideal candidate for this technology by serving as both memory and logic elements, thereby eliminating the need for frequent data transfer, significantly lowering the power consumption and latency. In our work, we are focusing on the implementation of basic logic gates (NAND, NOR, OR and AND) using Rf magnetron sputtered HfO_x-based memristors presented in [1] to facilitate its potential for scalable and energy efficient In-memory computing applications. There are various memristor based design styles in which IMPLY [5] and MAGIC [2, 4] are the most commonly used for performing stateful logic operations. IMPLY uses memristors to perform logic operations but requires an additional resistor. The input values are loaded in the memristor, and the output value is stored back into one of them, resulting in the destruction of one of the inputs during execution. In MAGIC (Memristor-Aided Logic), a separate memristor is dedicated to output, thus preserving all its inputs during the operations. Both the design styles can be used to implement the basic logic gates, but using MAGIC only NOR and NOT functions can be realized in the crossbar. Additionally, MAGIC offers the advantage of requiring fewer supply voltages compared to IMPLY and does not require an external resistor making it an efficient choice for realizing stateful logic operations.

HfO_x-based Memristor Modeling: This paper investigates the use of MAGIC design style for implementation of basic logic gates and major contribution of our work are summarized as follows.

The memristor is modeled using the VTEAM model in MATLAB, and optimized to replicate the switching behavior of the fabricated device.

Verilog-A Implementation: The optimized model is implemented in Cadence Virtuoso Circuit Simulator using Verilog-A, and its performance is evaluated based on the delay and energy required to read and write 1-bit data.

Logic Gate Implementation and Analysis: Basic logic gates such as NOR, OR, AND, and NAND are realized using the memristor model, and analyzed the delay and energy efficiency.

The paper is organized as Sect. 2 describes a memristor and modelling of a memristor Working principle and design of basic logic gates using MAGIC design styles are explained in Sect. 3. In Sect. 4 performance parameters like delay and energy associate with the HfO_x based memristor and stateful logic gates are evaluated. Finally, Sect. 5 concludes the work.

2 Background

2.1 Memristors

Memristors are passive two-terminal, non-linear resistive elements whose resistance depends on the amount and direction of charge that flows through it and retains this resistive state. Among different materials metal oxide-based devices are commonly used. In [1], the authors demonstrated a radio-frequency (RF) magnetron sputtered HfO_X-based memristor exhibiting reasonable bipolar resistive switching behaviour. The structure of the RF sputtered Hafnium Oxide ($Ag/HfO_X/Pt$)-based memristor device consists of three main layers: a platinum (Pt) bottom electrode, a hafnium oxide (HfO_X) switching layer, and a silver (Ag) top electrode as shown in Fig. 1. The platinum serves as the bottom electrode offering high conductivity and stability and providing efficient electrical contact with the HfO_X layer. The HfO_X layer of 65 nm thick, acts as the active switching medium in the device. It is chosen for its high dielectric constant, which facilitates resistive switching by enabling the formation and rupture of conductive filaments under applied voltage. Silver is used as the top electrode of 200 nm thick, and plays a crucial role in the switching behavior by forming metallic filaments through electrochemical migration during the device operation. When the voltage is applied the device exhibits bipolar resistive switching.

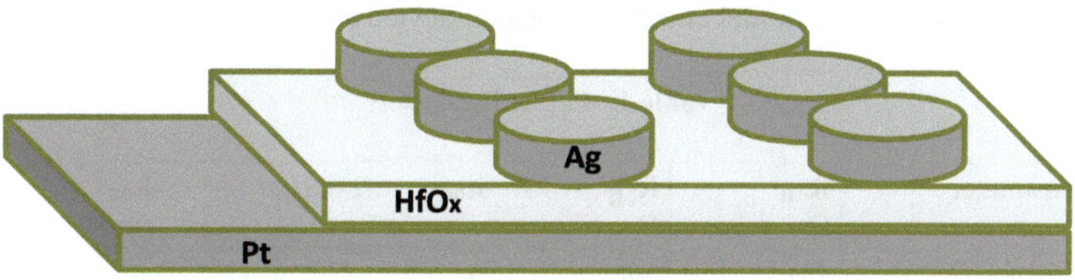

Fig. 1. Structure of Ag/HfOx/Pt Memristor

The device transitions between a high resistance state (HRS) of 2800 Ω and a low resistance state (LRS) of 181 Ω under the application of a SET voltage of 0.25 V and a RESET voltage of −0.2 V respectively. It also shows comparatively good retention of 12,000 s (LRS). In the subsequent sections this HfO_X-based memristor is employed for evaluation and implementation of logic circuits.

2.2 Modelling of Memristor

Mathematical models for the memristors are essential to understand, design, and integrate memristors into real-world circuits for memory, logic, or neuromorphic applications. Circuit designers need accurate models that can be implemented in SPICE simulators, Verilog-A, or MATLAB. Without reliable models, predicting circuit behavior and optimizing the performance of practical circuits becomes challenging. There are different mathematical memristive models based on accuracy and computational complexity. Here, we adopted the Voltage ThrEshold Adaptive Memristor -VTEAM model

[3] for developing the compact model used for simulating its electrical behavior in circuit applications. In the model, the derivative of the state variable, w is defined by (1) and the current–voltage relationship is given by (2)

$$\frac{dw(t)}{dt} = \begin{cases} k_{off} \cdot (\frac{v(t)}{v_{off}} - 1)^{\alpha_{off}} & 0 < v_{off} < v \\ 0, & v_{on} < v < v_{off} \\ k_{on} \cdot (\frac{v(t)}{v_{on}} - 1)^{\alpha_{on}} & v < v_{on} < 0 \end{cases} \tag{1}$$

$$i(t) = \left[R_{on} + \frac{Roff - Ron}{woff - won} .(w - w_{on}) \right]^{-1} \tag{2}$$

Where v_{on} and v_{off} are the ON and OFF threshold voltages, k_{off} and k_{on} represent the rates of change of the state variable w during switching. α_{off} and α_{on} are dimensionless coefficients that shape the non-linearity of the switching process, and R_{on} and R_{off} defines the resistances of the device when the limits of internal state variable w is in w_{on} and w_{off} respectively.

The VTEAM model is a general model that can be fit to the fabricated HfOx based memristor characteristics. Given the current–voltage characteristics of a specific memristor, a set of parameters is chosen to fit the VTEAM model to a reference I–V relationship. To fit the I–V curve, the relative RMS error is minimized using simulated annealing algorithms. The fitting parameters for HfO_x-based memristor are shown in Table 1. The I-V characteristics from the model and the fabricated device are shown in Fig. 2.

Table 1. Fitting Parameters

Parameter	**R_{on}** Ω	**R_{off}** Ω	**α_{on}**	**α_{off}**	**k_{on}** (m/s)
Values	**181**	2.8	9	2	1
Parameter	**k_{off}** (m/s)	**v_{on}** (V)	**v_{off}** (V)	**w_{on}** (m)	**w_{off}** (m)
Values	**−0.1 µ**	0.2	−0.2	65n	0

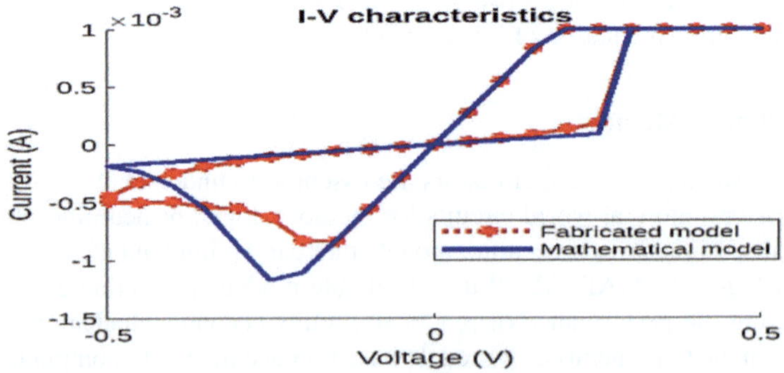

Fig. 2. I-V Characteristics

2.3 Realization of Basic Logic Functions

The MAGIC design style has dedicated memristors for inputs and outputs, so the input data is maintained during the entire computation. We utilized the MAGIC gates structure described in [6] and investigated its functionality and performance using our HfO_x–based memristor. Logical states are represented by resistance values, with logic *0* being represented by high resistance (R_{off}) and logic *1* by low resistance (R_{on}). The input memristors are denoted by A and B and output memristor by *O*. MAGIC based LOGIC gates are shown in Fig. 4.

It operates in two sequential steps. In the initialization phase, memristor *O* is set to an initial state. During the evaluation phase, the input memristors *A* and *B* are applied with a voltage *V0* with in the constrain limit [6]. Resulting current depending upon the value of the *V0* will determine the state of the output memristor *O*. Consider the NOR gate, evaluation memristor O is initialized to a low resistance state as shown in Fig. 3 and if needed the input memristors A and B with the desired input values. Evaluate voltage V0 is applied to the input memristors A and B. If both A and B are logic zero or in high resistance state, to ensure proper operation, voltage across the output memristor has to be below the switching threshold, such that memristor O remains in the initial state. If any one of the input A or B is in logic 1 state or in low resistance state, the voltage across the memristor O should be above the switching threshold for the memristor to switch from high resistance state to low resistance state.

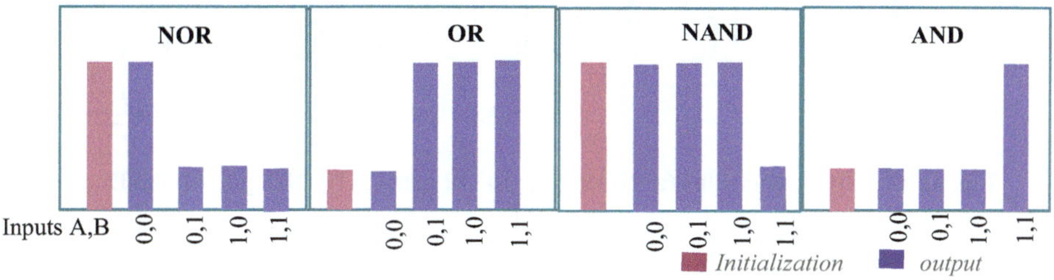

Fig. 3. Initialization and outputs of MAGIC NOR gate

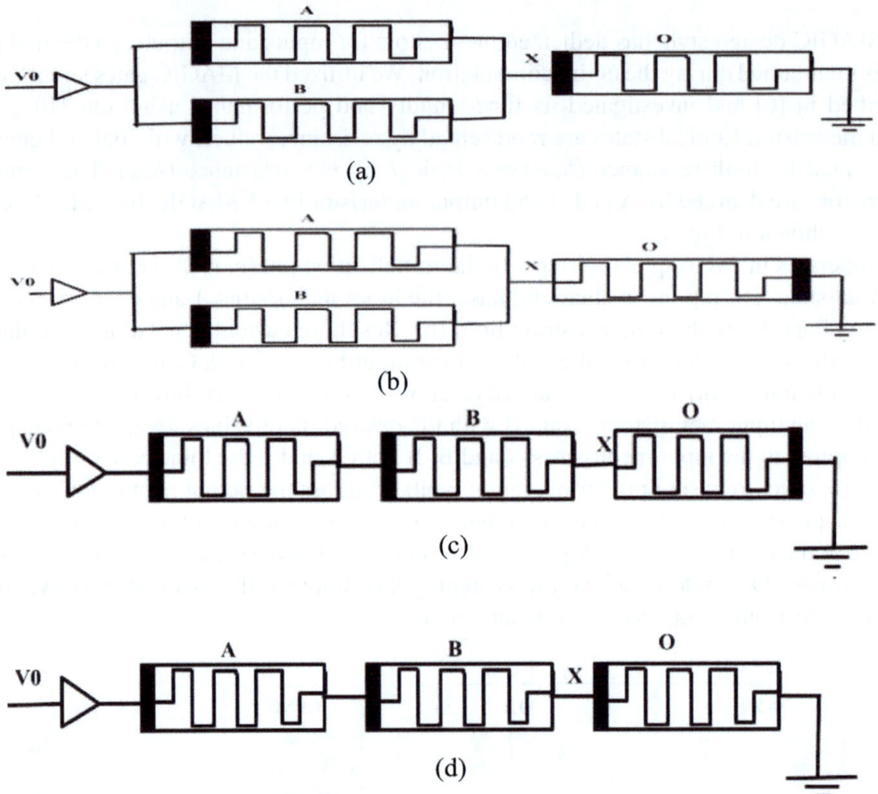

Fig. 4. MAGIC Logic gates (a) NOR gate (b) OR gate (c) NAND gate (d) AND gate

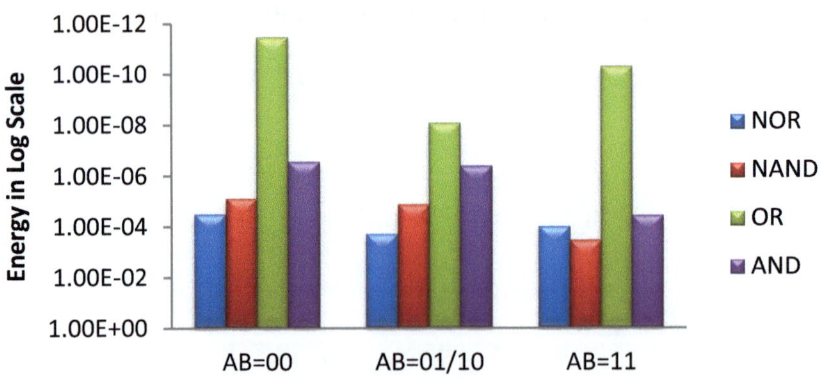

Fig. 5. Energy consumed by MAGIC gates using HfO_X- based memristor

The value of V0 must be carefully chosen to avoid the unintentional state change of the input memristor. The state of the input memristor A or B may be altered unintentionally. This happens when V0 is high and exceeds the switching threshold of input memristor A, it may change its resistance state, leading to incorrect logic operation. For example, if A and B was originally in the high-resistance state (logic 0) but V0 is high, it may switch to the low-resistance state (logic 1), which alters the intended logic function. Working of the all other gates are similar to that of the NOR gate expect the initialization value.

2.4 Evaluation and Results

A HfO_x-based memristor modeled using the VTEAM is implemented and the performance are analyzed for SET voltage of *400 mV* and RESET voltage of *−500 mV* using Cadence Virtuoso circuit simulator. The energy required to SET and RESET a memristor is *1.44 nJ* and *74.2 nJ* with delay of *6 μs* and *288 μs* respectively. All the logic gates using MAGIC design style are simulated using our model and verified for its functionality and performance. The delay associated with each logic gates are given in Table 2 and corresponding energy consumed by each logic gate is shown in Fig. 5 in logarithmic Scale. To satisfy the constraints of *V0* [6], the applied voltage values (*V0*) in mV are as follows: [*280, 500*], [*280, 700*], [*280, 500*], [*280, 830*] for NOR, NAND, OR and AND gates respectively.

Table 2. Delay analysis of MAGIC gates

AB	NOR	NAND	OR	AND
01/10	623 ms	–	12 ms	–
11	392.7 ms	552 ms	8 μs	30.09 ms

3 Conclusion

In this work, we presented the modeling and implementation of an HfO_x-based memristor for in-memory computing applications. A Verilog-A model is developed to accurately capture the electrical characteristics of the fabricated device. The designed model is then used to implement MAGIC (Memristor-Aided Logic) based logic gate structures. Our analysis demonstrates that the HfO_x-based memristor is a viable candidate for stateful logic applications, offering promising energy efficiency and delay characteristics. Future work will focus on optimizing the device parameters to further enhance performance and energy efficiency and also exploring the implementations of more complex computing tasks.

References

1. Sharon, A., Subin, P.S., Jayaraj, M.K., et al.: Fabrication and characterisation of memristor device using sputtered hafnium oxide. J. Mater. Sci. Mater. Electron. **34**, 1192 (2023). https://doi.org/10.1007/s10854-023-10613-5

2. Hoffer, B., Rana, V., Menzel, S., Waser, R., Kvatinsky, S.: Experimental demonstration of Memristor-Aided Logic (MAGIC) using Valence Change Memory (VCM). IEEE Trans. Electron Devices. **67**(8), 3115–3122 (2020)
3. Kvatinsky, S., et al.: VTEAM: A general model for voltage-controlled memristors. IEEE Trans Circuits Syst II Express Briefs. **62**(8), 786–790 (2015)
4. Thangkhiew, P.L., et al.: Efficient mapping of boolean functions to memristor crossbar using MAGIC NOR gates. IEEE Trans. Circuits Syst. I: Regul. Pap. **65**(8), 2466–2476 (2018)
5. Seiler, F., Taheri Nejad, N.: Efficient image processing via memristive-based approximate in-memory computing. IEEE Trans. Comput.-Aided Des. Integr. Circuits Syst. **43**(11), 3312–3323 (2024)
6. Kvatinsky, S., et al.: MAGIC—Memristor-aided logic. IEEE Trans Circuits Syst II Express Briefs. **61**(11), 895–899 (2014)

Amorphous WO_3 Thin Films for Energy-Efficient Electrochromic Smart Windows

P. Meenu[1], C. H. Sufeeda[2], and K. G. Deepa[1,2](✉)

[1] Department of Nanoscience and Nanotechnology, University of Kerala, Thiruvananthapuram, India
deepachaithanya@gmail.com
[2] Department of Physics, University of Kerala, Thiruvananthapuram 6955581, India

Abstract. Electrochromic smart windows are a way to develop energy-efficient buildings, changing their transmission properties with the application of electricity. In this study, amorphous WO_3 (a-WO_3) films are deposited using the potentiostatic electrodeposition method in a three-electrode system and demonstrated as switchable electrochromic smart windows. The deposition potential is varied between −0.6 to −0.9 V among which the film deposited at −0.8 V showed better electrochromic property. The deposition duration is also varied from 200 to 500 s to optimize thickness and morphology. Structural characterization studies are done with analysing XRD and Raman spectra revealed amorphous nature of the films. Optical analysis showed a decreasing bandgap from 3.4 to 3.3 eV with increasing deposition time, attributed to reduced quantum confinement in thicker films. Electrochemical evaluations, including cyclic voltammetry and chronoamperometry, demonstrated improved charge capacity and coloration efficiency for films deposited at −0.8 V for 400 s. These films exhibited rapid switching speed, and high transmittance modulation (>92%) with colouration efficiency of 65 cm^2/C.

Keywords: Electrodeposition · smart window · electrochromic property · amorphous

1 Introduction

Electrochromic smart windows are the most promising switchable window of today which changes the transmission property with the application of electricity. The energy efficient building has attracted significant interest in smart window technologies [1]. Electrochromic (EC) materials, particularly transition metal oxides, have emerged as promising candidates for such applications due to their reversible optical modulation in response to an applied voltage [2]. Tungsten oxide (WO_3) with numerous oxidation states is the most prominent among these due to its wide band gap, long-term cycling stability, and outstanding colouration efficiency [3]. WO_3 thin films can be synthesised using various techniques such as sputtering, sol–gel processing, and chemical vapour deposition [4]. However, these methods often involve processing high temperatures or

A. Antony et al. (Eds.): icONMAT 2025, SPPHY 328, pp. 241–251, 2025.
https://doi.org/10.1007/978-981-95-2697-0_30

complex fabrication steps, limiting their scalability and cost-effectiveness. Electrodeposition is an effective alternative with a low-cost, scalable, and low-temperature method for fabricating uniform and adherent WO_3 films on conductive substrates [5].

Several studies have shown the successful deposition of amorphous and nanocrystalline WO_3 films using aqueous and non-aqueous electrolytes. Zhang, Du & Yao used cathodic electrodeposition to prepare hydrated $WO_3.2H_2O$ films, yielding dual-band electrochromic behaviour with almost 92% optical modulation at 633 nm and 86% at 1100 nm and high colouration efficiency of 204 cm2/C at 1100 nm and 72 cm2/C at 633 nm [6]. Chithambararaj et al. (2022) explored electrodeposition of non-stoichiometric WO_3 films, showing improved electrochromic performance over stoichiometric counterparts. Variations in film composition led to enhanced optical modulation and faster switching behaviour, attributed to increased defect concentrations and porosity that facilitate ion insertion/extraction [7]. Zhang & Zhao (2022) developed a core–shell structure by electrodepositing WO_3 with polyaniline (PANI). This hybrid architecture boosted conductivity and electrochemical activity, resulting in faster response and improved cycling stability compared to pure WO_3 films [8].

Here we investigate the fabrication of amorphous WO_3 thin films by electrodeposition by varying the deposition potential and deposition time for electrochromic smart window applications. The structural, morphological, and electrochromic properties of the films were systematically characterised to analyse their performance. The results provide insights into optimising electrodeposition parameters, focusing on achieving high optical contrast, enhancing colouration efficiency, and reducing switching time.

2 Materials and Methods

Sodium tungstate dihydrate ($Na_2WO_4{\cdot}2H_2O$, $\geq$ 99%), hydrogen peroxide solution (H_2O_2, 30 wt % in H_2O), and lithium perchlorate ($LiClO_4$, $\geq$ 95%), Perchloric acid ($HClO_4$, 70%), Propylene carbonate ($C_4H_6O_3$, 99%) and Deionized water. All the chemicals were used without any further purification. The WO_3 films were electrochemically deposited on both FTO glass substrate (resistivity of 8 Ω sq-1) and stainless-steel substrates using a three-electrode system with FTO, Pt wire, and Ag/AgCl as the working, counter, and reference electrodes, respectively. Before use, the FTO glass/stainless-steel substrates were cleaned consecutively with soap solution, distilled water, acetone, and isopropyl alcohol under ultrasonication. Then cleaned by UV-Ozone cleaner for 10 min. The electrodeposition solution was prepared by dissolving 1.03 g of $Na_2WO_4{\cdot}2H_2O$ in 125 mL of DI water, and then 0.65 mL of H_2O_2 was added under magnetic stirring. Adding $HClO_4$ (70%) adjusted the pH to 1.0 (3 mL). WO3 films were deposited on FTO glass substrate (or, stainless-steel substrate) using conventional potentiostatic electrodeposition under voltages of −0.6 V, −0.7 V, −0.8 V and − 0.9 V (vs. Ag/AgCl) for 200 s. After electrodeposition, the films were washed with DI water and dried in air. XRD measurements are carried out on Bruker D8 ADVANCE with DAVINCI design operating at 40 kV and 40 mA to characterise the WO_3 films. In a θ-2θ arrangement, the diffraction pattern is recorded from 5 to 90°, and the x-ray source is the Cu Kα line with a wavelength of 1.5405 Å. Lab RAM HR Evolution (Horiba) spectrometer with 532 nm excitation was used to produce the Raman spectra. Absorption spectra of samples are recorded using a JASCO V-750 spectrophotometer. Carl Zeiss EVO 18 Research

Scanning electron microscope (SEM) was used to determine the surface morphology of the films. An electrochemical workstation Autolab, PGSTAT204 with the NOVA software was used to measure the cyclic voltammetry (CV), chronoamperometry (CA), and chronocoulometry (CC). The transmittance spectra of WO_3 thin films were captured with a UV–Vis Autolab spectrophotometer in the 400–800 nm range.

3 Results and Discussion

3.1 Structural and Morphological Studies

The XRD pattern of WO_3 prepared on stainless-steel (SS) substrate by electrodeposition is given in Fig. 1. All the as-deposited WO_3 films are amorphous in nature, as the XRD shows the absence of any sharp diffraction peaks other than the peaks of the substrate. Two sharp peaks observed in the XRD spectrum correspond to the stainless-steel substrate at around $2\theta \approx 44.5°$ and $74.7°$, which are associated with the (111) and (220) planes of austenitic stainless steel (typically Fe–Cr–Ni alloys) [9]. The amorphous nature of WO_3 films is common for films deposited at low temperatures without post-deposition annealing. This open, disordered structure improves both switching speed and colouration efficiency, which are key parameters in smart window performance, as it allows for easier ion transport during redox cycling [1].

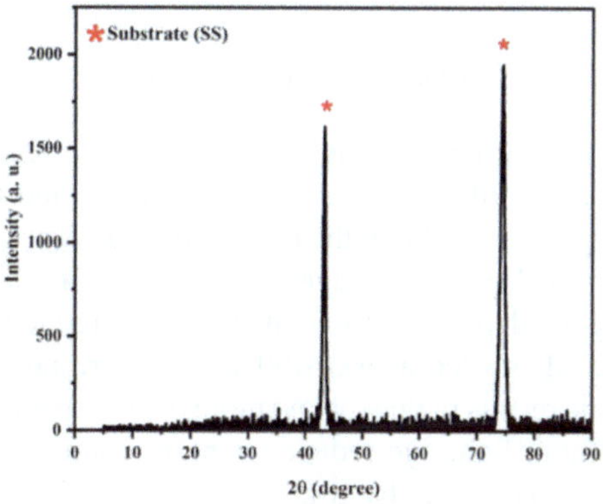

Fig. 1. XRD of the WO_3 samples electrodeposited on the stainless-steel substrate.

The structural analysis of the WO_3 thin films was further characterised by Raman spectroscopy. The Raman spectra of all the WO_3 samples are shown in Fig. 2. The Raman spectrum reveals broad and weak bands, which indicates the amorphous nature of the tungsten oxide structure. The broad feature around 650 cm^{-1} is attributed to the bridging W–O–W stretching modes within the WO_6 octahedra network. In contrast to crystalline WO_3, where these modes appear as sharp and well-defined peaks, the broadened nature of both peaks in this case reflects the lack of long-range order and a distribution of bond lengths and angles typical of amorphous structures. The broad band near 950 cm^{-1} corresponds to the terminal W=O stretching vibration, commonly found in disordered and hydrated tungsten oxide phases [10, 11].

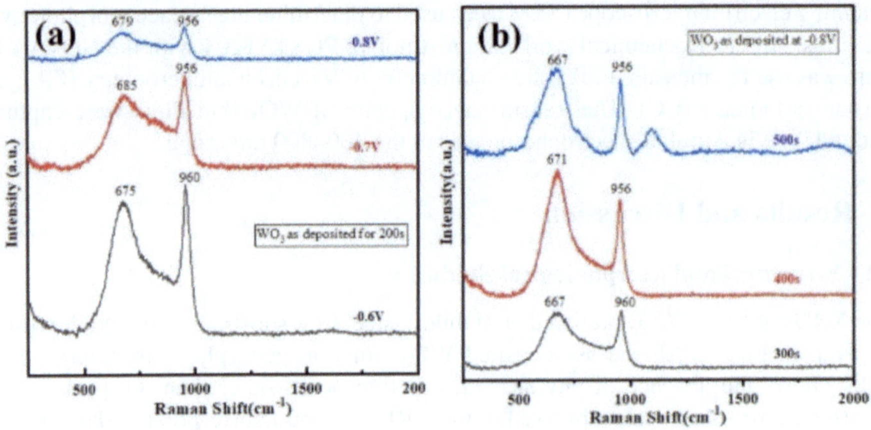

Fig. 2. Raman spectra of WO_3 samples at (a) different deposition voltages and (b) different deposition time.

The optical band gap (E_g) of the electrodeposited WO_3 thin films was determined using UV-Vis spectroscopy (Fig. 3). The absorbance spectra were recorded in the wavelength range of 300–800 nm, and the band gap was estimated using the Tauc plot method Eq. (1).

$$(\alpha h\upsilon)^2 = A\left(h\upsilon - E_g\right) \tag{1}$$

where, α is the absorption coefficient, $h\upsilon$ is the photon energy, E_g is the optical bandgap, and A is a constant. A plot of $(\alpha h\upsilon)^2$ vs. $h\upsilon$ was drawn, and the linear region was extrapolated to the energy axis to obtain the band gap value. The calculated energy gaps of thin films deposited for different voltages were about 3.46, 3.41, 3.37 and 3.55 eV (Fig. 3 (a)). On varying the deposition time, the band gap reduces slightly to 3.25, 3.27 and 3.19 eV for samples deposited in 300, 400 and 500 s respectively (Fig. 3 (b)). The increased deposition time results in thicker and more densely packed WO_3 films. Thicker films typically absorb more light and reduce quantum confinement effects, leading to band gap narrowing [12]. The wide band gap in this study is consistent with the high transparency observed in the bleached state, which is desirable for electrochromic smart window applications.

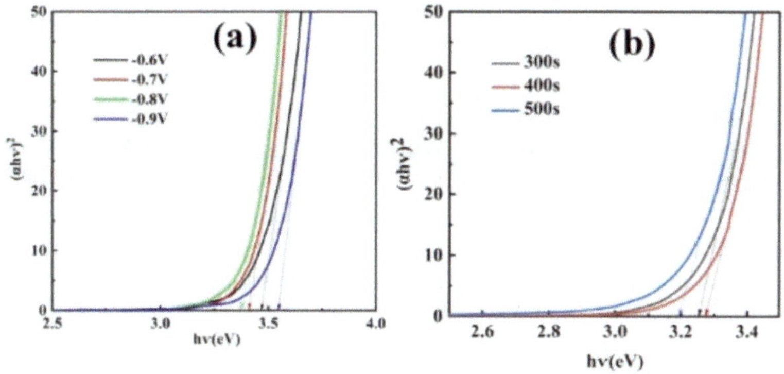

Fig. 3. Tauc plot of WO_3 samples at (a) different deposition voltages and (b) different deposition time

The surface morphology of the electrodeposited WO_3 thin films was examined using scanning electron microscopy (SEM). SEM image of as-deposited amorphous sample is shown in Fig. 4; the image shows clusters of particles on the surface of the film with a particle size of ~250 nm. The particles have spherical morphology.

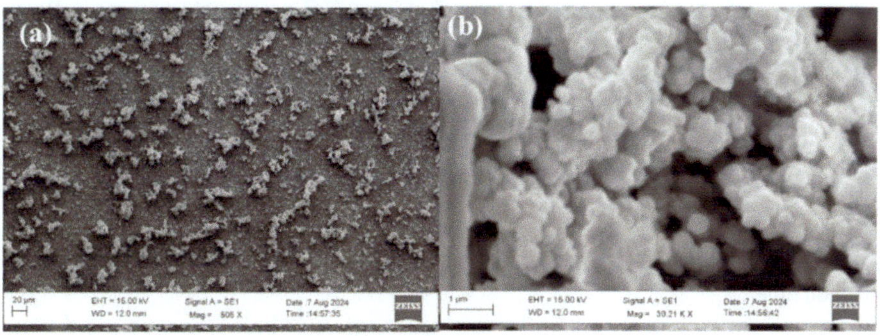

Fig. 4. SEM image of as-deposited amorphous sample

3.2 Electrochromic Studies

The electrochromic (EC) behaviour of the WO_3 films was investigated by cyclic voltammetry (CV), chrono amperometry (CA), chrono coulometry (CC) and in situ optical transmittance measurements at 633 nm in a standard three-electrode configuration, using WO_3 as the working electrode, a platinum counter electrode, and Ag/AgCl as the reference electrode. All measurements were performed in 1 M $LiClO_4$ in propylene carbonate electrolyte containing a Li^+ ion where electrochromism arises from the reversible redox reaction:

$$WO_3 + xM^+ + xe^- \leftrightarrow M_xWO_3, \tag{2}$$

Here, M^+ is the intercalating ion (Li^+).

The graph of CV in Fig. 5 depicts WO_3 films measured in a 1 M $LiClO_4$/PC electrolyte in the potential region of -1 V to 1 V (vs. Ag/AgCl) at a scan rate of 20 mV/s. The cathodic peak corresponds to the coloration process (insertion of ions/electrons), while the anodic peak represents the bleaching process (ion/electron extraction). The optical transmittance of the WO_3 thin films in the wavelength range 400–800 nm was recorded during electrochemical switching between bleached (oxidized) and coloured (reduced) states (Fig. 6 and Fig. 8). The chronoamperometry (CA) measurements (Fig. 7 and Fig. 9) were performed by alternately applying fixed potentials of -0.8 V (colouring) and + 1 V (bleaching) vs. Ag/AgCl in a 1 M $LiClO_4$/propylene carbonate electrolyte. The response time for bleaching and colouring of WO_3 films are calculated using chronoamperometry curves. The value for intercalated/deintercalated charge (Q) during the coloured/bleached state is determined using chrono coulometry (CC) plots (Fig. 10).

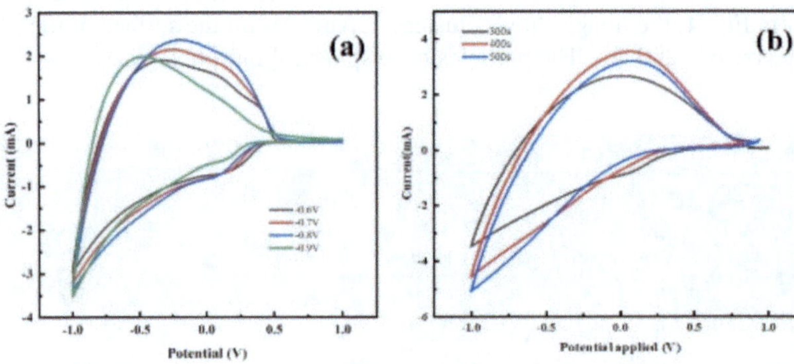

Fig. 5. Cyclic Voltammetry of WO_3 samples at (a) different deposition voltages and (b) different deposition time

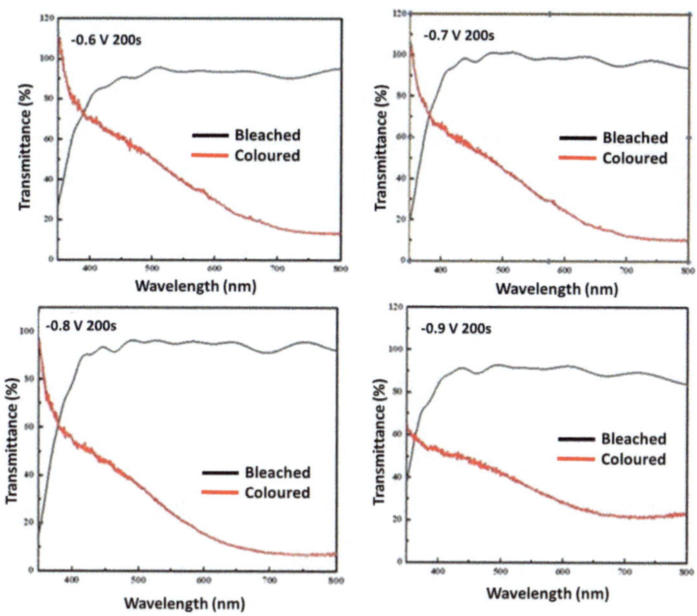

Fig. 6. (a) Transmittance of coloured and bleached state of WO_3 samples with different deposition voltage

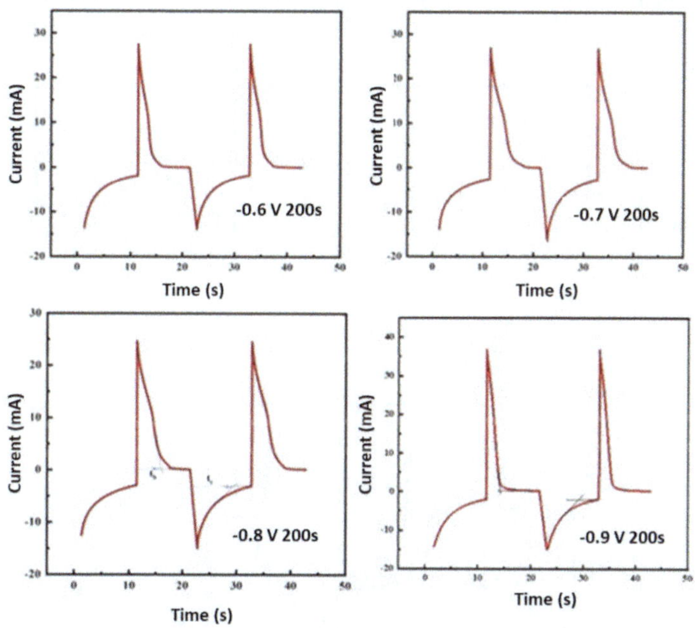

Fig. 7. Chronoamperometry of WO_3 samples with different deposition voltage

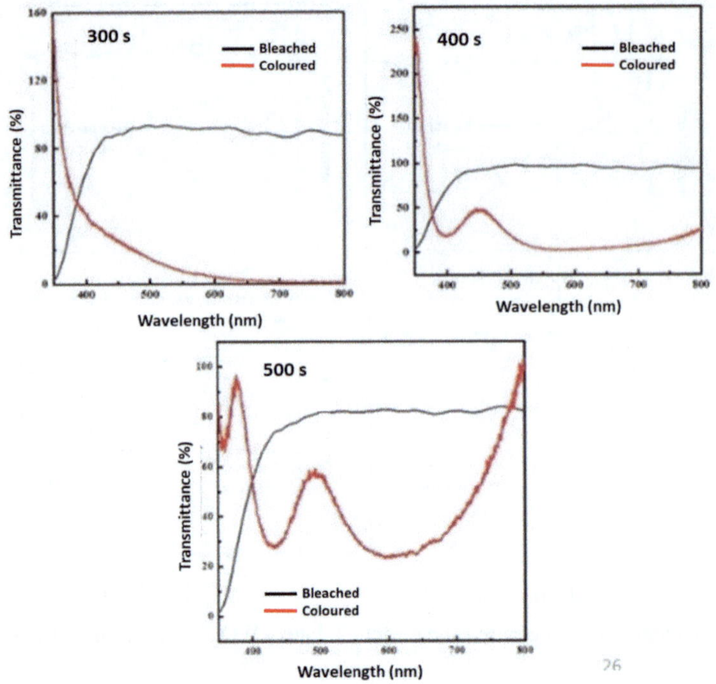

Fig. 8. Transmittance of coloured and bleached state of WO_3 samples with different deposition time

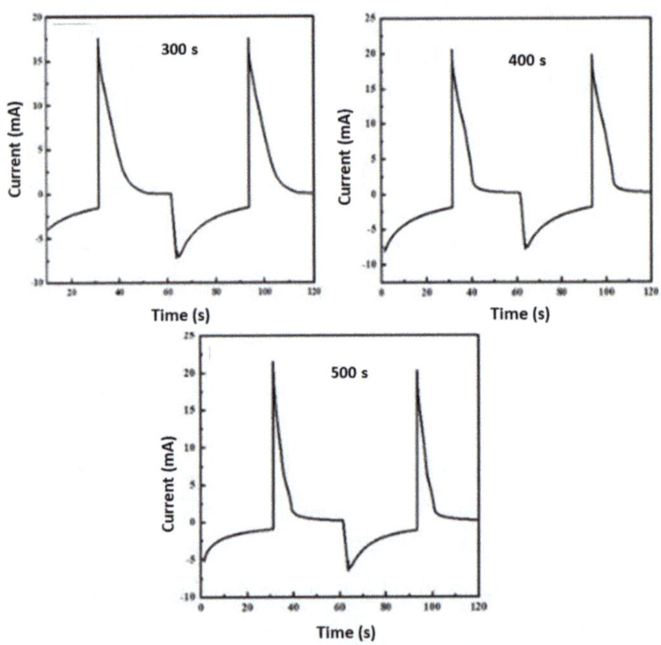

Fig. 9. Chronoamperometry of WO_3 samples with different deposition time

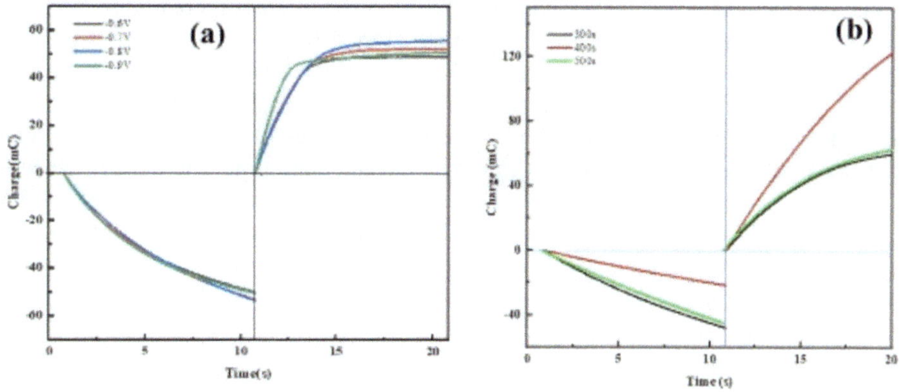

Fig. 10. Chrono coulometry of WO_3 samples at (a) different deposition voltages and (b) different deposition time.

The coloration efficiency was calculated using the equation:

$$CE = \frac{\Delta OD}{Q} = \frac{\log \frac{T_b}{T_c}}{Q} \quad (3)$$

where, ΔOD is the change in optical modulation, Q is the intercalated/deintercalated charge, T_b and T_c are transmittance in the bleached and coloured states, respectively.

The voltage optimisation is studied with the deposition potential varying between −0.6, −0.7, −0.8, and − 0.9 V for the fixed deposition time of 200 s. From the CV analysis, −0.8 V 200 s gives a high area under the CV curve indicates a high charge storage capacity and enhanced electrochemical activity of the material (Fig. 5(a)) [13]. In the bleached state, the films exhibited high transmittance values of above 90%, depending on film thickness and deposition voltage. In the coloured state, transmittance decreased significantly below 25%, showing a strong modulation in optical density due to ion intercalation and electronic transitions. The sample − 0.8 V 200 s shows the highest optical modulation of around 83%. The detailed study is shown in Fig. 6 and in Table 1. The coloration efficiencies are calculated and the film deposited at −0.8 V 200 s exhibited highest efficiency of 17.03 cm^2/C.

Table 1. Summary of electrochromic properties of a-WO_3 samples prepared at different deposition voltages

Deposition potential (V)	Response time (s)		Transmittance (%)		Optical modulation (ΔOD)	Intercalated charge (Q) (mC)	Colouration efficiency (CE) (cm^2/C)
	Bleaching time (t_b)	Colouration time (t_c)	Bleached (T_b)	Coloured (T_c)			
−0.6	3.17	6.64	93.62	22.83	70.79	−50.40	12.15
−0.7	4.61	7.74	99.44	17.79	81.65	−53.62	13.93
−0.8	4.65	7.82	95.60	11.67	83.93	−53.63	17.03
−0.9	2.55	7.21	90.48	25.19	65.29	−20.40	11.01

The voltage is fixed as −0.8 V and then the deposition time is increased to 300, 400, and 500 s from 200 s (Table 2). On analysing the Fig. 5(b), CV for different deposition time, −0.8 V 400 s gives a high area under the CV curve indicates a high charge storage capacity and enhanced electrochemical activity of the material [13]. The films showed transmittance values of over 90% for the deposition times of 300 s and 400 s, but it decreased to almost 82% for the 500 s in the bleached condition. The transmittance in the coloured state is below 25% for all the samples with deposition potential of −0.8 V irrespective of deposition time. The thorough investigation is displayed in Table 2 and Fig. 7. Also, the coloration efficiencies were calculated as in Table 2 and find that on analysing the coloration efficiencies, film deposited at −0.8 V 400 s have high efficiency of 65.06 cm^2/C along with highest optical modulation of around 92%.

Table 2. Summary of electrochromic properties of a- WO_3 samples with various deposition time

Samples	Response time (s)		Transmittance (%)		Optical modulation (ΔOD)	Intercalated charge (Q) (mC)	Colouration efficiency (CE) (cm^2/C)
	Bleaching time (t_b)	Colouration time (t_c)	Bleached (T_b)	Coloured (T_c)			
300	11.88	18.82	91.16	2.13	89.03	−48.22	33.83
400	9.18	22.18	95.95	3.75	92.20	−21.64	65.06
500	6.01	15.93	82.28	26.13	56.15	−44.59	11.17

4 Conclusion

a-WO_3 thin films were prepared using electrodeposition and the results demonstrate that deposition voltage and time plays a critical role in tuning film thickness, morphology, and electrochemical behaviour. Films deposited at −0.8 V for 400 s exhibited the most favourable properties, including high optical modulation (~92%), and fast switching times. The maximum coloration efficiency achieved was ~65 cm^2/C, with good charge capacity and stable cycling.

References

1. Granqvist, C.G.: Electrochromic materials: Out of a niche. Nat. Mater. **5**(2), 89–90 (2006)
2. Mortimer, R.J.: Electrochromic materials. Chem. Soc. Rev. **26**, 147–156 (1997)
3. Deb, S.K.: Opportunities and challenges in science and technology of WO_3 for electrochromic and related applications. Sol. Energy Mater. Sol. Cells. **92**(2), 245–258 (2008)
4. Niklasson, G.A., Granqvist, C.G.: Electrochromics for smart windows: thin films of tungsten oxide and nickel oxide, and devices based on these. J. Mater. Chem. **17**(2), 127–156 (2007)
5. Radhakrishnan, S., et al.: Electrodeposition of tungsten oxide thin films for electrochromic applications. Thin Solid Films. **593**, 101–107 (2015)
6. Zhang, Q., Du, R., Yao, A.: Cathodic electrodeposition of amorphous tungsten oxide dihydrate film for dual-band electrochromic modulation. RSC Adv. **15**(7), 5242–5251 (2025)

7. Li, Y., et al.: Enhanced electrochromic properties of WO_3 nanotree-like structures synthesized via a two-step solvothermal process showing promise for electrochromic window application. ACS Applied Nano Materials. **1**(6), 2552–2558 (2018)
8. Zhang, Y., Zhao, L.: The electrochromic properties of the film enhanced by forming WO_3 and PANI core–shell structure. J. Mater. Sci. Mater. Electron. **33**(26), 20802–20811 (2022)
9. Cullity, B.D., Stock, S.R.: Elements of X-ray Diffraction, 3rd edn. Pearson Education, Upper Saddle River (2001)
10. Bhatia, R., et al.: Raman and optical analysis of amorphous WO_3 thin films for electrochromic applications. Mater. Res. Bull. **48**, 2566–2572 (2013)
11. Gies, M., et al.: Electrochromic switching of tungsten oxide films grown by reactive ion-beam sputter deposition. J. Mater. Sci. **56**, 615–628 (2021)
12. Mohamedkhair, A.K., et al.: Tuning structural properties of WO_3 thin films for photoelectrocatalytic water oxidation. Catalysts. **11**(3), 381 (2021)
13. Zhen, Y., Jelle, B.P., Gao, T.: Electrochromic properties of WO_3 thin films: The role of film thickness. Anal. Sci. Adv. **1**(2), 124–131 (2020)

Author Index

A. Antony et al. (Eds.): icONMAT 2025, SPPHY 328, pp. 253–254, 2025.
https://doi.org/10.1007/978-981-95-2697-0

MIX
Papier aus verantwortungsvollen Quellen
Paper from responsible sources
FSC® C105338

If you have any concerns about our products,
you can contact us on
ProductSafety@springernature.com

In case Publisher is established outside the EU,
the EU authorized representative is:
Springer Nature Customer Service Center GmbH
Europaplatz 3, 69115 Heidelberg, Germany

Printed by Libri Plureos GmbH
in Hamburg, Germany